Formation and Control of Disinfection By-Products in Drinking Water

Formation and Control of Disinfection By-Products in Drinking Water

Dr. Philip C. Singer, Editor

Printed in the United States of America

American Water Works Association
6666 West Quincy Avenue
Denver, CO 80235

Library of Congress Cataloging-in-Publication Data

ISBN 0-89867-998-2

Formation and control of disinfection by-products.
p. cm.
Includes bibliographical references and index.
ISBN 0-89867-998-2
1. Water--Purification--Disinfection--By-products.
TD459.F67 1999
628.1'662--dc21 99-26056
CIP

Contents

Figures

Tables

Preface

Concern about disinfection by-products (DBPs) in drinking water dates back to the first published reports of the occurrence of trihalomethanes (THMs) in finished drinking water which appeared in 1994. Shortly thereafter, THMs were shown to be a by-product of the chlorination of water and were linked to increased cancers of the urinary and digestive tracts. These findings led to the initial regulation of THMs in finished US drinking waters, which occurred in 1979. Since that time, numerous studies have been conducted to identify other by-products arising from the disinfection of water; define water quality and treatment characteristics that lead to the formation of THMs and other DBPs; quantify the occurrence and distribution of DBPs in finished drinking water; develop and assess technologies and treatment modifications for the control of DBPs; evaluate health effects attributable to the presence of DBPs in drinking water; and determine regulatory approaches for balancing the risks between disinfection of drinking water and the production of DBPs arising from the disinfection processes. These studies ultimately led to the promulgation, by the US Environmental Protection Agency (USEPA), of the microbial/disinfection by-product (M/DBP) cluster; a series of rules consisting of the Information Collection Rule (1996), the Enhanced Surface Water Treatment Rule (1998), and the Disinfectants/Disinfection By-Products Rule (1998).

At the time of publication of this book, the water supply industry is in the midst of evaluating the data amassed from the Information Collection Rule, implementing the Interim Enhanced Surface Water Treatment Rule, and implementing Stage 1 of the Disinfectants/Disinfection By-Products (D/DBP) Rule. It can be safely assumed that these rules, and subsequent versions and stages of them, will form the framework for water treatment practice in the United States and elsewhere for many years to come.

This book evolved from a symposium held in March, 1997, on the occasion of the retirement of Dr. James M. Symons from the University of Houston. Dr. Symons served as the head of the Process Control Branch of the Drinking Water Research Division of the USEPA's Office of Research and Development for 20 years until he retired in 1982 to direct the environmental engineering program in the Department of Civil Engineering at the University of Houston. Because his tenure at USEPA overlapped with the discovery of THMs in drinking water, he became a central figure in USEPA's efforts to evaluate the formation, control, and regulation of THMs and other DBPs.

This book consists of chapters contributed by a number of individuals, most of whom presented papers at the Symons Symposium and all of whom are recognized experts in the area of formation, regulation, and control of disinfection by-products in drinking water. The book was edited by Dr. Philip C. Singer of the Department of Environmental Sciences and Engineering in the School of Public Health at the University of North Carolina at Chapel Hill. Dr. Singer, like Dr. Symons, has been intimately involved in many phases of DBP research since the THMs were first identified in drinking water in 1974 and has received numerous awards in recognition of his work. With the high level of interest in the subject of disinfection and disinfection by-products, the chapters in this book are expected to serve as excellent sources of reference material on this central theme of water purification.

Acknowledgments

The following experts in disinfection by-products authored or co-authored chapters in this book:

Chapter 1

James M. Symons, Bradenton, Fla.

Chapter 2

Stuart W. Krasner, Metropolitan Water District of Southern California, La Verne, Calif.

Chapter 3

Zaid K. Chowdhury, Malcolm Pirnie, Inc., Phoenix, Ariz.

Gary L. Amy, University of Colorado, Boulder, Colo.

Chapter 4

Jean-Philippe Croué, Laboratoire de Chimie de l'Eau et de l'Environnement, Université de Poitiers, France

Jean-François Debroux and **Gary L. Amy**, University of Colorado, Boulder, Colo.

George R. Aiken and **Jerry A. Leenheer**, United States Geological Survey, Boulder, Colo.

Chapter 5

Jennifer Orme Zavaleta and **Fred S. Hauchman**,
US Environmental Protection Agency, National Health and Environmental Effects Research Laboratory, Research Triangle Park, N.C.

Michael W. Cox, US Environmental Protection Agency, Office of Ground Water and Drinking Water, Washington, D.C.

Chapter 6

Patricia A. Murphy, US Environmental Protection Agency, Cincinnati, Ohio

Gunther F. Craun, Gunther F. Craun and Associates, Staunton, Va.

Chapter 7

Frederick W. Pontius, American Water Works Association, Denver, Colo.

Chapter 8

Gerald E. Speitel, Jr., University of Texas, Austin, Texas

Chapter 9

David A. Reckhow, University of Massachusetts, Amherst, Mass.

Chapter 10

Robert C. Hoehn, Virginia Tech., Blacksburg, Va.
Don J. Gates, Flora Vista, Calif.

Chapter 11

James P. Malley, Jr., University of New Hampshire, Durham, N.H.

Chapter 12

Stephen J. Randtke, University of Kansas, Lawrence, Kan.

Chapter 13

Vernon L. Snoeyink, Mary J. Kirisits, and **Costas Pelekani**, University of Illinois, Urbana, Ill.

Chapter 14

R. Rhodes Trussell and **Issam Najm**, Montgomery Watson, Pasadena, Calif.

Chapter 15

Joseph G. Jacangelo, Montgomery Watson, Herndon, Va.

Chapter 16

Raymond M. Hozalski, University of Minnesota, Minneapolis, Minn.

Edward J. Bouwer, The Johns Hopkins University, Baltimore, Md.

Chapter 17

Dennis Clifford, University of Houston, Houston, Texas

Philip Kim, Boyle Engineering Corporation, Bakersfield, Calif.

Paul Fu, CH2M-Hill, Santa Ana, Calif.

Chapter 18

Douglas M. Owen, Malcolm Pirnie, Inc., Carlsbad, Calif.

Chapter 19

Ronald F. Packham, Marlow Bucks, United Kingdom

CHAPTER · ONE

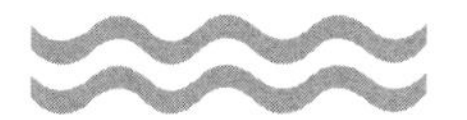

Disinfection By-Products: A Historical Perspective

JAMES M. SYMONS

The reactions of chlorine with organic material in water have been known for some time, e.g., chlorine and phenol produces chlorophenol tastes in water (Burttschell et al. 1959), chlorine with extracellular algal products causes off-flavors, and chlorine-substituted organic compounds are formed in chlorinated wastewater (Jolley 1973). The identification of "volatile" halogen-substituted organics eluded analysts, however, until the work of Rook (1971). This finding touched off a flurry of research and regulatory activity that has not ceased to this day, and likely will continue well into the future.

Although occasionally some more current work is included, this historical review stops just before the development of the Strawman Rule for Disinfectants and Disinfection By-Products (USEPA 1989). The release of that proposed rule was the beginning of "round two" of the US Environmental Protection Agency's (USEPA's) attempt to control the formation of disinfection by-products (DBPs). Further discussion of USEPA's more recent regulatory efforts are presented in chapter 7.

This book presents state-of-the-art knowledge on particular subjects relating to DBP formation in water. This initial chapter gives a general perspective of the overall subject, with the details on a given subject to follow in subsequent chapters.

EFFECT OF EARLY ANALYTICAL METHODOLOGY

To understand why chloroform and related compounds were not frequently reported in drinking water prior to the publications of Rook (1974) and Bellar and Lichtenberg (1974), a review of the analytical techniques of the late 1960s and early 1970s for the measurement and detection of organics in water is necessary. In most cases, then as now, concentrating the organics before analysis was required. Two concentrating methods were common. The carbon adsorption method (CAM) involved the adsorption of the organics onto granular activated carbon (GAC), followed by desorption with chloroform, evaporating the chloroform, and weighing the extract (Anon. 1962). The liquid–liquid extraction method (LLE) involved mixing the water sample with an immiscible solvent, withdrawing the solvent layer, and evaporating the solvent to a very small volume to further concentrate the organics prior to gas chromatographic (GC) analysis (American Society of Testing and Materials [ASTM] 1973). In the first method, the extract was contaminated with the extraction solvent, chloroform, and in the second method, any chloroform in the original water sample was lost while heating to evaporate the solvent. Thus, neither analytical method could readily detect chloroform in the original sample.

A pioneer in trihalomethane (THM) analysis and study, Rook started working at the Rotterdam Waterworks in 1963. Rook used his experience from a previous position in a brewery to adapt a headspace analysis method used to measure off-flavors in beer to study taste-and-odor problems in Rotterdam drinking water. This work resulted in a 1971 Dutch paper describing the method (Rook 1971). The chloroform peak appeared as the largest peak after chlorination of the water (Rook 1997).

The first paper in English concerning this work was in 1972 (Rook 1972). In the headspace analysis technique, the sample was heated to 60°C and the gas obtained in a 12-hour period was adsorbed on a small amount of activated silica. This concentrate could then be separated by GC.

In 1971, Bellar, working at the USEPA research laboratory in Cincinnati, Ohio, was given the assignment of measuring the concentration of volatile organic contaminants (VOCs) in wastewater. The direct aqueous injection method for GC analysis was not adequate (ASTM 1973). Because his background was in air pollution, Bellar adapted an air sampling method (Bellar and Sigsby 1970) called the *purge-and-trap method* (Bellar and Lichtenberg 1974). Because no solvents were involved, volatile contaminants such as chloroform could be detected.

In mid-1973, Bellar decided to sample other drinking waters and to perform the key experiment of sampling several points at the Cincinnati water treatment plant. The analysis of the Cincinnati samples demonstrated that chloroform was not present in the source water, the Ohio River, but increased steadily in concentration following source water chlorination.

Bellar was not the first to find chloroform in US drinking water. McAuliffe (1971), using headspace analysis, reported chloroform as one contaminant he found in tap water. Further, in 1972, USEPA used a CAM sampling technique to analyze tap water in New Orleans, La. (Friloux 1971; USEPA 1972). Here, however, instead of desorbing the organics off the GAC with chloroform, heat was used. Thus, the extract was not contaminated with chloroform, and chloroform was found in the finished water. Because the source water was not tested in either of these cases, the formation of chloroform in the treatment plant was not recognized. Instead, the chloroform was assumed to have been present in the source water.

INFLUENCE OF EARLY STUDIES AND SURVEYS

Rook (1974) believed that natural organic matter (NOM), as measured by color, was the precursor to the formation of THMs. His work showed a strong relationship between source water color and chloroform levels when source water chlorination was practiced. Rook found that both colored natural water free of industrial pollution and peat soaked in distilled water produced chloroform on chlorination. Previous work (Booth and Saunders

1950) on the chlorination of substituted quinones—moieties Rook expected were similar to those found in natural color—greatly influenced his thinking. The salient idea was to link the resorcinol structure to the still-not-too-well-defined structure of humic substance (Rook 1997).

Based on Rook's findings, USEPA in Cincinnati decided that THM formation was an important issue that needed prompt study. In November 1974, USEPA announced the start of a nationwide survey of source and finished waters. The study would eventually be called the National Organics Reconnaissance Survey (NORS). In the study, 80 locations were selected that represented a wide variety of source water qualities and treatment practices. Of these 80 locations, only one used ozone; the other locations used either free chlorine or chloramines. The plan was to analyze source and finished water (point of entry to the distribution system) for the four THMs and two other VOCs that were suspected to form during disinfection. The other two VOCs were carbon tetrachloride and 1,2-dichloroethane. The samples were analyzed in Cincinnati using the purge-and-trap method, followed by GC.

The NORS started in January 1975, and the results were available Apr. 18, 1975. What made this timeliness remarkable was that largely uncharted territory was being explored, both analytically and in understanding the problem. Among other results, the data showed the following five major points (Symons et al. 1975):

1. All 79 locations that chlorinated or chloraminated had THMs in various concentrations in their tap water. Also, where carbon tetrachloride or 1,2-dichloroethane or both were present, their concentrations did not increase during treatment.

2. Chloramination produced lower concentrations of THMs than free chlorination.

3. In most cases, the most abundant THM present was chloroform. A frequency distribution plot of all of the data produced median values of 21 µg/L for chloroform, 6 µg/L for bromodichloromethane, 1.2 µg/L for dibromochloromethane, and below the detection limit (BDL) for bromoform.

4. In some locations, the THM concentration pattern noted in item 3 above did not hold. For example, Brownsville, Texas, had the following concentrations: chloroform—12 μg/L, bromodichloromethane—37 μg/L, dibromochloromethane—100 μg/L, and bromoform—92 μg/L.
5. Although Miami, Fla., had the highest chloroform concentration—311 μg/L, Huron, S.D., had the highest total trihalomethane (TTHM) concentration—482 μg/L, compared to Miami's 427 μg/L. Note: the original publication summed the THMs correctly as μmol/L.

In addition to the basic data and their relationship to disinfection practice, attempts were made to correlate TTHMs to various group parameters that had been measured. The NORS data in Figure 1-1 show the relationship of TTHM concentration to nonpurgeable organic carbon (NPOC) concentration.

USEPA in Cincinnati used a two-pronged approach to further study this issue. One approach was for the chemists to attempt to understand the THM formation reaction more fully. The second approach was for the engineers to investigate methods of control. The chemists, headed by Alan Stevens, began to revisit some of Rook's work and attempted to determine how the THM formation reaction was influenced by (1) precursor type and precursor concentration, (2) disinfection temperature, (3) form of the disinfectant, (4) pH of disinfection, and (5) bromide ion. Tom Love, who headed the control studies, developed the following conceptional model of THM formation:

$$\text{Chlorine} + \text{Precursors} \rightarrow \text{THMs} \qquad (1\text{-}1)$$

as a simple method of visualizing the three approaches for controlling THM concentrations, i.e., (1) remove the THMs, (2) remove the precursors, or (3) change disinfectants (Love et al. 1975).

Based in part on the work of the USEPA chemists, Stevens and his colleagues concluded that the other important aspect of these experiments was the dramatic change in the rate of chloroform formation when the results of source and settled water chlorination were compared (Figure 1-2). Conventional alum coagulation caused the removal of most of the precursor material from the source water (Stevens et al. 1976).

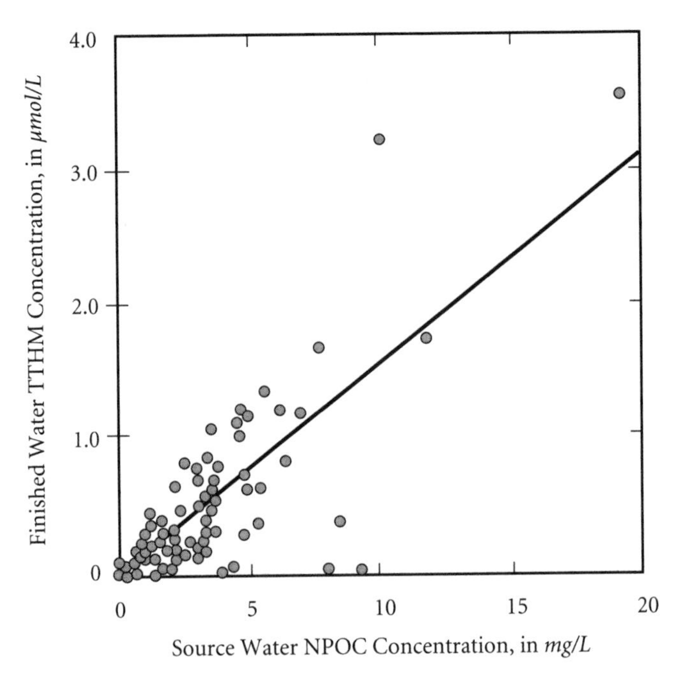

Source: Symons et al. (1981).

Figure 1-1 Raw water nonpurgeable organic carbon versus finished water total trihalomethanes (Data obtained from the National Organics Reconnaissance Survey)

Because the treatment practice at the Cincinnati Waterworks at that time was to add chlorine and alum coagulant to Ohio River water just prior to entering 2 to 3 days of storage in off-stream reservoirs, the utility decided to delay chlorination until after the storage reservoirs to allow precursors to be removed prior to chlorination (Bolton 1977). The point of chlorination was moved on July 14, 1975, and this change had a dramatic effect on chloroform concentrations in the distributed water (Figure 1-3) (Symons et al. 1981).

Although the elimination of source water chlorination in Cincinnati significantly lowered chloroform formation, this experiment was the first demonstration of a commonly seen pattern during THM control experiments, i.e., the limited control of bromine-substituted THMs as compared to chloroform when THM precursors were removed.

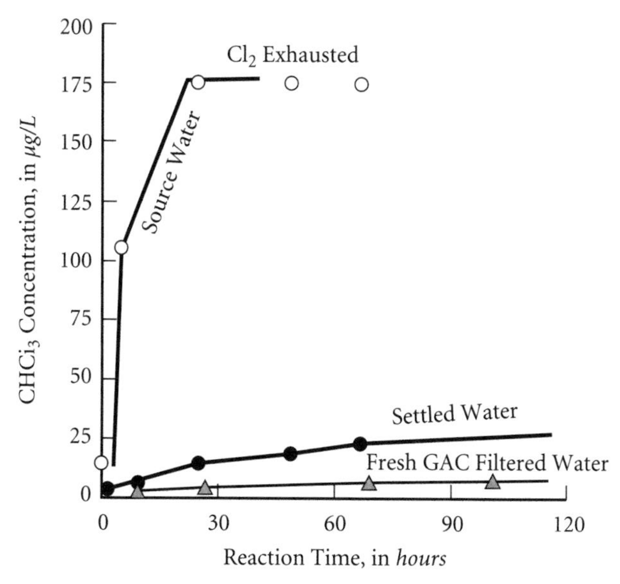

Source: Stevens et al. (1976).

Figure 1-2 Effect of treatments on chloroform production
8 mg/L chlorine dose, 25°C, pH 7

Because the change in chloroform concentration was so dramatic, many expected that the simple and inexpensive solution to THM formation during conventional water treatment would be to move the point of first chlorination to after coagulation and settling, as was demonstrated in the laboratory by Babcock and Singer (1979) and on a plant scale in Durham, N.C. (Young and Singer 1979). The success of this change in treatment practice did vary, however, depending on the amount and nature of the precursors, detention times in the coagulation basin, and coagulation conditions (Symons et al. 1981).

In the fall of 1975, the first Water Chlorination—Environmental Impact and Health Effects conference convened (Oct. 22–24, 1975). This and the five conferences that followed affectionately became known as the "Jolley Conferences" after their founder Robert L. Jolley of Oak Ridge National Laboratory. They were *the* place to present one's research on various aspects of water chlorination down through the years. Researchers both inside and outside USEPA made ample use of this technical forum, the last one being held in 1987.

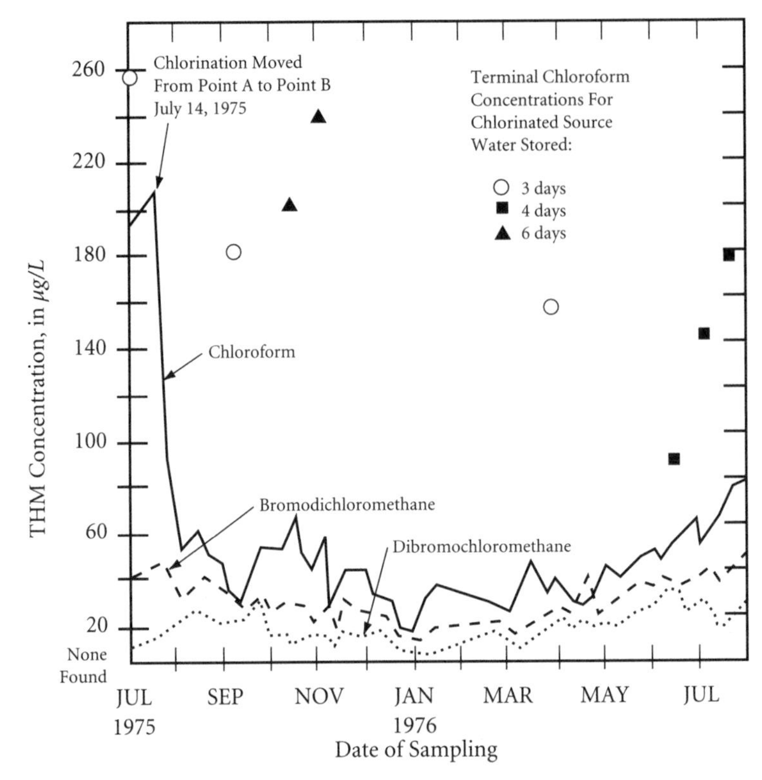

Source: Symons et al. (1981).

Figure 1-3 Trihalomethane concentrations in Cincinnati, Ohio, tap water

The reader is referred to the proceedings of these conferences as a storehouse of research findings.

On Dec. 24, 1975, a notice appeared in the *Federal Register* (1975) mandating a special study. This special study was subsequently named the National Organics Monitoring Survey (NOMS). NOMS repeated and expanded the NORS work, by taking several samples over time at each location to obtain some insight on the seasonal variation of organic contaminants (Brass et al. 1977).

The USEPA chemists in Cincinnati performed another in-house study involving bromide ion. Here, several solutions of commercial humic acid were chlorinated with the same concentration of chlorine, but different concentrations of spiked bromide

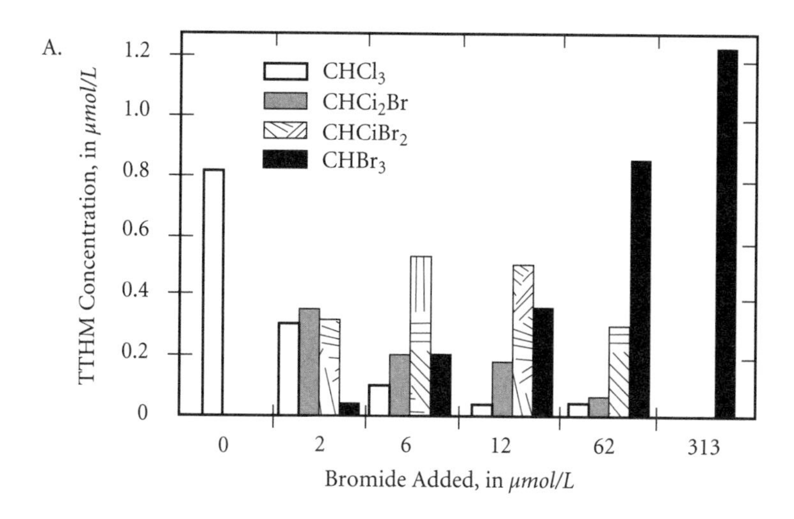

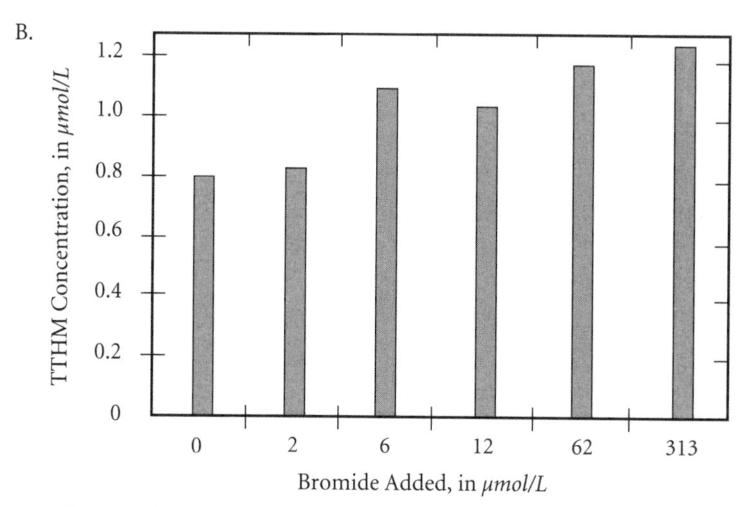

Source: Symons et al. (1981).

Figure 1-4 Trihalomethanes formed by reaction of humic acid with aqueous chlorine in the presence of varying bromide ion concentrations using a chlorine dose of 282 microequivalents per litre

ion. Not only did they observe the same enrichment of bromine in the resulting THMs as had Rook (1974), but they also showed a shift in speciation (Figure 1-4A) and an *increase* in the μmol/L concentration of TTHM as the added bromide ion concentration was increased (Figure 1-4B) (Symons et al. 1981). This provided new insight into the importance of bromide ion in THM formation. More recently, a similar effect has been demonstrated in another class of DBPs, the haloacetic acids (HAAs) (Pourmoghaddas et al. 1993; Cowman and Singer 1996).

Following the Love model (Eq 1-1), one method to control THM concentrations was to remove precursors. To assess the performance of any unit process being studied for possible precursor removal, however, the precursors themselves had to be measured. Precursors could not be measured directly, because in any given sample, it is an unknown mixture of the organic matter present in that sample. Surrogates were investigated (e.g., NPOC, ultraviolet [UV] absorbance, fluorescence, and chlorine demand) (Symons et al. 1975), and although higher values of the surrogates generally predicted higher values of THMs resulting from chlorination, the scatter in the data was too great for surrogates to be any more than a qualitative predictor of resultant TTHMs (see Figure 1-1 for NPOC data).

In considering this problem, Stevens had the idea of measuring THM precursors by determining the *potential* for THM formation in any given sample by chlorinating it, allowing the precursors to be converted to THMs, and then measuring the resulting THM concentrations (Stevens and Symons 1977). The research that led to the development of this test, the trihalomethane formation potential (THMFP) test, showed that in any given sample, a small percentage of the carbon present in the organic matter participated in the THM reaction. Furthermore, this percentage was strongly influenced by the pH, temperature, and reaction time of the test. Studies on the influence of the chlorine residual at the end of the test and the bromide ion concentration came later and were summarized by Symons et al. (1993). The Stevens and Symons (1977) paper was the basis for Standard Method 5710B (APHA, AWWA, and WEF 1995).

A modification of Standard Method 5710B (APHA 1995) called the uniform formation conditions (UFC) test was proposed by Summers et al. (1996). In this test, the incubation conditions were prescribed for all samples, as in 5710B, but the conditions were milder, i.e., 24-hour free chlorine residual of 1 ± 0.4 mg/L, pH 8 ± 0.2, temperature 20 ± 1°C, and incubation time 24 ± 1 hour.

Standard Method 5710 (APHA 1995) also contains a modification of the precursor test, the simulated distribution system (SDS) test (5710C) (Miller and Hartman 1982). Here, rather than adding chlorine to the sample as in the THMFP test, a finished water sample, either actual or ersatz, is held at the temperature of the distribution system being investigated for the time of travel to the end of the distribution system, or to some intermediate point being studied. The resulting DBP concentrations then predict the DBP concentrations in the tap water at the site under study (Koch et al. 1991).

In 1978, the first effort to determine a kinetic model for the THM reaction was started (Trussell and Umphres 1978). Good fits with experimental data were obtained, but the authors concluded that more work needed to be done.

DEVELOPMENT OF THE THM RULE

On Mar. 1, 1976, a National Cancer Institute (NCI) report on the carcinogenesis bioassay of chloroform was released (NCI 1976). In this report, chloroform was classified as a suspected human carcinogen. Although the study on which this conclusion was based has not held up over time, it had a big impact in 1976, and chloroform is still regulated as a human carcinogen.

Based on the information at hand and the NCI report on carcinogenesis, on July 19, 1976, USEPA put an Advanced Notice for Proposed Rule Making (ANPRM) in the *Federal Register* (1976). This was an announcement that USEPA was considering regulating THMs, was proposing some optional approaches, and was asking for public comment prior to drafting a proposed rule. The proposed options were: "I. Maximum Contaminant Level (MCL); A. Establish MCLs for Specific Organic Chemicals, B. Establish MCLs for general organic indicators, C. Combinations of MCLs for specific compounds and general organic indicators, or II. Designated Treatment Techniques to Control Either Specific Contaminants (e.g. Trihalomethanes) or Total Organics [as measured by total organic carbon (TOC)]." The ANPRM started the DBP regulatory process.

Two years later, the proposed THM regulation was published in the *Federal Register* (1978). The proposed THM regulation on organic contaminant control was to establish an MCL for TTHMs. Total trihalomethanes were defined as the arithmetic sum of the concentrations of the four common THMs on a mass basis, rather than on a molar basis. The MCL was 0.10 mg/L (incorrectly equated to "100 parts per billion" in the proposed regulation) and it would apply to water utilities serving 75,000 people or more. The use of micromolar concentrations for reporting THMs was not accepted by USEPA's upper management because of the difficulty of explaining micromolar concentrations to lay people. In the proposed rule, utilities serving from 10,000 to 75,000 people would have to monitor for one year. Utilities serving a population of less than 10,000 would be exempt from this regulation and would be regulated at the discretion of the primacy agency.

The proposed TTHM MCL was selected as a compromise. It was a concentration that was expected to be achievable at a reasonable cost and that would provide some protection against the perceived health risk in locations where TTHM concentrations were relatively high, but would not force utilities to alter their treatment practices too much. As experience with meeting this regulation was developed, many expected future MCL values to be lowered. The smaller locations were proposed to be excluded for the following reason. The USEPA was concerned that if the smaller utilities tried to alter their disinfection practice to lessen TTHM concentrations, because of a lack of technical expertise, an increased risk of microbial contamination in the finished water might result.

In 1979, to support the forthcoming TTHM regulation, USEPA in Cincinnati was gathering all of the experimental data on THMs that it and others had collected (Symons et al. 1981). This peer-reviewed book for a long time served as the single main repository for THM information.

Following the Love model (Eq 1-1), this book contained studies on (1) removal of THMs by air stripping, adsorption on GAC and polymeric resin, and oxidation; (2) removal of precursors by clarification, adsorption on GAC, oxidation, and oxidation and biodegradation; and (3) the behavior of alternate disinfectants, including chlorine dioxide, chloramines, ozone, and bromine chloride. In addition, chapters were included on the formation reaction, the THMFP test, the effect THM control had on microbiological quality, and the costs of various treatments.

As the first decade of the DBP era ended, the final rule on THM control was promulgated on Nov. 29, 1979 (*Federal Register* 1979). The summary noted that USEPA had received 391 written comments on the proposed THM regulation (*Federal Register* 1978) and had heard from 157 witnesses during the public hearings. Of the comments received, 306 out of the 391 expressed concern "...for the basis of health effects data that support the proposed THM regulations."

The major change USEPA made in response to the comments about the proposed THM regulation (*Federal Register* 1978) related to applicability. The final rule was to cover all water utilities serving 10,000 or more, although the effective date was two years from promulgation for systems serving more than 75,000 and four years for the smaller utilities. The printed document was 83 pages long (*Federal Register* 1979), but the actual THM regulation was only two of those 83 pages.

The second decade of the DBP era started with the American Water Works Association (AWWA); the City of Englewood, Colo.; and the Capital City Water Company filing a petition for review, asking the court for "a review of the final rule" on Jan. 11, 1980. The outcome of this court challenge was to modify the THM rule (*Federal Register* 1983), including, for the first time, a codification of best available technology (BAT) for THM control. The BAT consideration is now included in all USEPA drinking water regulations.

DEVELOPMENT OF THE DBP RULE

The next set of technical developments were in two areas. The first area was the development of analytical techniques used to show the formation of many more DBPs than the four THMs, and the second area involved studies to show the effectiveness and impact of water treatment modifications on drinking water quality that occurred after the THM regulation went into effect.

The Search for Additional DBPs

Individual compounds. Research to look for additional by-products from chlorination other than THMs began with Dr. Russell L. Christman of the University of North Carolina and his team's use of the most powerful analytical techniques available at that time. The work, involving derivatization of polar compounds and subsequent GC/mass spectrometry (MS) analysis, identified another group of DBPs, the HAAs, that formed during chlorination. The HAAs often formed in concentrations close to or greater than the THMs (Christman et al. 1979; Norwood et al. 1980). Using Standard Method 6251 (APHA, AWWA, and WEF 1995)—based on the early work of Quimby et al. (1980) and modified by others since—these analytes are commonly measured, and, based on subsequent animal toxicology studies (see chapter 5), are being regulated (*Federal Register* 1998). This same team expanded the list of halogen-substituted by-products even further (Christman et al. 1983). That same year, Uden and Miller (1983) sampled two water supplies in an attempt to identify chlorinated acids and choral using microwave plasma emission as a chlorine-selective detector following GC.

Using LLE with pentane, Trehy and Beiber (1980) were able to demonstrate the formation of haloacetonitriles (HANs) from the chlorination of several amino acids and natural water. They speculated that the structures of humic substances include amino acid residues related to these amino acids. Trehy and Beiber (1980) also found that HANs were destroyed by the dechlorination agents used for THM samples. Currently, to make a "complete" analysis of the known DBPs, several samples must be collected, each preserved in a different manner.

Hemming et al. (1986) identified another disinfection by-product in chlorinated water. This compound, 3-chloro-4-(dichloromethyl)-5-hydroxy-2(5H)-furanone, was nicknamed MX as it is an extremely potent mutagen in bacterial systems and appears to be responsible for a substantial portion (30 to 60 percent) of the mutagenicity of chlorinated drinking water (Meier, Blasak, and Konhl 1987).

In 1987, Stevens presented the results of intensive sampling of finished water at 10 water utilities (Stevens et al. 1990). These samples were analyzed for 22 DBPs, including four THMs, four HANs, two HAAs, two haloaldehydes, five haloketones (HKs), three chlorophenols, chloropicrin, and cyanogen chloride (CNCl). In addition, mass spectra that were obtained from these samples were compared with their library of 782 mass spectra derived from the chlorination of commercial fulvic acid. They concluded that approximately one half of the compounds in the test list of 22 appear to be of significance in the water supplies tested. Furthermore, they noted that 196 compounds that can be attributed to the chlorination process have been found in one or more of the 10 utilities' finished water.

A group parameter. As the list of halogen-substituted DBPs grew longer, interest developed in a group parameter that would measure as many of the halogen-substituted DBPs as possible in a single measurement. One possibility was a method of measuring the quantity of organic halogen (OX) adsorbed on granular activated carbon (adsorbable organic halogen—AOX) (Kühn and Sontheimer 1973).

This work evolved into an analytical method for measuring total organic chlorine (TOCl) in water by adsorbing the TOCl on powdered activated carbon, pyrolyzing the activated carbon, and measuring the chloride ion in condensed combustion vapors (Kühn and Sontheimer 1975; Dressman, McFarren, and Symons 1977; Takahashi 1979; Dressman, Najar, and Redzikowski 1979). This became Standard Method 5320 (APHA, AWWA, and WEF 1995) called *dissolved organic halogen* (DOX). The DOX is a group parameter that is a surrogate measurement of all of the halogen-substituted compounds in a sample that can be recovered by this technique. Because no single standard is available, accuracy cannot be determined and the method reports all halogens as chloride. Additionally, recovery of known compounds show that some are rather poorly recovered.

Over the years, many researchers became interested in the problem of how much of the total halogen-substituted DBPs formed as measured by Standard Method 5320 were being measured as individual analytes. Christman et al. (1983) and Reckhow and Singer (1984) demonstrated that the halogen-substituted DBPs that could be measured did not account for all the OX in a sample as measured by Standard Method 5320 (termed TOX, or total organic halogen).

Based on their work at six utilities, Singer and Chang (1989) concluded that the TTHMs accounted for only about 26 percent of the TOX, the remainder of the halogen-substituted DBPs being compounds other than THMs. Six years later, Singer and his co-workers sampled eight water utilities in North Carolina (Singer, Obelensky, and Greiner 1995). This time they compared the chlorine-equivalent concentrations of the total of three THMs (no bromoform was detected), four HAAs, two HANs, two HKs, and chloropicrin to TOX. Again, only about one third of the TOX could be accounted for by summing these 12 different individual DBPs. In 1990 studies, Stevens et al. (1990) and Reckhow and Singer (1990) identified approximately 30 to 60 percent of the TOX with the target halogen-substituted DBPs they measured.

Many of the by-products that result from the chlorination reaction are still unknown as of this writing. This is probably because they are too polar or too high in molecular weight to be detected by conventional GC techniques. In the Disinfectants and Disinfection By-Products (D/DBP) rule (*Federal Register* 1998), USEPA is planning to control the concentrations of all halogen-substituted DBPs by regulating the concentrations of TTHM, HAA5, and controlling the concentration of precursor as measured by TOC (see chapter 7).

Effect of the THM Rule on Water Quality

After the THM rule became effective in 1979, some water utilities had to make changes in their practices to come into compliance. To determine the effect on finished water quality, the American Water Works Association Research Foundation (AWWARF) sponsored a questionnaire survey sent to 1,255 of the 3,081 utilities serving 10,000 customers or more. Seventy-three percent of the surveys were completed and returned (McGuire and Meadow 1988). The results showed that enactment of the THM regulation resulted (on average) in a 40 to 50 percent lessening in TTHM concentrations for the larger utilities surveyed. The median TTHM concentration among all the respondents was 38 μg/L, a value not much different than the results of USEPA's NORS (Symons et al. 1975) or NOMS (Brass et al. 1977). Although the median concentration was not influenced much, utilities with high TTHM levels were able to lessen their TTHM concentrations substantially.

Of those systems that implemented THM control measures, the majority did one or more of the following: (1) modified their point(s) of chlorine application, (2) changed their chlorine dosages, and (3) adopted the use of chloramines to ensure compliance with the MCL for TTHMs. A significant number of utilities changed from free chlorine to chloramines as part of their disinfection practices. Compliance with the 0.10 mg/L MCL was not found to be particularly costly. Nevertheless, USEPA was concerned that these changes in disinfection practice might be influencing the microbial safety of tap water. These concerns were partly responsible for the development of the Surface Water Treatment Rule (SWTR) (*Federal Register* 1987a, 1989a) and the Total Coliform Rule (TCR) (*Federal Register* 1987b, 1989b).

The following year, the Metropolitan Water District of Southern California conducted a survey of 35 water utilities (10 in California and 25 in the rest of the United States), measuring DBPs in finished water quarterly for a year, starting in the spring of 1988 (Krasner et al. 1989). They tested for four THMs, five HAAs, four HANs, two HKs, chloropicrin, chloral hydrate, CNCl, formaldehyde, acetaldehyde, and 2,4,6-trichlorophenol. In addition, TOX concentrations were determined. The median TTHM concentration was 39 µg/L, very similar to the three surveys cited above. The median of the sum of the five HAA concentrations on a mass basis was 19 µg/L. HAA5 represented the second largest class of DBPs in these waters, confirming the earlier work (Christman et al. 1979; Norwood et al. 1980), and influencing USEPA's future decision to propose regulating this class of DBPs (*Federal Register* 1994).

Another important finding was that in locations that chloraminated, CNCl concentrations were significantly higher than for systems using free chlorine. Although the National Academy of Sciences (NAS) had reviewed the DBPs formed from chlorine dioxide and ozone (NAS 1980), and Sontheimer (1979) had shown some halogen substitution from chloramines, the finding of Krasner et al. (1989) was the first to show the formation of a specific DBP from chloramination, a popular substitute for free chlorine when THM control was required.

In addition to the organic ozone by-products cited by NAS (1980), Haag and Hoigné (1983) demonstrated the formation of bromate ion when ozonating waters containing bromide ion. Kurokawa et al. (1990) demonstrated the carcinogenicity of bromate ion, and USEPA is regulating it (*Federal Register* 1998). These findings and those that followed, as detailed in subsequent chapters in this book, raised concerns that using an alternative to free chlorine for disinfection might be just trading one problem for another.

Krasner et al. (1989) also found that certain DBPs (i.e., HANs, HKs, chloral hydrate, and CNCl) were not stable in a distribution system where the pH was relatively high (e.g., pH 9). Finally, the molar sum of the 19 halogen-substituted DBPs measured was correlated to the molar concentration of TOX. Although no line of best fit was presented in the paper, an estimated line appears to have a slope of about 0.3. This confirmed the previously cited work of Singer and co-workers (Singer and Chang 1989; Singer, Obelensky, and Greiner 1995) and, to some degree, Stevens et al. (1990) and Reckhow and Singer (1990) who showed that measurable halogen-substituted DBP concentrations accounted for 30 to 60 percent of the measured TOX concentration.

RELATED USEPA REGULATIONS

Several recent USEPA regulations are linked and as water utilities attempt to comply with them all, their efforts are likely to influence the resulting DBP concentrations in finished and distributed water. As mentioned previously and noted by Singer (1993), the SWTR, which was first proposed in 1987 (*Federal Register* 1987a) and promulgated in 1989 (*Federal Register* 1989a), arose from growing concerns about the presence of pathogenic microorganisms in surface water supplies and the inadequacy of existing treatment and disinfection practices to properly remove and inactivate many of these organisms, particularly the cysts of *Giardia lamblia*. Also, as noted previously, concerns arose that many of the practices adopted by utilities to control THM levels may have compromised disinfection effectiveness.

A major part of the SWTR is the concept of controlling disinfection by the $C \times T$ product. This technique specifies that a sufficient concentration (C) of disinfectant must be present in the water for a sufficient contact time (T) to ensure an adequate degree of inactivation of the organism in question. To achieve the degree of inactivation specified in the rule, in many water treatment plants either C, T, or both had to be increased, with a corresponding effect on DBPs. Another key component of the SWTR is the requirement that the residual disinfectant concentration in the water entering the distribution system not be less than 0.2 mg/L for more than 4 hours, and that a measurable amount must be present in at least 95 percent of all samples collected in the distribution system. This applies to both filtered and unfiltered systems. Subsequently, USEPA proposed an Enhanced SWTR (ESWTR) with potentially more stringent disinfection requirements for *Giardia*, as well as requirements for the control of *Cryptosporidium* (*Federal Register* 1994), see chapter 7. The disinfection requirements in these two rules will influence the extent of DBP formation that will occur in a given location.

A third rule, the Total Coliform Rule (*Federal Register* 1987b, 1989b), also mandated greater attention to disinfection, with likely corresponding implications for DBP formation. A fourth rule, the Lead and Copper Rule (*Federal Register* 1991), required that utilities distribute "noncorrosive" water. Frequently this has meant that the pH of distributed water needed to be increased. This change in pH affects the chemistry of DBP formation (see chapters 2 and 3), as well as the effectiveness of some disinfectants. The final additional rule that will have an effect on DBP formation is the impending Groundwater Disinfection Rule (*Federal Register* 1992a).

THE RELATIVE RISK DILEMMA

By now, the interrelationship between microbial safety and chemical safety was becoming apparent to USEPA. No relaxation of microbial quality could be tolerated and better microbial quality was being sought. Any mandated increase in disinfection level might, however, decrease chemical water quality. This trade-off is discussed in detail in chapter 6.

The USEPA's attempt to balance the microbial and chemical risks in drinking water proved to be beyond their capability in the early 1990s. Its inability to accomplish this balance led the agency to adopt a negotiated-rulemaking (reg neg) approach to controlling DBP concentrations in US drinking water (*Federal Register* 1992b). Details of the reg-neg process are presented in chapter 7.

SUMMARY

This chapter reviews the research and surveys that provided the groundwork for the intensive research activities on DBPs in the 1990s. In more recent years, great accomplishments have been made in the areas of analytical capabilities, both of the DBPs themselves and their precursors; health effects of the DBPs; regulatory approaches; the chemistry of various alternative disinfectants; treatment for precursor removal; and the costs of the various control approaches. Each subsequent chapter in this book reviews a separate topic related to the DBP issue. The background information provided in this chapter offers a framework for those topics.

ACKNOWLEDGMENTS

To gather information beyond my own files for this chapter, I had contact with many of the individuals who were involved with the work cited herein. I wish to thank all of them who gave generously of their time to help me. In particular, thanks to J.J. Rook, PhD, and Tom Bellar, who generously helped me fill in some of the details; Fred Pontius and Herb Brass, who helped me with some of the references; and Philip C. Singer, PhD, and Stuart Krasner for their editorial guidance and suggestions. I also want to thank the University of Houston Environmental Engineering Program graduate students Anthony Tripp, Timothy Nedwed, PhD, and Julie M. Davis who gave generously of their time to review the original manuscript.

REFERENCES

Anon. 1962. Tentative Method for Carbon Chloroform Extract (CCE) in Water. *Jour. AWWA*, 54(2):223–227.

American Public Health Association, American Water Works Association, and Water Environment Federation. 1995. *Standard Methods for the Examination of Water and Wastewater*, 19th ed. Washington D.C.: APHA.

ASTM. 1973. Tentative Recommended Practice for Measuring Volatile Organic Matter in Water by Aqueous-Injection Gas Chromatography. *Annual Book of ASTM Standards*, Pt. 23, Water ASTM D 2908-70T. American Society for Testing and Materials.

Babcock, D.B., and Singer, P.C. 1979. Chlorination and Coagulation of Humic and Fulvic Acids. *Jour. AWWA*, 71(3):149–152.

Bellar, T.A., and J.J. Lichtenberg. 1974. Determining Volatile Organics at Microgram-per-Litre Levels by Gas Chromatography. *Jour. AWWA*, 66(12):739–744.

Bellar, T.A., and J.E. Sigsby. 1970. Non-Cryogenic Trapping Techniques for Gas Chromatography. Internal Report. Research Triangle Park, N.C.: US Environmental Protection Agency, Div. of Chem. and Physics.

Bellar, T.A., J.J. Lichtenberg, and R.C. Kroner. 1974. The Occurrence of Organohalides in Chlorinated Drinking Water. *Jour. AWWA*, 66(12):703–706.

Bolton, C.M. 1977. Cincinnati Research in Organics. *Jour. AWWA*, 69(7):405–406.

Booth, H., and B.C. Saunders. 1950. *Chemy Ind.*, p. 824.

Brass, H.J., M.A. Feige, T. Halloran, J.W. Mello, D. Munch, and R.F. Thomas. 1977. The National Organic Monitoring Survey: Sampling and Analyses for Purgeable Organic Compounds. In *Drinking Water Quality Enhancement Through Source Protection*. Pojesek, R.B., ed. Ann Arbor, Mich.: Ann Arbor Science Publishers.

Burttschell, R.H., A.A. Rosen, F.M. Middleton, and M.B. Ettinger. 1959. Chlorine Derivatives of Phenol Causing Taste and Odor. *Jour. AWWA*, 51(2):205–214.

Christman, R.F., D.S. Norwood, D.S. Millington, J.D. Johnson, and A.A. Stevens. 1983. Identity and Yields of Major Halogenated Products of Aquatic Fulvic Acid Chlorination. *Environ. Sci. & Tech.*, 17(10):625–628.

Christman, R.L., J.D. Johnson, D.L. Norwood, W.T. Liao, J.R. Hass, F.K. Pfaender, M.R. Webb, and M.J. Bobenreith. 1979. Chlorination of Aquatic Humic Substances. Final Report of USEPA Project R-804430, EPA 600/2-81-016, NTIS Accession No. PB 81-161952. Cincinnati, Ohio: US Environmental Protection Agency.

Dressman, R.C., B.A. Najar, and R. Redzikowski. 1979. The Analysis of Organohalides (OX) in Water as a Group Parameter. In *Proc.* 1979 *AWWA Water Quality Technology Conference*. Denver, Colo.: American Water Works Association.

Dressman, R.C., E.F. McFarren, and J.M. Symons. 1977. An Evaluation of the Determination of Total Organic Chlorine (TOCl) in Water By Adsorption onto Ground Granular Activated Carbon, Pyrohydrolysis, and Chorine (sic) Ion Measurement. In *Proc.* 1977 *AWWA Water Quality Technology Conference*. Denver, Colo.: American Water Works Association.

Federal Register. 1975. National Interim Drinking Water Regulations: Subpart E - Special Monitoring Regulations for Organic Chemical. *Fed. Reg.*, 40(248):59587–59588.

———. 1976. Organic Chemical Contaminants; Control Options in Drinking Water. *Fed. Reg.*, 41(136):28991–28998.

———. 1978. Interim Primary Drinking Water Regulations; Control of Organic Chemical Contaminants in Drinking Water; Proposed Rule. *Fed. Reg.*, 43(28):5756–5780.

———. 1979. National Interim Primary Drinking Water Regulations; Control of Trihalomethanes in Drinking Water: Final Rule. *Fed. Reg.*, 44(231):68624–68707.

———. 1983. National Interim Primary Drinking Water Regulations; Trihalomethanes; Final Rule. *Fed. Reg.*, 48(40):8406–8414.

———. 1987a. National Primary Drinking Water Regulations; Filtration, Disinfection; Turbidity; *Giardia lamblia*, Viruses, *Legionella*, and Heterotrophic Bacteria; Proposed Rule. *Fed. Reg.*, 52(212):42178–42222.

———. 1987b. National Primary Drinking Water Regulations; Total Coliforms; Proposed Rule. *Fed. Reg.*, 52(212):42224–42245.

———. 1989a. National Primary Drinking Water Regulations; Filtration, Disinfection; Turbidity; *Giardia lamblia*, Viruses, *Legionella*, and Heterotrophic Bacteria; Final Rule. *Fed. Reg.*, 54(124):27486–27541.

———. 1989b. National Primary Drinking Water Regulations; Total Coliforms (including Fecal Coliforms and E. Coli); Final Rule. *Fed. Reg.*, 54(124):27544–27568.

———. 1991. Maximum Contaminant Level Goal and National Primary Drinking Water Regulation for Lead and Copper; Final Rule. *Fed. Reg.*, 56(110):26460–26564.

———. 1992a. Draft Ground Water Disinfection Rule Available for Public Comment. *Fed. Reg.*, 57(148):33960.

———. 1992b. Notice of Intent to Form an Advisory Committee to Negotiate the Drinking Water Disinfection By-products Rule and Announcement of a Public Meeting. *Fed. Reg.*, 57(179):41533.

———. 1994. National Primary Drinking Water Regulations; Enhanced Surface Water Treatment Requirements; Proposed Rule. *Fed. Reg.*, 59(145):38668–38829.

———. 1998. National Primary Drinking Water Regulations: Disinfectants and Disinfection By-Products; Notice of Data Availability; Proposed Rule. *Fed. Reg.*, 62(212):59388–59484.

Friloux, J. 1971. Petrochemical Wastes as a Pollution Problem in the Lower Mississippi River. USEPA Report. Baton Rouge, La.: Lower Mississippi Basin Office, Water Quality Office.

Haag, W.R., and J. Hoigné. 1983. Ozonation of Bromide-Containing Waters: Kinetics of Formation of Hypobromous Acid and Bromate. *Environ. Sci. & Tech.*, 17(5):261.

Health, Education, and Welfare. 1972. *Toxic Substances List.* Christensen, H.E., ed. Health Services and Mental Health Administration. Rockville, Md.: NIOSH.

Hemming, J., et al. 1986. Determination of the Strong Mutagen 3-Chloro-4-(dichloromethyl)-5-hydroxy-2(5H)-furanone in Chlorinated Drinking Water. *Chemosphere*, 15:549.

Jolley, R.L. 1973. Chlorination Effects on Organic Constituents in Effluents from Domestic Sanitary Sewage Treatment Plants. ORNL/TM-4290. Oak Ridge, Tenn.: Oak Ridge National Laboratory.

Koch, B., S.W. Krasner, M.J. Sclimenti, and W.K. Schimpff. 1991. Predicting the Formation of DBPs by the Simulated Distribution System. *Jour. AWWA*, 83(10):62–70.

Krasner, S.W., M.J. McGuire, J.G. Jacangelo, N.L. Patania, K.M. Reagan, and E.M. Aieta. 1989. The Occurrence of Disinfection By-Products in US Drinking Water. *Jour. AWWA*, 81(8):41–53.

Kühn, W., and H. Sontheimer. 1973. Einige Untersuchungen Zur Bestimmung von organische Clorverbundungen auf Aktivkohlen. *Vom Wasser*, 41:65–79.

———. 1975. Zur analytischen Erfassung organischer Chlorverbundungen mit der temperaturprogrammierten Pyrohydrolyse. *Vom Wasser*, 43:327–341.

Kurokawa, Y., et al. 1990. Toxicity and Carcinogenicity of Potassium Bromate. *Envir. Health Perspectives*, 87:309.

Love, O.T., Jr., J.K. Carswell, A.A. Stevens, and J.M. Symons. 1975. Treatment of Drinking Water for Prevention and Removal of Halogenated Organic Compounds (An EPA Progress Report). Presented at 1975 AWWA Annual Conference.

McAuliffe, C. 1971. Gas Chromatographic Determination of Solutes by Multiple Phase Equilibrium. *Chemical Technology*, 1:46.

McGuire, M.J., and R.G. Meadow. 1988. AWWARF Trihalomethane Survey. *Jour. AWWA*, 80(1):61–68.

Meier, J.R., W.F. Blasak, and R.B. Konhl. 1987. Mutagenic and Clastogenic Properties of 3-Chloro-4-(dichloromethyl)-5-hydroxy-2(5H)-furanone: A Potent Bacterial Mutagen in Drinking Water. *Enviro. & Molecular Mutagenesis*, 10:411.

Miller, R., and D.J. Hartman. 1982. Feasibility Study of Granular Activated Carbon Adsorption and On-Site Regeneration. EPA-600/2-82-087A, Cincinnati, Ohio.: US Environmental Protection Agency.

NAS. 1980. The Chemistry of Disinfectants in Water: Reactions and Products. In *Drinking Water and Health*. Washington, D.C.: National Academy of Sciences.

NCI. 1976. Report on Carcinogenesis Bioassay of Chloroform, Bethesda, Md., Technical Information Service No. PB 264018/AS.

Norwood, D.L., J.D. Johnson, R.L. Christman, J.R. Hass, and M.J. Bobenreith. 1980. Reactions of Chlorine with Selected Aromatic Models of Aquatic Humic Material. *Environ. Sci. & Tech.*, 14(2):187.

Pourmoghaddas, H., A.A. Stevens, R.N. Kinman, R.C. Dressman, L.A. Moore, and J.C. Ireland. 1993. Effect of Bromide Ion on the Formation of HAAs During Chlorination. *Jour. AWWA*, 85(1):82–87.

Quimby, B.D., M.F. Delaney, P.C. Uden, and R.M. Barnes. 1980. Determination of the Aqueous Chlorination Products of Humic Substances by Gas Chromatography with Microwave Emission Detection. *Anal. Chem.*, 52:259–263.

Reckhow, D.A., and P.C. Singer. 1984. The Removal of Organic Halide Precursors by Preozonation and Alum Coagulation. *Jour. AWWA*, 76(4):151–157.

———. 1990. Chlorination By-products in Drinking Waters: From Formation Potentials to Finished Water Concentrations. *Jour. AWWA*, 82(4):113–180.

Rook, J.J. 1971. Headspace Analysis in Water (translated title). H_2O, 4(17):385–387.

———. 1972. Production of Potable Water From a Highly Polluted Source. *Water Treatment and Examination*, 21:259–271.

———. 1974. Formation of Haloforms During the Chlorination of Natural Water. *Water Treatment and Examination*, 23:234–243.

———. 1997. Personal communication.

Singer, P.C. 1993. Trihalomethanes and Other By-Products Formed by Chlorination of Drinking Water. In *Keeping Pace with Science and Engineering*. Washington, D.C.: National Academy Press.

Singer, P.C., and S.D. Chang. 1989. Correlations Between Trihalomethanes and Total Organic Halides Formed During Water Treatment. *Jour. AWWA*, 81(8):61–65.

Singer, P.C., A. Obelensky, and A. Greiner. 1995. DBPs in Chlorinated North Carolina Drinking Waters. *Jour. AWWA*, 87(10):83–92.

Sontheimer, H. 1979. Effectiveness of Granular Activated Carbon for Organics Removal. In *Proc.* 1978 *AWWA Annual Conference*. Denver, Colo.: American Water Works Association.

Stevens, A.A., L.A. Moore, C.J. Slocum, B.K. Smith, D.R. Seeger, and J.C. Ireland. 1990. By-Products of Chlorination at Ten Operating Utilities. In *Water Chlorination - Chemistry, Environmental Impact and Health Effects*, Vol. 6. Jolley, R.L., et al., eds.

Stevens, A.A., and J.M. Symons. 1977. Measurement of Trihalomethane and Precursor Concentrations Changes Occurring During Water Treatment and Distribution. *Jour. AWWA*, 69(10):546–554.

Stevens, A.A., C.J. Slocum, D.R. Seeger, and G.G. Robeck. 1976. Chlorination of Organics in Drinking Water. *Jour. AWWA*, 68(11):615–620.

Summers, R.S., S.M. Hooper, H.M. Shukairy, G. Solarik, and D. Owen. 1996. Assessing DBP Yield: Uniform Formation Conditions. *Jour. AWWA*, 88(6):80–93.

Symons, J.M., T.A. Bellar, J.K. Carswell, J. DeMarco, K.L. Kropp, G.G. Robeck, D.R. Seeger, C.J. Slocum, B.L. Smith, and A.A. Stevens. 1975. National Organics Reconnaissance Survey for Halogenated Organics. *Jour. AWWA*, 76(11):634–647.

Symons, J.M., S.W. Krasner, L.A. Simms, and M. Sclimenti. 1993. Measurement of THM and Precursor Concentrations Revisited: The Effect of Bromide Ion. *Jour. AWWA*, 85(1):51–62.

Symons, J.M., A.A. Stevens, R.M. Clark, E.E. Geldreich, O.T. Love, Jr., and J. DeMarco. 1981. *Treatment Techniques for Controlling Trihalomethanes in Drinking Water*. EPA-600/2-81/156, NTIS Accession Number, PB 82163197.

Takahashi, Y. 1979. Analysis Techniques for Organic Carbon and Organic Halogen. In *Proc. EPA/NATO-CCMS Conference on Adsorption Techniques*. EPA 510/9-84-005. Washington, D.C.: USEPA.

Trehy, M.L., and T.I. Bieber. 1980. Effects of Commonly Used Water Treatment Processes on the Formation of THMs and DHANs. In *Proc. 1980 AWWA Annual Conference*. Denver, Colo.: American Water Works Association.

Trussell, R.R., and M.D. Umphres. 1978. The Formation of Trihalomethanes. *Jour. AWWA*, 70(11):604–612.

Uden, P.C., and J.W. Miller. 1983. Chlorinated Acids and Chloral in Drinking Water. *Jour. AWWA*, 75(10):524–527.

USEPA. 1972. Industrial Pollution of the Lower Mississippi River in Louisiana. Dallas, Texas: Surveillance and Analysis Division, Region VI.

———. 1989. Discussion of Strawman Rule for Disinfectant and Disinfection By-Products. Washington, D.C.: Criteria and Standards Division, USEPA, Office of Drinking Water.

Young, J.S., Jr., and P.C. Singer. 1979. Chloroform Formation in Public Water Supplies: A Case Study. *Jour. AWWA*, 71(2):87–95.

CHAPTER · TWO

Chemistry of Disinfection By-Product Formation

STUART W. KRASNER

INTRODUCTION

Disinfection by-product (DBP) formation can be summarized by the following equation:

$$\text{Disinfectant} + \text{Precursor} \rightarrow \text{DBPs} \qquad (2\text{-}1)$$

All chemical disinfectants (chlorine [Cl_2], monochloramine [NH_2Cl], ozone [O_3], and chlorine dioxide [ClO_2]) are known to form various types of DBPs. Disinfection by-product precursors include natural organic matter (NOM) (e.g., algae) and bromide. Disinfection by-product formation is affected by various parameters (e.g., pH, temperature, time, disinfectant dose, and residual) (Singer 1994).

Some DBPs represent halogen substitution by-products, whereas others are oxidation by-products. In addition, disinfectant combinations (e.g., from pre- and postdisinfection) can form secondary by-products.

Table 2-1 lists some of the major DBPs that have been identified as a result of chlorination (Stevens et al. 1989; Krasner et al. 1989; Cowman and Singer 1996), chloramination (Krasner et al. 1989; Cowman and Singer 1996; Krasner et al. 1996c), ozonation (Glaze and Weinberg 1993; Xie and Reckhow 1992; Najm and Krasner 1995; Kuo et al. 1996), and disinfection with chlorine dioxide (Aieta and Berg 1986; Richardson et al. 1994).

During 1988–89, a DBP study of 35 utilities nationwide was conducted (Krasner et al. 1989). In this study, treatment plant effluents were analyzed for trihalomethanes (THMs), five haloacetic acids (HAAs), haloacetonitriles (HANs), haloketones, chloral hydrate, chloropicrin, cyanogen chloride (CNCl), 2,4,6-trichlorophenol, formaldehyde, and acetaldehyde. (The three mixed bromochloro HAA species in Table 2-1 and tribromoacetic acid were not measured in this study, as analytical methods and commercial standards were not available for those species at that time.)

On a weight basis, THMs were the largest class of DBPs detected in the 35-utility study; the second largest fraction was the HAAs. The data indicated that the median level of total THMs (TTHMs) was approximately twice that of the sum of the five HAAs (HAA5). The third largest fraction was the aldehydes (i.e., formaldehyde and acetaldehyde). Although these two low-molecular-weight aldehydes are typically considered to be by-products of ozonation (Glaze and Weinberg 1993), they also appear to be by-products of chlorination. Every target-compound DBP in the 35-utility DBP study was detected at some time in some utility's water during the study; however, 2,4,6-trichlorophenol was only detected at low levels at a few utilities during the first sampling quarter and was not detected in subsequent samplings.

While THMs tended to be present in higher concentration than HAAs in the 35-utility DBP study, the HAA concentration consistently exceeded the concentration of THMs in a six-utility DBP study in North Carolina (Singer et al. 1995). In 1996, the American Water Works Association (AWWA) Water Industry Database (WIDB now known as WATERSTATS) was updated with available HAA data. The cumulative frequency distribution of the ratio of TTHMs to HAAs for WATERSTATS ranged from ~0.75:1 (10th percentile) to ~2.5:1 (90th percentile); the median ratio of TTHMs to HAAs was ~1.45:1.

Table 2-1 Major disinfection by-products formed during disinfection of drinking water

DBP Class	Individual DBPs	Chemical Formula
Trihalomethanes	Chloroform	$CHCl_3$
	Bromodichloromethane	$CHCl_2Br$
	Dibromochloromethane	$CHClBr_2$
	Bromoform	$CHBr_3$
Haloacetic acids	Monochloroacetic acid	$CH_2ClCOOH$
	Dichloroacetic acid	$CHCl_2COOH$
	Trichloroacetic acid	CCl_3COOH
	Bromochloroacetic acid	$CHBrClCOOH$
	Bromodichloroacetic acid	$CBrCl_2COOH$
	Dibromochloroacetic acid	$CBr_2ClCOOH$
	Monobromoacetic acid	$CH_2BrCOOH$
	Dibromoacetic acid	$CHBr_2COOH$
	Tribromoacetic acid	CBr_3COOH
Haloacetonitriles	Trichloroacetonitrile	$CCl_3C{\equiv}N$
	Dichloroacetonitrile	$CHCl_2C{\equiv}N$
	Bromochloroacetonitrile	$CHBrClC{\equiv}N$
	Dibromoacetonitrile	$CHBr_2C{\equiv}N$
Haloketones	1,1-Dichloroacetone	$CHCl_2COCH_3$
	1,1,1-Trichloroacetone	CCl_3COCH_3
Miscellaneous chlorinated organic compounds	Chloral hydrate	$CCl_3CH(OH)_2$
	Chloropicrin	CCl_3NO_2
Cyanogen halides	Cyanogen chloride	$ClC{\equiv}N$
	Cyanogen bromide	$BrC{\equiv}N$
Oxyhalides	Chlorite	ClO_2^-
	Chlorate	ClO_3^-
	Bromate	BrO_3^-
Aldehydes	Formaldehyde	$HCHO$
	Acetaldehyde	CH_3CHO
	Glyoxal	$OHCCHO$
	Methyl glyoxal	CH_3COCHO
Aldoketoacids	Glyoxylic acid	$OHCCOOH$
	Pyruvic acid	$CH_3COCOOH$
	Ketomalonic acid	$HOOCCOCOOH$
Carboxylic acids	Formate	$HCOO^-$
	Acetate	CH_3COO^-
	Oxalate	$OOCCOO^{-2}$
Maleic acids	2-*tert*-Butylmaleic acid	$HOOCC(C(CH_3)_3)$: $CHCOOH$

In 1985, the US Environmental Protection Agency (USEPA) measured chlorination DBPs at 10 operating utilities, using both target-compound and broad-screen analyses (Stevens et al. 1989b). In this study, 196 compounds that could be attributed to the chlorination process were found in one or more of the 10 utilities' finished waters. Approximately half of these compounds contained chlorine, but many of the halogenated and nonhalogenated compounds were of unknown structure. Furthermore, the halogenated compounds, cumulatively, accounted for between 30 and 60 percent of the total organic halogen (TOX) found in these samples.

Although chloramines can minimize THM formation (see chapter 8), TOX can form, albeit at a lower concentration than is the case with free chlorine (Stevens, Moore, and Miltner 1989). In the 35-utility DBP study, CNCl was observed to be preferentially formed in chloraminated waters as compared to chlorinated waters (Krasner et al. 1989). However, typically only 5 to 20 percent of the measured TOX during chloramination could be accounted for when compared to the molar sum of the TTHMs, six HAAs (HAA5 plus bromochloroacetic acid), and the cyanogen halides (CNCl plus cyanogen bromide) (Symons et al. 1996).

Ozone can convert NOM in water to aldehydes, aldoketoacids, carboxylic acids (see chapter 9), and other biodegradable organic material (BOM) (Krasner et al. 1996b). Carboxylic acids represent the largest fraction (e.g., ~30 percent) of the BOM. However, a large fraction (e.g., ~60 percent) of the BOM has not been identified. Although most ozonation by-products are oxygenated species, the presence of bromide will result in the formation of bromate and brominated organic by-products (Doré et al. 1988; Glaze and Weinberg 1993). However, only a fraction of the dissolved organic bromide has been measured as targeted brominated organic DBPs (Glaze et al. 1993).

Studies show that the TOX formed by chlorine dioxide is from 1 to 25 percent of the TOX formed by chlorine under the same reaction conditions (Symons et al. 1981; Aieta and Berg 1986). However, when chlorine dioxide is used, the inorganic by-products chlorite and chlorate are produced (see chapter 10). In addition, Richardson et al. (1994) identified maleic acids and a number of other by-products of chlorine dioxide treatment.

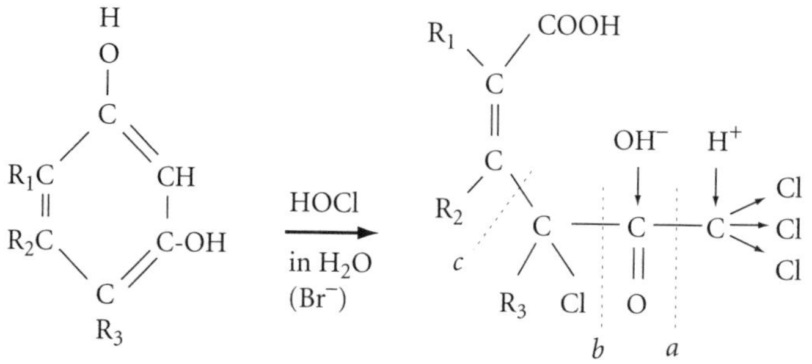

Source: Adapted from Rook (1977).

Figure 2-1 Haloform reaction with fulvic acids and resorcinol

CHEMISTRY OF DBP FORMATION

Haloform Reaction

Rook (1977) postulated that the haloform reaction occurred with the resorcinol-type moiety of fulvic acids (components of NOM). The proposed pathway first involved a fast chlorination of the carbon atoms that are activated by ortho hydroxide (OH) substituents or phenoxide ions in an alkaline environment. (Hypochlorous acid [HOCl] is the typical source of the electrophilic halogenating species Cl^+.) The reaction initially gives the intermediate carbanion, which is rapidly halogenated to the product shown in Figure 2-1 (the intermediate structures are not illustrated). After the aromatic structure has been halogenated and opened, cleavage at *a* will result in the formation of THMs (e.g., chloroform [$CHCl_3$]). Alternatively, oxidative and hydrolytic cleavages at *b* will yield an HAA (e.g., trichloroacetic acid, TCAA [Cl_3CCOOH]) or chloral hydrate ($Cl_3CCH(OH)_2$). In addition, cleavage at *c* will result in haloketone formation. If bromide is present, mixed bromochloro by-products will be formed.

Norwood et al. (1980) studied the reactions of chlorine with a series of selected aromatic compounds designed to model the monomeric components of aquatic humic material. All of the compounds studied produced measurable amounts of chloroform, with resorcinol derivatives showing the greatest yields. The resorcinol reaction was found to proceed through several chlorinated intermediates, which included 3,5,5-trichlorocyclopent-3-ene-1,2-dione, to chloroform and chlorinated acids. Alternatively, chlorination of 3-methoxy-4-hydroxycinnamic acid gave a low chloroform yield; instead, it produced chlorinated substitution products and chlorophenols, which broke down on further reaction to chloroacetic acids.

Christman et al. (1978) noted that the configuration of the cyclopentene derivative produced during the chlorination of resorcinol tended to confirm Rook's hypothesis that the carbon between the hydroxyl groups is the center for chloroform production. In Rook's proposed haloform reaction, fast chlorination of the activated carbon atoms of a substituted resorcinol yields a straight-chain ketoacid after ring rupture (Figure 2-1). However, the stable cyclopentene derivative observed by Christman and co-workers (1978) is inconsistent with Rook's hypothetical mechanism.

Norwood, Christman, and Hatcher (1987) found that although phenolic moieties are present in fulvic acids, they account for only a minor fraction of the total carbon. In addition, these researchers demonstrated that although phenolic ring rupture mechanisms appear to be important in organic halogen formation, other aqueous chlorination mechanisms involving aliphatic and other types of aromatic structures need to be considered. Because the aqueous chlorine species are electrophiles, they tend to react with electron-rich sites in organic structures (Rook 1977; Norwood et al. 1980; Reckhow and Singer 1985). Activated aromatic rings, aliphatic ß-dicarbonyls, and amino nitrogen are examples of electron-rich organic structures that react strongly with chlorine (Harrington et al. 1996).

Morris and Baum (1978) observed that the base-catalyzed series of halogenation and hydrolysis reactions in the haloform reaction typically occur with methyl ketones or compounds oxidizable to that structure. However, these researchers noted that acetone or other simple methyl ketones react too slowly to account for the formation of chloroform under the usual water chlorination conditions. Alternatively, ß-diketones are one group of compounds that form carbanions relatively quickly within the contact time of most water treatment plants and distribution systems (Morris and Baum 1978). In addition, the pyrrole ring, which occurs in chlorophylls and other natural products, is another structure in which active carbanion formation occurs

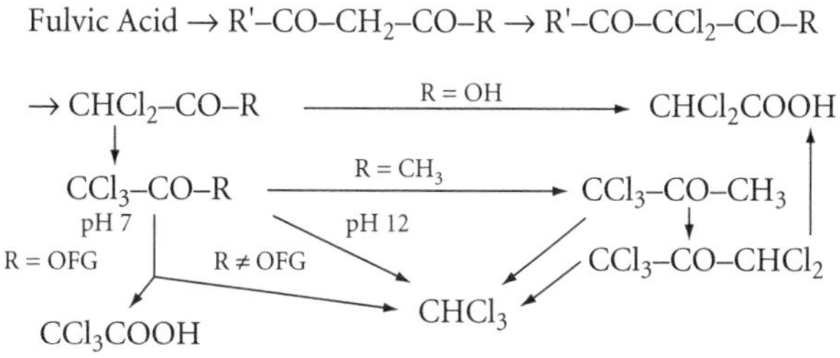

Fulvic Acid → R'–CO–CH_2–CO–R → R'–CO–CCl_2–CO–R

→ $CHCl_2$–CO–R —(R = OH)→ $CHCl_2COOH$

CCl_3–CO–R —(R = CH_3)→ CCl_3–CO–CH_3

pH 7 pH 12

R = OFG R ≠ OFG CCl_3–CO–$CHCl_2$

$CHCl_3$

CCl_3COOH

NOTE: OFG = oxidizable functional group.
Source: Adapted from Reckhow and Singer (1985).

Figure 2-2 Generalized conceptual model for the formation of major organic halide products from fulvic acid

(Morris and Baum 1978). With the pyrrole ring, the hydrogen atoms in the ortho position to the nitrogen atom are activated like those in phenol and provide sites for chlorination and subsequent haloform formation.

Reckhow and Singer (1985) developed a generalized conceptual model for the formation of the major organic halide products from fulvic acids (Figure 2-2), in which fulvic acids can be considered as polyfunctional unsaturated organic molecules with both aliphatic and aromatic components. Activated ionic substitution reactions play an important role in the formation of chloroform and HAAs from humic substances. Typically, these highly activated compounds contain ß-diketone moieties or structures that can readily be oxidized by chlorine to give ß-diketone moieties (R'-CO-CH_2-CO-R). Because most of the reactions between chlorine and humic materials result in oxidation rather than chlorine incorporation, oxygenated functional groups (e.g., ß-diketone groups) should be formed.

In the Reckhow and Singer (1985) model, with the formation of a ß-diketone moiety, the activated carbon will quickly become fully substituted with chlorine. Hydrolysis then occurs rapidly, yielding a monoketone group. If the remaining "R" group is a hydroxyl group, the reaction will stop, yielding dichloroacetic acid (DCAA [$CHCl_2COOH$]). Otherwise, the structure will be further chlorinated to a trichloromethyl species. This intermediate species is base-hydrolyzable to chloroform. At neutral pH, if the R group is an oxidizable functional group capable of readily

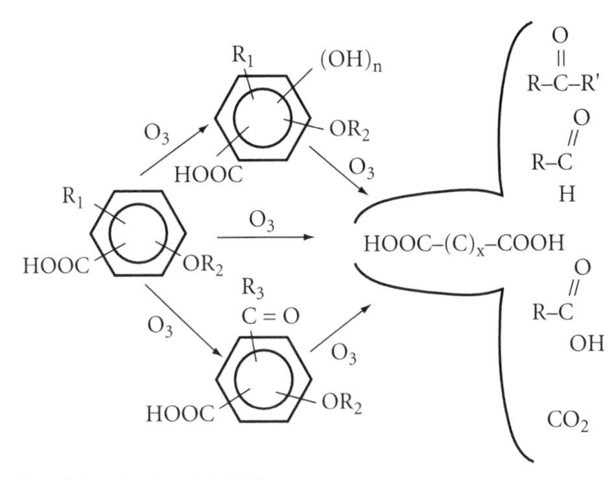

Source: Adapted from Doré et al. (1988).

Figure 2-3 A summary of the action of ozone in disinfection by-product formation

donating an electron pair to the rest of the molecule, TCAA is expected to form. In the absence of such an oxidative cleavage, hydrolysis will prevail, yielding chloroform.

Oxidation Reactions

Figure 2-3 shows a summary of the action of ozone in DBP formation (after Doré et al. 1988). Substitution reactions (top left portion of figure) lead to hydroxylation of the aromatic ring, whereas reactions on the aliphatic chains (especially with unsaturated groups) lead to the formation of carbonyl functionalities (bottom left portion of figure). Subsequent reactions lead to the opening of the aromatic ring and to the formation (right-hand portion of figure) of ketones (R-CO-R'), aldehydes (RCHO), organic acids (RCOOH), aliphatic compounds (HOOC-$(C)_x$-COOH), and carbon dioxide (CO_2).

Oxidation reactions can occur when any strong oxidant–disinfectant (e.g., ozone or chlorine) is used. For example, Le Cloirec and Martin (1985) demonstrated that chlorine can oxidize amino acids to yield aldehydes (Figure 2-4). Initially, there is substitution of a chlorine atom to produce an N-chloramine. (HOOC-CH(CH_3)-NHCl is formed when α-alanine is chlorinated.)

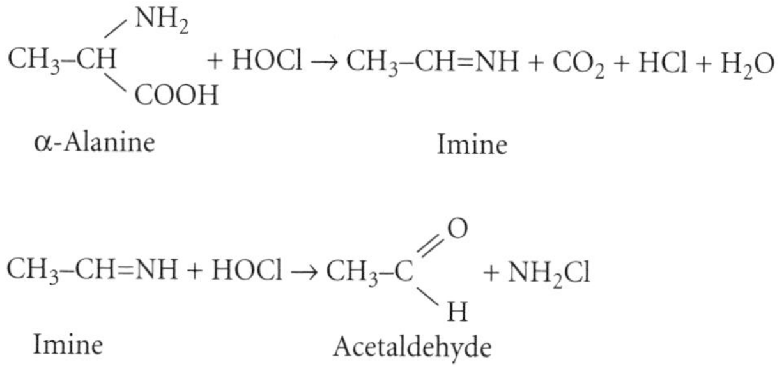

$$CH_3\text{–}CH(NH_2)(COOH) + HOCl \rightarrow CH_3\text{–}CH{=}NH + CO_2 + HCl + H_2O$$

α-Alanine Imine

$$CH_3\text{–}CH{=}NH + HOCl \rightarrow CH_3\text{–}C({=}O)H + NH_2Cl$$

Imine Acetaldehyde

Source: Adapted from LeCloirec and Martin (1985).

Figure 2-4 Oxidation of an amino acid by chlorine to form an aldehyde

After chlorine substitution, an electronic rearrangement of the N-chloramine leads to the formation of the corresponding imine. The imine, which is unstable, reacts with additional HOCl (chlorine substitution followed by rearrangement) to yield an aldehyde. Although aldehydes are typically associated with ozonation (Jacangelo et al. 1989), Krasner et al. (1989) found aldehyde formation during chlorination (typically at lower concentrations than with ozonation).

Chlorine dioxide (another strong oxidant–disinfectant) reacts with phenols (or phenolic groups attached to humic acids) to form dicarboxylic acids, including maleic acid (Wajon et al. 1982; Aieta and Berg 1986; Richardson et al. 1994). The organic products identified from the reaction between chlorine dioxide and phenols include chlorophenols, *p*-benzoquinone, maleic acid, and oxalic acid. Wajon et al. (1982) studied the reaction of chlorine dioxide with phenol and with hydroquinone. These researchers found that the removal of an electron from the substrate by chlorine dioxide to form a phenoxyl radical and chlorite was the rate-determining step. In the case of hydroquinone, chlorine dioxide removes another electron from the phenoxy radical, forming *p*-benzoquinone and another chlorite ion. In the case of phenol, chlorine dioxide adds to the phenoxy radical in the para position to the oxygen atom, forming *p*-benzoquinone and resulting in the release of hypochlorous acid. Products of further oxidation include maleic and oxalic acid (Masschelein 1979; Richardson et al. 1994).

Secondary Effects of Ozonation

The direct action of the ozone molecule and the addition and substitution reactions of chlorine both proceed, in part, through electrophilic attack by these oxidizing agents on precursor sites having strong densities of electronic charge (e.g., on aromatic rings) (Doré et al. 1988). Thus, preozonation can degrade some of the molecular sites reactive to chlorine. Reckhow and Singer (1984) found that preozonation destroyed a certain percentage of the precursors for THMs, TOX, TCAA, and dichloroacetonitrile (DCAN). However, ozonation resulted in no net effect on the precursors of DCAA and an increase in the precursors for 1,1,1-trichloroacetone (1,1,1-TCA). Increases in precursor levels may be caused by the transitory formation of polyhydroxylated aromatic compounds or by the accumulation of methylketone functions (Figure 2-2) that are only slightly reactive with ozone (Doré et al. 1988).

In addition to the effects of ozonation on DBP precursors, ozonation by-products can lead to the formation of secondary by-products from postchlorination of preozonated waters. McKnight and Reckhow (1992) studied secondary by-products formed during ozonation–chlorination. Acetaldehyde produced by ozonation can react with chlorine to form chloroacetaldehyde. When acetaldehyde undergoes an initial chlorine substitution reaction, the reaction proceeds rapidly to form the trichlorinated product chloral hydrate.

According to Scully (1990), the following reaction between formaldehyde (H–CHO)—another aldehyde produced during ozonation—and monochloramine may occur under acidic conditions to form CNCl:

$$\text{H–CHO} + \text{NH}_2\text{Cl} \xrightarrow[-\text{H}_2\text{O}]{\text{H}^+} \text{H–CH=N–Cl} \xrightarrow[-\text{HCl}]{} \text{H–CN} \xrightarrow{\text{NH}_2\text{Cl}} \text{CNCl} \quad (2\text{-}2)$$

In this reaction, N-chloraldimines are likely to be intermediates, whereas nitriles are the products. Pedersen et al. (1995) performed a study that demonstrated that formaldehyde can react with chloramines to form CNCl. Because aldehydes (e.g., formaldehyde and acetaldehyde) can be removed in biologically active filters (Weinberg et al. 1993), the formation of secondary by-products will depend on the mode of filter operation (whether the filters are prechlorinated or postchlorinated, media type, etc.) (Coffey et al. 1996).

THE EFFECTS OF DBP PRECURSORS ON DBP FORMATION

The Effects of NOM on DBP Formation

Natural organic material, a mixture of humic and nonhumic substances, contributes to the DBP precursor levels in drinking water (Owen et al. 1993). The total organic carbon (TOC) concentration of a water is generally a good indication of the amount of THMs and other DBP precursors present (Singer and Chang 1989). The specific ultraviolet absorbance (SUVA) of a water, which is defined as the UV absorbance (measured in cm^{-1}) times 100 divided by the dissolved organic carbon (DOC) concentration (mg/L), has been found to be a good surrogate for the water's humic content (Edzwald and Van Benschoten 1990). Humic substances have higher SUVAs and higher DBP formation potentials (DBPFPs) than the nonhumic fraction (Owen, Amy, and Chowdhury 1993).

Other researchers (as described in Krasner et al. 1996a) found that the higher the SUVA of isolated fractions of NOM, the higher the THM formation potential (THMFP) and TOX formation potential (TOXFP) (Table 2-2). Krasner et al. (1996a) observed a similar relationship in bulk waters (Table 2-3). Thus, the UV-to-DOC ratio, which is a reflection of the aromatic content of the NOM, provides an indication of the possible reactivity of the organic matter to form THMs and other DBPs. This is consistent with the haloform reaction of Rook (1977), in which the resorcinol-type moiety is reactive in forming DBPs.

Reckhow, Singer, and Malcolm (1990) isolated 10 aquatic humic and fulvic acids and studied their reactions with chlorine. These researchers found that UV absorbance was greater for the humic acids, reflecting their higher aromatic content and greater molecular size. In addition, the yields of TOX, THMs, HAAs, and HANs were greater for the humic acids than for the corresponding fulvic acids from the same source. However, the yield of 1,1,1-TCA was higher for the fulvic fractions than for the corresponding humic fractions, which may result from a higher methyl

Table 2-2 Chlorine reactivity of the NOM fractions isolated from Apremont reservoir (France)

Fraction	Chlorine Demand (*mg Cl_2/mg C*)	TOXFP (*μg Cl/mg C*)	THMFP (*μg $CHCl_3$/mg C*)	SUVA (*L/mg·m*)
Humic acids*	3 ± 0.2	277 ± 34	51 ± 2	4.6
Fulvic acids*	1.4 ± 0.12	140 ± 5	26 ± 2	3.1
Hydrophilic acids*	1.2 ± 0.2	101 ± 12	21 ± 1.4	2.0
Hydrophobic neutral fraction	0.27	40	12	2.0

*Average values for chlorine demand and DBPFP calculated from four to six experiments.
Source: Adapted from Krasner et al. (1996a).

Table 2-3 Chlorine reactivity of some representative bulk surface waters from California

Sample	SUVA (*L/mg·m*)	THMFP/C (*μmol/μmol*)
Mandeville Island drainage	5.0	0.91 percent
State Project water (SPW)	3.0	0.80 percent
Sacramento River water	2.2	0.75 percent
Colorado River water	1.5	0.50 percent

Source: Adapted from Krasner et al. (1996a).

ketone content in the unreacted fulvic fractions. Furthermore, these researchers observed a significant positive correlation between the TCAA/THM ratio and the SUVA. Ultraviolet absorbance is also a measure of the degree of conjugation of a compound, which would be consistent with the proposed haloform reaction of Reckhow and Singer (1985) in which a conjugated system (R group) capable of donating electrons to the carbonyl group in the intermediate species CCl_3-CO-R will preferentially lead to TCAA formation over chloroform formation (Figure 2-2).

The Effects of Algae on DBP Formation

In addition to humic substances, algae can be a source of DBP precursors. Hoehn et al. (1990) found that algae—both their biomass and their extracellular products—react readily with chlorine to produce THMs. These researchers observed that the algal extracellular products, on reaction with chlorine, generally yielded greater quantities of chloroform from the available TOC than did the algal biomass. Furthermore, they observed that high-yielding THM precursors were liberated by algae in greater abundance during the late exponential phase of growth than at any other time during the algal life cycle.

Algae are a source of amino acids. Trehy and Bieber (1981) found that the chlorination of certain amino acids, as well as humic acid, resulted in the formation of HANs (e.g., DCAN):

$$\underset{\displaystyle NH_2}{R\text{–}\underset{|}{CH}\text{–}COOH} + Cl_2 \rightarrow Cl_2CHC{\equiv}N \qquad (2\text{-}3)$$

For example, halogenation of aspartic acid (R group shown above in the amino acid equals CH_2COOH) is expected to yield cyanoacetic acid ($HOOCCH_2CN$) as one of the products of this reaction. This product contains a highly activated methylene group, which should be rapidly halogenated. Halogenation would transform CH_2 to CX_2, which would be followed by decarboxylation leading to a dihaloacetonitrile.

The Effects of Bromide on DBP Formation

Sources of bromide include saltwater intrusion, connate water (ancient, geologically trapped seawater), oil-field brines, and industrial and agricultural chemicals (APHA 1995; Krasner, Sclimenti, and Means 1994). Bromine chemistry in DBP formation first involves the oxidation of bromide to hypobromous acid (HOBr) by aqueous chlorine (HOCl):

$$HOCl + Br^- \rightarrow HOBr + Cl^- \qquad (2\text{-}4)$$

$$HOCl + HOBr + NOM \rightarrow DBPs \qquad (2\text{-}5)$$

Trussell and Umphres (1978) found that the concentration (in millimoles) of THMs produced per mole of TOC present for the formation reaction was related to the ratio of the concentration (in millimoles) of bromide incorporated in THMs (THM-Br) and the moles of TOC present. In addition, these researchers observed that the presence of bromide in a source water influenced the rate of the THM reaction as well as the THM yield;

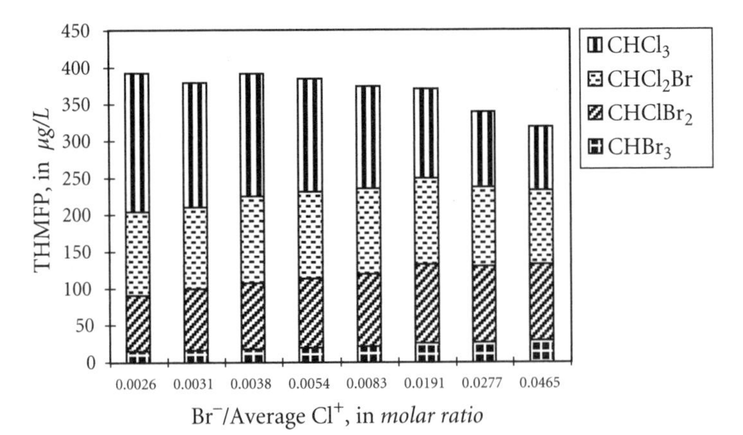

Source: Symons et al. (1993).

Figure 2-5 Effect of Br^-/average Cl^+ in THMFP testing of SPW (TOC = 2.7 mg/L, Br^- = 0.34 mg/L, Cl_2 = 10–120 mg/L)

i.e., the rate of THM formation was higher in waters with increased concentrations of bromide.

Amy, Tan, and Davis (1991) found that HOCl functions as a more effective oxidant, whereas HOBr behaves as a more efficient halogen substitution agent. These researchers performed THMFP tests and observed that, in general, less than 5 to 10 percent of the HOCl became incorporated in the THMs (THM-Cl), whereas as much as 50 percent or more of the bromide became incorporated in THM-Br. In addition, these researchers found that as the ratio of bromide to TOC increased, the percentage of brominated DBPs increased. This can occur when there is either an increase in bromide concentration or a decrease in TOC concentration.

Symons et al. (1993) found that the ratio of bromide to the average free available chlorine (Cl^+) controls bromine substitution. Hypobromous acid attacks more sites in the precursor than HOCl and reacts with them faster than HOCl (Stevens 1984). However, in THMFP tests, very large chlorine doses can "outcompete" the HOBr. As shown in Figure 2-5, as the ratio of bromide to Cl^+ decreased (i.e., because of an increase in the chlorine dose) in THMFP testing of California state project water (SPW), the concentration of brominated THMs tended to decrease, whereas the amount of chloroform increased.

Cowman and Singer (1996) found that bromine is more reactive than chlorine in substitution and addition reactions that form HAAs. These researchers observed that HAA speciation for any class of haloacetic acids is dependent on the ratio of bromide to chlorine. However, the distribution among tri-, di-, and monohalogenated HAA classes appeared to be independent of bromide concentration over the range of concentrations studied.

THE EFFECTS OF WATER QUALITY PARAMETERS ON DBP FORMATION

The Effects of pH and Reaction Time on DBP Formation

Morris and Baum (1978) found that the rate of the haloform reaction is highly affected by the first step of the reaction, which involves ionization of the methyl ketone moiety. This proceeds at a faster rate at higher pH levels. These researchers found that chlorination of most of the organic substances they studied exhibited greatly increased production of chloroform at higher pH values.

Stevens et al. (1976) observed that the rate of chloroform formation increased with an increase in pH. Fleischacker and Randtke (1983) found that the formation of chloroform increased with an increasing pH of chlorination, whereas the formation of nonpurgeable organic chlorine decreased with increasing pH. Reckhow and Singer (1984) observed a similar effect of pH on the formation of chloroform and TOX. In addition, these researchers found that TCAA formation significantly decreased at alkaline pH levels. Likewise, 1,1,1-TCA and DCAN showed reduced concentrations at high pH (Reckhow and Singer 1985).

Stevens, Moore, and Miltner (1989) evaluated the effects of pH and reaction time on DBP formation (Table 2-4). They evaluated chlorination pH levels of 5, 7, and 9.4 for reaction times of 4, 48, and 144 hours. For the THMs, formation increased over time, especially within the first 48 hours. Moreover, THM formation was higher with increasing pH. The HAA formation increased over time, with pH (in the range of 5 to 9.4) having no significant effect on DCAA formation, whereas TCAA formation was lower at pH 9.4 than at the lower pH levels.

Table 2-4 Summary of effects of pH and reaction time on formation of DBPs

DBP	pH 5	pH 7	pH 9.4
THMs	Lower formation		Higher formation
TCAA	Similar formation		Lower formation
DCAA	Similar formation, perhaps slightly higher at pH 7		
Chloral hydrate	Similar formation		Forms within 4 h; decays over time
DCAN	Higher formation	Forms within 4 h; then decays over time	Lower formation
1,1,1-TCA	Higher formation	Lower formation	Not detected

Source: Adapted from Stevens, Moore, and Miltner (1989).

Chloral hydrate formation increased over time at pH 5 and 7, whereas chloral hydrate that had formed within 4 hours at pH 9.4 decayed over time at the elevated pH. Dichloroacetonitrile formation only increased over time at pH 5, whereas DCAN formed within 4 hours at pH 7, decayed over time at the neutral pH, and was formed to a low extent at all reaction times at pH 9.4. Finally, 1,1,1-TCA formed at a higher concentration at pH 5 than at pH 7 and was not detected at pH 9.4.

The phenomenon observed by these researchers is caused, in part, by base-catalyzed hydrolysis of some non-THM DBPs. Some DBPs, which are electrophiles, can be destroyed by nucleophiles, such as hydroxide. Croué and Reckhow (1989) measured the rate constants of hydrolysis for a number of DBPs at pH levels of 6.1, 7.2, and 8.5 (Table 2-5). Only a slight loss of chloropicrin was observed (less than 10 percent after 3 hours), regardless of pH. The hydrolysis rate for this DBP was insignificant under these conditions. On the other hand, trichloroacetonitrile (TCAN) showed a significant rate of hydrolysis, which was base-catalyzed. Dichloroacetonitrile is also known to undergo a base-catalyzed hydrolysis (Trehy and Bieber 1981), although the hydrolysis rates measured by Croué and Reckhow were small. Bieber and Trehy (1983) found a somewhat higher hydrolysis rate constant for DCAN at pH 8.3

Table 2-5 Rate constants ($\times 10^{-6}\ s^{-1}$) for hydrolysis of selected DBPs

DBP	pH 6.1	pH 7.2	pH 8.5
Chloropicrin	10	13	10
TCAN	6	17	174
DCAN	NA*	3.1	4.2
1,1,1-TCA	NA	NA	73

*NA = Not available, not determined.
Source: Adapted from Croué and Reckhow (1989).

(i.e., $6.4 \times 10^{-6}\ s^{-1}$) in their study. Finally, Croué and Reckhow measured a significant decomposition rate for 1,1,1-TCA at pH 8.5.

The electron-withdrawing nature of the CCl_3 group (Betterton, Erel, and Hoffmann 1988) of 1,1,1-TCA is consistent with a high rate of destruction of this compound in the presence of hydroxide ions. Likewise, electron-withdrawing substituents (e.g., chlorine) on the alpha carbon of a nitrile should facilitate nucleophilic attack on the nitrile function in DCAN (Bieber and Trehy 1983). Moreover, the hydrolysis of trihaloacetaldehydes (e.g., chloral hydrate) and trihalomethyl ketones (e.g., 1,1,1-TCA) results in the release of haloforms (e.g., chloroform) over time, especially at high pH levels (Trehy and Bieber 1981).

Krasner et al. (1989) studied the stability of DBPs in two chloraminated distribution systems, where postchloramination minimized further DBP formation after the use of prechlorination within the plant. Plant 1 operated a distribution system with a pH level of 8. All of the DBPs presented in Table 2-6 were stable in this distribution system. Alternatively, plant 2 treated the water at the plant at pH 7.3 and distributed the water at pH 9.0. The THMs, HAAs, and chloropicrin were stable in this distribution system, whereas the HANs and chloral hydrate degraded over time, and 1,1,1-TCA was not detected once the pH had been raised to 9.0. These full-scale data are consistent with the findings of Stevens, Moore, and Miltner (1989) and Croué and Reckhow (1989).

Table 2-6 Effects of pH and time on disinfection by-products (μg/L) in two chloraminated distribution systems

	Plant 1*		Plant 2†		
Parameter	Plant Effluent	Distribution System	Filter Influent	Plant Effluent	Distribution System
pH	8.0	7.9	7.3	9.0	9.0
THMs	44	50	38	39	44
HAAs	40	46	34	39	44
HANs	6.3	7.7	5.0	1.5	0.5
Chloral hydrate	4.5	5.2	4.0	1.2	0.5
1,1,1-TCA	2.3	2.2	1.5	ND‡	ND
Chloropicrin	0.44	0.59	0.09	0.17	0.18

*Plant 1 was a conventional treatment plant with prechlorination–postchloramination. The raw- and finished-water TOC concentrations were 4.2 and 2.8 mg/L, respectively. The raw- and finished-water UV absorbances were 0.178 and 0.102 cm^{-1}, respectively. The raw-water bromide concentration was 0.05 mg/L. The raw- and finished-water temperatures were 9 and 13°C, respectively. The total chlorine residuals in the plant effluent and the distribution system were 3.5 and >2.0 mg/L, respectively. The detention time in the distribution system was >7 h.

†Plant 2 was a conventional treatment plant with prechloramination. The raw- and finished-water TOC concentrations were 7.7 and 4.8 mg/L, respectively. The raw- and finished-water UV absorbances were 0.255 and 0.095 cm^{-1}, respectively. The raw-water bromide concentration was 0.22 mg/L. The water temperatures were in the range of 26 to 30°C. The total chlorine residuals in the plant effluent and the distribution system were both 3.5 mg/L. The detention time in the distribution system was in the range of 18 to 20 h.

‡ND = Not detected.

Source: Source: Adapted from Krasner et al. (1989).

The Effects of Temperature and Seasonal Variability on DBP Formation

Stevens et al. (1976) studied the effects of temperature on the rate of reaction of precursors present in Ohio River water. Chloroform production—after 96 hours at pH 7—at 3, 25, and 40°C was <50, ~100, and >200 μg/L, respectively.

In the 35-utility DBP study, the median THM formation was highest in the summer and lowest in the winter. Because seasonal variations can result from changes in the concentration of DBP precursors (i.e., NOM and bromide) as well as temperature fluctuations, Krasner et al. (1990) sorted the data by dividing the results into different temperature ranges. The median THM concentration of the high-temperature (24 to 31°C) group was significantly higher than the median THM concentration of the other temperature groups. Most of the other DBPs, except for the haloketones, followed the same trend as the THMs. When the data for 1,1,1-TCA were sorted into different temperature ranges, the median concentration (1.4 μg/L) in the cold-temperature (1.1 to 8.5°C) group was significantly higher than the median concentration (0.2 μg/L) in the warm-temperature (16 to 23°C) group. This DBP is a reactive intermediate rather than a stable end product. The lower temperatures in the winter suggest less reactivity and a lower rate of production of final end products.

Seasonal variability must also include consideration of the nature of the organic precursors, which may vary in composition with the season (Singer 1994). Zumstein and Buffle (1989) studied the seasonal evolution of NOM in a river and lake system. Increases in the DOC and UV values were observed in the Grenet River (Switzerland) after heavy rainfalls. These increases can be attributed to increased leaching of soil organic matter during high river discharges. Zumstein and Buffle also found that the DOC concentrations in the upper layers of Lake Bret (Switzerland), which is a eutrophic lake, were at a maximum during the period from May to September. The epilimnetic DOC increases in spring and summer were ascribed to the biological activity of the lake.

Moreover, the concentration of bromide can be seasonally related (e.g., wet-weather versus dry-weather conditions) (Singer 1994). Krasner, Sclimenti, and Means (1994) found significant variations in bromide concentrations in SPW that were related to the extent of saltwater intrusion. Such a phenomenon was highly affected by seasonal variability in rainfall and was most pronounced during drought years.

The Effects of Chlorine Dose and Residual on DBP Formation

The rate, extent, and distribution of DBP formation is affected by the chlorine dose and residual (Singer 1994). For example, higher doses and residuals favor the formation of HAAs over THMs. In addition, higher chlorine doses result in a higher proportion of trihalogenated HAAs as compared to di- and monohalogenated HAAs. Reckhow and Singer (1985) found that the concentration of TCAA increased more than that of DCAA with increasing chlorine dose. These researchers also observed a reduction in the concentration of 1,1,1-TCA and DCAN with increasing chlorine dose, which may be the result of a direct reaction between chlorine and these two intermediate DBPs. Furthermore, increases in the chlorine dose shift the formation of DBPs to the less bromine-substituted species (Symons et al. 1993).

The Effects of Water Quality Parameters on DBP Formation Testing

Disinfection by-product formation can be evaluated at the bench scale using either THMFP (or DBPFP) methods (Stevens and Symons 1977; Symons et al. 1993), simulated distribution system (SDS) testing (Koch et al. 1991), and/or uniform formation condition (UFC) tests (Summers et al. 1996). Each test method has been optimized to yield specific information because the water quality parameters used in each procedure highly influence the yield and speciation of DBP formation.

The DBPFP method is an indirect measurement of the amount of DBP precursors in a water (APHA 1995). The DBPFP is a useful measure of the precursor material remaining at any stage of treatment (Stevens and Symons 1977). For example, Reckhow and Singer (1984) used DBPFP testing to evaluate the removal of various DBP precursors by ozonation and coagulation.

Typically, DBPFP tests are conducted for incubation periods up to seven days to ensure that the formation curve has plateaued out. It is critical that a positive chlorine residual be maintained throughout the entire storage period, as the continued reaction of precursor with chlorine to yield THMs and many other DBPs depends on the maintenance of a chlorine residual (Stevens and Symons 1977). Thus, chlorine doses higher than those used in the actual treatment plant may be required in DBPFP testing. The DBPFP tests are usually conducted at a warm temperature (e.g., 20 or 25°C) and at a neutral pH (e.g., 7.0) or the pH of the raw or distributed water (e.g., 8.0). As discussed earlier in this chapter, each of these water quality parameters can affect the yield and speciation of the DBPs formed in the FP test.

The intent of the DBPFP test is not to predict the concentration and types of DBPs that would form in an actual distribution system; rather, this method is used to determine the potential to form DBPs based on the amount of precursors in the raw or treated water. Moreover, because of the relatively high chlorine doses used in the procedure, the speciation of DBPs formed typically tends to have a higher ratio of chlorinated to brominated species than would be found in the actual distribution system. Strictly speaking, the majority of precursors for chloroform and bromoform are the same, whereas the chlorine-to-bromide ratio affects the relative proportion of the individual species (Symons et al. 1993).

Alternatively, SDS tests are used to predict the actual concentration and speciation of DBPs that would form in a distribution system (Koch et al. 1991). Simulated distribution system conditions (reaction time, pH, and temperature) are customized for each system to be evaluated. The SDS tests use chlorine doses and residuals that match those used in the actual treatment plant and distribution system, respectively; therefore, the speciation of chlorinated and brominated species formed in SDS tests has been found to parallel that formed in the actual distribution system (Koch et al. 1991; Symons et al. 1993).

Because SDS conditions are site-specific, Summers et al. (1996) developed the UFC test to enable direct comparisons of DBP formation in different waters using conditions representative of many US distribution systems. Uniform formation condition tests are conducted at a pH of 8.0 ± 0.2, a temperature of 20.0 ± 1.0°C, and an incubation time of 24 ± 1 hour, with a chlorine residual after 24 hours of 1.0 ± 0.4 mg/L (Summers et al. 1996). The storage time was selected, in part, because the average mean detention time in the distribution systems surveyed in the WIDB was 1.3 days. The pH value was chosen to reflect the influence of the Lead and Copper Rule on distribution system pH. The temperature chosen reflected the average temperature used in SDS testing according to a survey conducted by Summers and colleagues (1996). The chlorine residual goal was based, in part, on the average mean chlorine residual (i.e., 0.9 mg/L) in the distribution systems in the WIDB.

Summers et al. (1996) performed a sensitivity analysis to evaluate the effects of variations in the water quality conditions used in the UFC test on DBP formation. These researchers found that most of the three-day DBP formation (for THMs, HAAs, and TOX) occurred in the first 24 hours of reaction for the three waters that were evaluated. Disinfection by-product formation (especially for THMs and HAAs) was highly influenced by temperature, thus confirming the need for a standard temperature for UFC testing. A reaction pH of 8.0 was used to represent a pH level at which the base-catalyzed formation of THMs was not overemphasized at the expense of TCAA formation, which is lower at higher pH values. The effect of chlorine residual—in the range of low residuals ($\leq$ 3 mg/L) that are commonly found in distribution systems—did not significantly affect the UFC formation of THMs and TOX, whereas HAA production tended to increase with increasing residual.

SUMMARY AND CONCLUSIONS

Disinfection by-product formation involves halogen substitution and/or oxidation reactions. Disinfection by-product formation and degradation are affected by numerous parameters (TOC, UV, bromide, chlorine dose, pH, time, and temperature). The UV-to-DOC ratio (SUVA) is an indication of the humic content of a water, as well as an indication of the possible reactivity of the NOM in that water to form THMs and other DBPs.

The different classes of DBPs do not all follow the same relationships with respect to these parameters. The ratio of bromide to Cl^+ affects DBP speciation within a given class. In addition, some non-THM DBPs are degraded by base-catalyzed hydrolysis. A complex set of equilibria and kinetics is necessary to understand DBP formation and degradation.

For chlorination, THMs and HAAs represent the two largest classes of DBPs detected on a weight basis. Other halogenated by-products (e.g., HANs, chloral hydrate, and haloketones) form to a lesser extent, especially at alkaline pH levels, in part because of base-catalyzed hydrolysis. In addition, oxidation reactions from chlorine can form nonhalogenated by-products (e.g., aldehydes).

REFERENCES

Aieta, E.M., and J.D. Berg. 1986. A Review of Chlorine Dioxide in Drinking Water Treatment. *Jour. AWWA*, 78(6):62.

American Public Health Association, American Water Works Association, and Water Environment Federation. 1995. *Standard Methods for the Examination of Water and Wastewater*. Washington, D.C.: APHA.

Amy, G.L., L. Tan, and M.K. Davis. 1991. The Effects of Ozonation and Activated Carbon Adsorption on Trihalomethane Speciation. *Water Res.*, 25(2):191.

Betterton, E.A., Y. Erel, and M.R. Hoffmann. 1988. Aldehyde–Bisulfite Adducts: Prediction of Some of Their Thermodynamic and Kinetic Properties. *Environ. Sci. & Tech.*, 22(1):92.

Bieber, T.I., and M.L. Trehy. 1983. Dihaloacetonitriles in Chlorinated Natural Waters. In *Water Chlorination: Environmental Impact and Health Effects*, Vol. 4. Ann Arbor, Mich.: Ann Arbor Science Publishers.

Christman, R.F., J.D. Johnson, J.R. Hass, F.K. Pfaender, W.T. Liao, D.L. Norwood, and H.J. Alexander. 1978. Natural and Model Aquatic Humics: Reactions with Chlorine. In *Water Chlorination: Environmental Impact and Health Effects*, Vol. 2. Ann Arbor, Mich.: Ann Arbor Science Publishers.

Coffey, B.M., S.W. Krasner, M.J. Sclimenti, P.A. Hacker, and J.T. Gramith. 1996. A Comparison of Biologically Active Filters for the Removal of Ozone By-Products, Turbidity, and Particles. In *Proc. AWWA Water Quality Technology Conference*. Denver, Colo.: American Water Works Association.

Cowman, G.A., and P.C. Singer. 1996. Effect of Bromide Ion on Haloacetic Acid Speciation Resulting from Chlorination and Chloramination of Aquatic Humic Substances. *Environ. Sci. & Tech.*, 30(1):16.

Croué, J-P, and D.A. Reckhow. 1989. Destruction of Chlorination Byproducts with Sulfite. *Environ. Sci. & Tech.*, 23(11):1412.

Doré, M., N. Merlet, B. Legube, and J-P Croué. 1988. Interactions Between Ozone, Halogens, and Organic Compounds. *Ozone Science and Engineering*, 10:153.

Edzwald, J.K., and J.E. Van Benschoten. 1990. Aluminum Coagulation of Natural Organic Matter. In *Proc. Fourth Int'l Gothenburg Symposium on Chemical Treatment*, Madrid, Spain.

Fleischacker, S.J., and S.J. Randtke. 1983. Formation of Organic Chlorine in Public Water Supplies. *Jour. AWWA*, 75(3):132.

Glaze, W.H., H.S. Weinberg, and J.E. Cavanagh. 1993. Evaluating the Formation of Brominated DBPs During Ozonation. *Jour. AWWA*, 85(1):96.

Glaze, W.H., and H.S. Weinberg. 1993. *Identification and Occurrence of Ozonation By-Products in Drinking Water.* Denver, Colo.: American Water Works Association Research Foundation and American Water Works Association.

Harrington, G.W., A. Bruchet, D. Rybacki, and P.C. Singer. 1996. Characterization of Natural Organic Matter and Its Reactivity with Chlorine. In *Water Disinfection and Natural Organic Matter: Characterization and Control.* Minear, R.A., and G.L. Amy, eds. Washington, D.C.: American Chemical Society.

Hoehn, R.C., D.B. Barnes, B.C. Thompson, C.W. Randall, T.J. Grizzard, and P.T.B. Shaffer. 1980. Algae as Sources of Trihalomethane Precursors. *Jour. AWWA*, 72(6):344.

Jacangelo, J.G., N.L. Patania, K.M. Reagan, E.M. Aieta, S.W. Krasner, and M.J. McGuire. 1989. Ozonation: Assessing Its Role in the Formation and Control of Disinfection By-products. *Jour. AWWA*, 81(8):74.

Koch, B., S.W. Krasner, M.J. Sclimenti, and W.K. Schimpff. 1991. Predicting the Formation of DBPs by the Simulated Distribution System. *Jour. AWWA*, 83(10):62.

Krasner, S.W., M.J. McGuire, J.G. Jacangelo, N.L. Patania, K.M. Reagan, and E.M. Aieta. 1989. The Occurrence of Disinfection By-products in US Drinking Water. *Jour. AWWA*, 81(8):41.

Krasner, S.W., K.M. Reagan, J.G. Jacangelo, N.L. Patania, E.M. Aieta, and K.M. Gramith. 1990. Relationships Between Disinfection By-Products and Water Quality Parameters: Implications for Formation and Control. In *Proc. AWWA Annual Conference.* Denver, Colo.: American Water Works Association.

Krasner, S.W., M.J. Sclimenti, and E.G. Means. 1994. Quality Degradation: Implications for DBP Formation. *Jour. AWWA*, 86(6):34.

Krasner, S.W., J-P Croué, J. Buffle, and E.M. Perdue. 1996a. Three Approaches for Characterizing NOM. *Jour. AWWA*, 88(6):66.

Krasner, S.W., B.M. Coffey, P.A. Hacker, C.J. Hwang, C.-Y. Kuo, A.A. Mofidi, and M.J. Sclimenti. 1996b. Characterization of the Components of BOM. In *Proc. 4th Internat'l BOM Conf.*, Waterloo, Ont.

Krasner, S.W., J.M. Symons, G.E. Speitel, Jr., A.C. Diehl, C.J. Hwang, R. Xia, and S.E. Barrett. 1996c. Effects of Water Quality Parameters on DBP Formation During Chloramination. In *Proc. AWWA Annual Conference.* Denver, Colo.: American Water Works Association.

Kuo, C.-Y., H.-C. Wang, S.W. Krasner, and M.K. Davis. 1996. Ion-Chromatographic Determination of Three Short-Chain Carboxylic Acids in Ozonated Drinking Water. In *Water Disinfection and Natural Organic Matter: Characterization and Control.* Washington, D.C.: American Chemical Society.

Le Cloirec, C., and G. Martin. 1985. Evolution of Amino Acids in Water Treatment Plants and the Effect of Chlorination on Amino Acids. In *Water Chlorination: Chemistry, Environmental Impact and Health Effects*, Vol. 5. Chelsea, Mich.: Lewis Publishers.

Masschelein, W.J. 1979. *Chlorine Dioxide: Chemistry and Environmental Impact of Oxychlorine Compounds.* Ann Arbor, Mich.: Ann Arbor Science Publishers.

McKnight, A., and D.A. Reckhow. 1992. Reactions of Ozonation By-Products With Chlorine and Chloramines. In *Proc. AWWA Annual Conference.* Denver, Colo.: American Water Works Association.

Morris, J.C., and B. Baum. 1978. Precursors and Mechanisms of Haloform Formation in the Chlorination of Water Supplies. In *Water Chlorination: Environmental Impact and Health Effects*, Vol. 2. Ann Arbor, Mich.: Ann Arbor Science Publishers.

Najm, I.N., and S.W. Krasner. 1995. Effects of Bromide and NOM on By-Product Formation. *Jour. AWWA*, 87(1):106.

Norwood, D.L., J.D. Johnson, R.F. Christman, J.R. Hass, and M.J. Bobenrieth. 1980. Reactions of Chlorine with Selected Aromatic Models of Aquatic Humic Material. *Environ. Sci. & Tech.*, 14(2):187.

Norwood, D.L., R.F. Christman, and P.G. Hatcher. 1987. Structural Characterization of Aquatic Humic Material. 2. Phenolic Content and Its Relationship to Chlorination Mechanism in an Isolated Aquatic Fulvic Acid. *Environ. Sci. & Tech.*, 21(8):791.

Owen, D.M., G.L. Amy, and Z.K. Chowdhury. 1993. *Characterization of Natural Organic Matter and Its Relationship to Treatability*. Denver, Colo.: American Water Works Association Research Foundation and American Water Works Association.

Pedersen, E.J., III, E.T. Urbansky, B.J. Mariñas, and D.W. Margerum. 1995. Formation of Cyanogen Chloride from the Reaction of Monochloramine and Formaldehyde. Presented at the 210th Nat'l Mtg, American Chemical Society, Chicago, Ill.

Reckhow, D.A., and P.C. Singer. 1984. The Removal of Organic Halide Precursors by Preozonation and Alum Coagulation. *Jour. AWWA*, 76(4):151.

———. 1985. Mechanisms of Organic Halide Formation During Fulvic Acid Chlorination and Implications with Respect to Preozonation. In *Water Chlorination: Chemistry, Environmental Impact and Health Effects*, Vol. 5. Chelsea, Mich.: Lewis Publishers.

Reckhow, D.A., P.C. Singer, and R.L. Malcolm. 1990. Chlorination of Humic Materials: Byproduct Formation and Chemical Interpretations. *Environ. Sci. & Tech.*, 24(11):1655.

Richardson, S.D., A.D. Thruston, Jr., T.W. Collette, K.S. Patterson, B.W. Lykins, Jr., G. Majetich, and Y. Zhang. 1994. Multispectral Identification of Chlorine Dioxide Disinfection Byproducts in Drinking Water. *Environ. Sci. & Tech.*, 28(4):592.

Rook, J.J. 1977. Chlorination Reactions of Fulvic Acids in Natural Waters. *Environ. Sci. & Tech.*, 11(5):478.

Scully, F.E., Jr. 1990. Reaction Chemistry of Inorganic Monochloramine: Products and Implications for Drinking Water Disinfection. Presented at the 200th Nat'l Mtg, American Chemical Society, Washington, D.C.

Singer, P.C., and S.D. Chang. 1989. Correlations Between Trihalomethanes and Total Organic Halides Formed During Water Treatment. *Jour. AWWA*, 81(8):61.

Singer, P.C. 1994. Control of Disinfection By-Products in Drinking Water. *Jour. of Env. Eng.*, 120(4):727.

Singer, P.C., A. Obolensky, and A. Greiner. 1995. DBPs in Chlorinated North Carolina Drinking Waters. *Jour. AWWA*, 87(10):83.

Stevens, A.A., C.J. Slocum, D.R. Seeger, and G.G. Robeck. 1976. Chlorination of Organics in Drinking Water. *Jour. AWWA*, 68(11):615.

Stevens, A.A., and J.M. Symons. 1977. Measurement of Trihalomethane and Precursor Concentration Changes. *Jour. AWWA*, 69(10):546.

Stevens, A.A. 1984. Personal communication.

Stevens, A.A., L.A. Moore, and R.J. Miltner. 1989. Formation and Control of Non-Trihalomethane Disinfection By-products. *Jour. AWWA*, 81(8):54.

Stevens, A.A., L.A. Moore, C.J. Slocum, B.L. Smith, D.R. Seeger, and J.C. Ireland. 1989. By-Products of Chlorination at Ten Operating Utilities. In *Disinfection By-Products: Current Perspectives*. Denver, Colo.: American Water Works Association.

Summers, R.S., S.M. Hooper, H.M. Shukairy, G. Solarik, and D. Owen. 1996. Assessing DBP Yield: Uniform Formation Conditions. *Jour. AWWA*, 88(6):80.

Symons, J.M., A.A. Stevens, R.M. Clark, E.E. Geldreich, O.T. Love, Jr., and J. DeMarco. 1981. Treatment Techniques for Controlling Trihalomethanes in Drinking Water. EPA-600/2-81-156. Cincinnati, Ohio: USEPA, Municipal Environmental Research Laboratory.

Symons, J.M., S.W. Krasner, L.A. Simms, and M.J. Sclimenti. 1993. Measurement of THM and Precursor Concentrations Revisited: The Effect of Bromide Ion. *Jour. AWWA*, 85(1):51.

Symons, J.M., R. Xia, A.C. Diehl, G.E. Speitel, Jr., C.J. Hwang, S.W. Krasner, and S.E. Barrett. 1996. The Influence of Operational Variables on the Formation of Dissolved Organic Halogen During Chloramination. In *Water Disinfection and Natural Organic Matter: Characterization and Control*. Washington, D.C.: American Chemical Society.

Trehy, M.L., and T.I. Bieber. 1981. Detection, Identification and Quantitative Analysis of Dihaloacetonitriles in Chlorinated Natural Waters. In *Advances in the Identification & Analysis of Organic Pollutants in Water*, Vol. 2. Ann Arbor, Mich.: Ann Arbor Science Publishers.

Trussell, R.R., and M.D. Umphres. 1978. The Formation of Trihalomethanes. *Jour. AWWA*, 70(11):604.

Wajon, J.E., D.H. Rosenblatt, and E.P. Burrows. 1982. Oxidation of Phenol and Hydroquinone by Chlorine Dioxide. *Environ. Sci. & Tech.*, 16(7):396.

Weinberg, H.S., W.H. Glaze, S.W. Krasner, and M.J. Sclimenti. 1993. Formation and Removal of Aldehydes in Plants That Use Ozonation. *Jour. AWWA*, 85(5):72.

Xie, Y., and D.A. Reckhow. 1992. A New Class of Ozonation By-Products: The Ketoacids. In *Proc. 1992 AWWA Annual Conference*. Denver, Colo.: American Water Works Association.

Zumstein, J., and J. Buffle. 1989. Circulation of Pedogenic and Aquagenic Organic Matter in an Eutrophic Lake. *Water Res.*, 23(2):229.

CHAPTER · THREE

Modeling Disinfection By-Product Formation

ZAID K. CHOWDHURY
GARY L. AMY

Natural organic matter (NOM) found in surface and groundwaters serves as the primary precursor for the formation of disinfection by-products (DBPs). Naturally occurring bromide (Br^-) in the presence of NOM also serves as a DBP precursor. The levels of DBPs formed as a result of disinfection depend on the concentration of precursor materials, disinfectant types and dosages, and water quality conditions. As such, the DBPs formed during water treatment can be classified as halogenated DBPs (resulting from chlorination and ozonation), oxidation DBPs (resulting from ozonation), and inorganic DBPs (primarily resulting from chlorine dioxide addition and ozonation). The halogenated DBPs primarily include trihalomethanes (THMs), haloacetic acids (HAAs), haloacetonitriles (HANs), haloketones (HKs), and chloral hydrate (CH). Oxidation DBPs include aldehydes, aldo and keto acids, carboxylic acids, and bromate (BrO_3^-). The by-products of chlorine dioxide addition are chlorite (ClO_2^-) and chlorate (ClO_3^-). The modeling efforts described in this chapter primarily address chlorination and ozonation by-products.

Due to the complex nature of DBP precursors and their corresponding reactions with disinfectants, models for quantification of DBPs have largely been developed using empirical approaches. This modeling approach involves statistical analysis of data derived from controlled laboratory or field experiments to develop functional relationships to predict the concentration of DBPs based on water quality and reaction conditions. Some efforts at developing mechanistic models for predicting DBPs have also been undertaken. This chapter provides a summary of various modeling efforts for DBP prediction during water treatment.

FACTORS AFFECTING DBP FORMATION

The extent of DBP formation depends on the dosages of disinfectants, the nature and concentration of precursors, and various water quality parameters. Organic precursors for DBP formation are often characterized by the concentration of total organic carbon (TOC) or dissolved organic carbon (DOC) and ultraviolet light absorbance at 254 nm (UV_{254}). Naturally occurring bromide is an inorganic DBP precursor. The type and dosage of the disinfectant determines the types and amounts of DBPs formed during disinfection. Some water quality constituents, such as ammonia, reduce the effective disinfectant dosage and, consequently, reduce the formation of DBPs. Other water quality parameters, such as pH and temperature (θ), also affect the formation of DBPs. One of the most important factors in determining DBP formation is the length of time (reaction time, *t*) a particular disinfectant remains in contact with the precursors.

The rates of DBP formation change as the reactions proceed, reflecting the complex nature of the reactions involved. Most DBP formation occurs at a fast initial rate followed by a slower rate of formation that often continues for days in the presence of residual disinfectants. Researchers indicate that the DBP formation reaction starts to slow down after several hours of contact time (AWWARF 1991; Koch et al. 1991). The same group of researchers also indicate that formation of brominated DBPs is a relatively fast reaction compared to the formation of chlorinated by-products. The general kinetic trends described above are relevant for THM formation, where most research has been performed to date. There are exceptions, however, in the case of certain DBPs (such as chloral hydrate) that undergo hydrolysis reactions resulting in a decrease in concentration with increasing reaction times. There is still another group of DBPs (e.g., the HANs) that show a decrease in concentration due to continuing reaction with chlorine (Reckhow and Singer 1985; Krasner et al. 1989).

A number of researchers (e.g., Stevens et al. 1976; Lange and Kawczynski 1978; Trussell and Umphres 1978) have observed a positive correlation between the pH of the water and the extent of THM formation. Since those studies, the impact of pH on a number of other chlorination by-products has been reported (e.g., Miller and Uden 1983; Oliver and Thurman 1983; Reckhow and Singer 1985). The formation of haloacetic acids is found to be favored at a lower pH of chlorination, whereas ozonation at a low pH can reduce bromate formation. The level of bromide has also been associated with an increase in molar THM yields and a shift in speciation towards the more brominated forms (Amy et al. 1985); this same trend has been observed for the HAAs (Cowman and Singer 1996).

EMPIRICAL MODELS FOR CHLORINATED DBP FORMATION

Moore, Tuthhill, and Polakoff (1979) obtained raw and treated water quality information from 19 operating water treatment facilities in Massachusetts (low bromide waters) to study the formation of THMs. They proposed the following simple chloroform formation model based on the applied chlorine dose, reflecting the fact that the extent of chloroform formation parallels that of chlorine consumption:

$$CHCl_3 \text{ in } mg/L = -11.24 + 22.23 \cdot (Cl_2 Dose \text{ in } mg/L) \quad (3\text{-}1)$$

Trussell and Umphres (1978) also included a term for organic carbon in their proposed rate equation

$$\frac{d(TTHM)}{dt} = k_n \cdot (\mathrm{Cl_2}) \cdot (\mathrm{C})^m \qquad (3\text{-}2)$$

Where:

m = the order of the reaction with respect to the concentration of organic carbon precursor, C
k_n = reaction rate constant

Kavanaugh et al. (1980) suggested an alternative form of the THM rate equation by including a THM yield parameter (f) as shown in the following equation:

$$\frac{d(TTHM)}{dt} = k_n \cdot (TOC) \cdot \left(\mathrm{Cl_2}Dose - \frac{3(TTHM)}{f}\right)^m \qquad (3\text{-}3)$$

After these early efforts in developing simple THM predictive models based on stoichiometry and rate equations, a shift towards empirical modeling occurred. This shift was primarily caused by a need for developing relatively accurate quantitative predictions of THM concentrations within applicable boundary conditions while recognizing the complexity of NOM characteristics. Various empirical models were developed by Engerholm and Amy (1983), Urano, Wada, and Takemosa (1983), Amy, Chadik, and Chowdhury (1987), Morrow and Minear (1987), and Hutton and Chung (1992). Early efforts in empirical modeling of THM formation concentrated mainly on developing nonlinear power function type relationships for predicting total THM concentrations. For example, the model proposed by Amy, Chadik, and Chowdhury (1987) was

$$\begin{aligned} TTHM = {} & 0.00309 \cdot (UVABS \cdot TOC)^{0.440} \\ & \cdot (CLDOSE)^{0.409} \cdot (RXNTM)^{0.265} \\ & \cdot (TEMP)^{1.06} \cdot (pH - 2.6)^{0.715} \cdot (BR + 1)^{0.0358} \end{aligned} \qquad (3\text{-}4)$$

Where:

$TTHM$ = total THM concentration, in μmoles/L
$UVABS$ = UV absorbance at 254 nm, in $\mathrm{cm^{-1}}$
TOC = total organic carbon concentration, in mg/L

CLDOSE = chlorine dose at the beginning of the formation reaction, in mg/L
RXNTM = reaction time, in hours
TEMP = reaction temperature, in °C
pH = reaction pH
BR = concentration of bromide, in mg/L

Most experiments from which these empirical models were developed were designed to maximize the formation of THMs (i.e., formation potential conditions). As a result, the distribution of THM species was different than what would be expected under more realistic chlorination conditions in a water treatment plant. The database used for developing these equations included TOC values ranging from 3.0 to 13.8 mg/L. The reaction time ranged from 0.1 to 168 hours; reaction temperature ranged from 10 to 30°C; pH ranged from 4.6 to 9.8; UV absorbance ranged from 0.063 to 0.489 cm^{-1}; bromide levels ranged from 0.010 to 0.245 mg/L, and chlorine dosages ranged from 1.5 to 41.4 mg/L. More recently, Amy et al. (1991) augmented the above database with additional data and proposed a set of empirical equations for individual THM species. In this effort, the database was segmented with respect to bromide concentrations and different constants were developed for each species. The form of the equations are shown below. The values of the fitting constants for the three ranges of bromide concentrations are shown in Table 3-1.

$$[\mathrm{CHCl_3}] = a[UV \cdot TOC]^b[\mathrm{Cl_2}Dose]^c[t]^d \cdot [Temp]^e[pH - 2.6]^f[\mathrm{Br} + 1]^g \tag{3-5}$$

$$[\mathrm{CHCl_2Br}] = a_1[UV \cdot TOC]^{b_1}[\mathrm{Cl_2}Dose]^{c_1}[t]^{d_1} \cdot [Temp]^{e_1}[pH - 2.6]^{f_1}[\mathrm{Br}]^{g_1} \tag{3-6}$$

$$[\mathrm{CHClBr_2}] = a_2[UV/TOC]^{b_2}[\mathrm{Cl_2}Dose]^{c_2}[t]^{d_2} \cdot [Temp]^{e_2}[pH - 2.6]^{f_2}[\mathrm{Br}]^{g_2} \tag{3-7}$$

$$[\mathrm{CHBr_3}] = a_3[UV]^{b_3}[\mathrm{Cl_2}Dose]^{c_3}[t]^{d_3} \cdot [Temp]^{e_3}[pH - 2.6]^{f_3}[\mathrm{Br}/TOC]^{g_3} \tag{3-8}$$

$$[TTHM] = [\mathrm{CHCl_3}] + [\mathrm{CHCl_2Br}] + [\mathrm{CHClBr_2}] + [\mathrm{CHBr_3}] \tag{3-9}$$

Table 3-1 Coefficients for THM predictive equations 3-5 through 3-8

Coefficients	Br < 0.25 mg/L	0.25 mg/L < Br < 0.75 mg/L	Br > 0.75 mg/L
a	0.381	0.00776	28.47
b	0.533	0.988	1.0837
c	0.457	0.1829	0.1718
d	0.251	0.2675	0.2121
e	0.986	2.6131	1.8227
f	0.978	–0.7367	–0.2576
g	0.533	–2.6614	–8.0572
a1	0.835	0.066	0.0615
b1	0.074	0.3519	0.4845
c1	0.381	0.8515	1.1501
d1	–0.258	0.2939	0.2976
e1	0.681	1.3478	1.4203
f1	0.886	–0.2774	–0.7399
g1	0.710	0.0821	–2.111
a2	3.043	0.1246	0.3654
b2	–0.0199	–0.2930	0.140
c2	–0.0105	0.7073	1.3343
d2	0.2499	0.2576	0.2748
e2	0.5285	0.7275	0.6448
f2	1.5339	0.6492	–0.2901
g2	2.0455	1.2451	0.7757
a3	11.82	29.28	686663
b3	0.3407	–1.453	0.8122
c3	–0.2468	3.338	5.4964
d3	0.1256	0.2052	0.2466
e3	–0.7812	–2.802	–5.9748
f3	2.0894	0.9105	2.566
g3	1.1332	2.7728	6.4774

Where:

$[TTHM]$ = concentration of THMs, in μmoles/L
$[CHCl_3]$ = concentration of chloroform, in μmoles/L
$[CHCl_2Br]$ = concentration bromodichloromethane, in μmoles/L
$[CHClBr_2]$ = concentration of dibromochloromethane, in μmoles/L
$[CHBr_3]$ = concentration of bromoform, in μmoles/L

Empirical models for predicting haloacetic acid formation as a result of chlorination have also been reported (AWWARF 1991). The following are the reported empirical equations for various HAA species:

$$MCAA = 1.634(TOC)^{0.753}(Br + 0.01)^{-0.085} \cdot (pH)^{-1.124}(\mathrm{Cl_2}Dose)^{0.509}(t)^{0.300} \tag{3-10}$$

$$DCAA = 0.605(TOC)^{0.291}(UV)^{0.726} \cdot (\mathrm{Br} + 0.01)^{-0.568}(\mathrm{Cl_2}Dose)^{0.480} \cdot (t)^{0.239}(Temp)^{0.665} \tag{3-11}$$

$$TCAA = 87.182(TOC)^{0.355}(UV)^{0.901} \cdot (\mathrm{Br} + 0.01)^{-0.679}(pH)^{-1.732} \cdot (\mathrm{Cl_2}Dose)^{0.881}(t)^{0.264} \tag{3-12}$$

$$MBAA = 0.176(TOC)^{1.664}(UV)^{-0.624} \cdot (\mathrm{Br})^{0.795}(pH)^{-0.927}(t)^{0.145}(Temp)^{0.450} \tag{3-13}$$

$$DBAA = 84.94(TOC)^{-0.620}(UV)^{0.651} \cdot (\mathrm{Br})^{1.073}(\mathrm{Cl_2}Dose)^{-0.200}(t)^{0.120}(Temp)^{0.657} \tag{3-14}$$

Where:

$MCAA$ = concentration of monochloroacetic acid, in μg/L
$DCAA$ = concentration of dichloroacetic acid, in μg/L
$TCAA$ = concentration of trichloroacetic acid, in μg/L
$MBAA$ = concentration of monobromoacetic acid, in μg/L
$DBAA$ = concentration of dibromoacetic acid, in μg/L

The database used for developing these equations include TOC values ranging from 2.8 to 11 mg/L. The reaction time ranged from 0.1 to 105 hours; reaction temperature ranged from 13 to 20°C; pH ranged from 5.6 to 9.0; UV absorbance ranged from 0.05 to 0.38 cm^{-1}; bromide levels ranged from 0.005 to 0.430 mg/L; and chlorine dosages ranged from 3.0 to 25.0 mg/L.

Equation 3-4 and Eq 3-10 through 3-14 were incorporated into the Water Treatment Plant Simulation computer program developed for the US Environmental Protection Agency (USEPA) (Harrington, Owen, and Chowdhury 1992). This program was used by USEPA to evaluate the cost effect of alternative maximum contaminant levels (MCLs) for the THMs and HAAs. Although these equations represented the best predictive equations at the time of regulatory assessment for the DBP Rule, they only represent the central tendency of the databases from which the equations were developed and as a result suffer from several limitations. One major limitation is that these equations were based on raw water chlorination and not chlorination of treated waters (e.g., coagulated–settled waters or granular activated carbon [GAC]-treated waters), and thus are most appropriate for prechlorination. Consequently, the water treatment plant (WTP) simulation program that embodies these equations could only account for the impact of a reduction in the amount of precursor material but not a change in characteristics of the precursors. The other principal limitation is the restrictive boundary conditions. Due to the empirical nature of these equations, they are not expected to perform well outside of the water quality conditions included in the database from which the equations were originally derived. Efforts are currently underway to develop predictive equations based on reaction kinetics of DBP formation that will be appropriate to coagulated–settled waters and may remove some of the restrictions regarding boundary conditions.

EMPIRICAL MODELS FOR OZONATED DBP FORMATION

Ozone dosage and residual, bromide concentration, and pH levels have been reported to positively affect the formation of bromate, while the concentration of organic matter appears to negatively affect bromate formation. Various empirical power function type relationships have been proposed by Amy et al. (1998), Song et al. (1996), and Ozekin (1994). The following are the proposed equations by Amy, Song, and Ozekin, respectively:

$$[\mathrm{BrO_3^-}] = 2.32 \times 10^{-6}[DOC]^{-1.47}[\mathrm{O_3}Dose]^{1.65}[Br^-]^{0.81} \cdot [\mathrm{NH_3\text{-}N}]^{-0.121}[pH]^{5.35}[t]^{0.28} \quad (3\text{-}15)$$

$$[\mathrm{BrO_3^-}] = 10^{-6.11}[DOC]^{-1.18}[\mathrm{O_3}Dose]^{1.42}[Br^-]^{0.88} \cdot [\mathrm{NH_3\text{-}N}]^{-0.18}[pH]^{5.11}[t]^{0.27}[IC]^{0.18} \quad (3\text{-}16)$$

$$[\mathrm{BrO_3^-}] = 1.63 \times 10^{-6}[DOC]^{-1.26}[\mathrm{O_3}Dose]^{1.57} \cdot [\mathrm{Br^-}]^{0.73}[pH]^{5.82}[t]^{0.28} \quad (3\text{-}17)$$

Where:

$[BrO_3^-]$ = concentration of bromate, in μg/L
DOC = dissolved organic carbon concentration, in mg/L
O_3Dose = transferred ozone dose at the beginning of the formation reaction, in mg/L
t = reaction time, in hours
$NH_3\text{-}N$ = ammonia nitrogen concentration, in mg/L
IC = inorganic carbon concentration, in mg/L as $CaCO_3$
pH = reaction pH
Br = concentration of bromide, in mg/L

The above equations were based on batch ozonation experiments of raw waters. The ranges of water quality conditions included in the database were bromide from 0.07 to 0.44 mg/L, reaction time of 1 to 120 min, pH of 6.5 to 8.5, ozone dosages from 1 to 10 mg/L (transferred dose), inorganic carbon concentrations from 1 to 216 mg/L as $CaCO_3$, temperature from 20 to 30°C, and DOC concentration from 1.1 to 8.4 mg/L.

Ozonation also produces other by-products, such as aldehydes, biodegradable organic carbon (or assimilable organic carbon), and bromoform (formed when ozonating bromide-containing waters). The biodegradable by-products are often removed by biologically active granular media filtration after the ozonation process. In the absence of such processes, the biodegradable component may pass on into the distribution system and may result in regrowth within the distribution system infrastructure. Models to predict the concentration of ozonation by-products, other than bromate, are scarce.

MECHANISTIC MODELS FOR DBP FORMATION

Due to the complex nature of NOM, most researchers have historically used empirical approaches for developing models to predict DBP concentrations during water treatment. As discussed earlier, one of the significant shortcomings of these empirical models is that they do not perform very well under conditions other than those from which they were developed. On the other hand, mechanistic models are expected to be applicable under a much broader range of conditions. For this reason, there has always been a desire to develop mechanistic models for DBP formation. Several mechanistic modeling formats have been proposed for developing models for THM formation (Adin et al. 1991), bromate formation (Haag and Hoigné 1983), and distribution of HAA species (Cowman and Singer 1996). These models are still in their rudimentary forms, and research is continuing to refine them.

SUMMARY

Several empirical models for predicting various DBPs of concern during chlorination and ozonation have been developed. These models have proven useful for predicting DBP concentrations; however, caution should be exercised when using these models outside of the boundary conditions from which they were developed.

Even within the boundary conditions, these empirical models only predict central tendency. Consequently, the accuracy of the predictions will depend on the specific water quality and treatment conditions for which the models are applied. Mechanistic models are expected to be more robust and able to predict concentrations with better accuracy for a wide range of waters. Adequate mechanistic models are not yet available, however, and research is continuing on this front.

REFERENCES

Adin, A., J. Katzhendler, D. Alkaslassy, and Ch. Rav-Acha. 1991. Trihalomethane Formation in Chlorinated Drinking Water: A Kinetic Model. *Water Res.*, 25(7):707–805.

Amy, G.L., M. Siddiqui, K. Ozekin, H. Wei, and X. Wang. 1997. Empirically-Based Models For Predicting Chlorination and Ozonation By-Products; Haloacetic Acids, Chloral Hydrate, and Bromate. USEPA Report, CX 819579.

Amy, G.L., P.A. Chadik, and Z.K. Chowdhury. 1987. Developing Models for Predicting THM Formation Potential and Kinetics. *Jour. AWWA*, 79(7).

Amy, G.L., P.A. Chadik, Z.K. Chowdhury, and P.H. King. 1985. Factors Affecting Incorporation of Bromide Into Brominated Trihalomethanes During Chlorination. In *Water Chlorination: Environmental Impact and Health Effects*, Vol. 5. Jolley, R.L., ed. Chelsea, Mich.: Lewis Publishers.

AWWARF. 1991. Disinfection By-Products Database and Model Project. Denver, Colo.: American Water Works Association Research Foundation.

Cowman, G.A., and P.C. Singer. 1996. Effect of Bromide Ion on Haloacetic Acid Speciation Resulting From Chlorination and Chloramination of Aquatic Humic Substances. *Environ. Sci. & Tech.*, 30(1):16.

Engerholm, B.A., and G.L. Amy. 1983. A Predictive Model for Chloroform Formation with Chlorination of Humic Substances. *Water Res.*, 17:1797.

Haag, W., and J. Hoigné. 1983. Ozonation of Bromide Containing Water Kinetics of Formation of Hypobromous Acid and Bromate. *Environ. Sci. & Tech.*, 17:261.

Harrington, G., D. Owen, and Z.K. Chowdhury. 1992. Developing Models to Simulate DBP Formation During Water Treatment. *Jour. AWWA*, 84(11):78.

Hutton, P.H., and F.I. Chung. 1992. Simulating THM Formation Potential in Sacramento Delta. *Jour. Water Resources Planning, ASCE*, 118(5).

Kavanaugh, M.C., A.R. Trussell, J. Cromer, and R.R. Trussell. 1980. An Empirical Kinetic Model for Trihalomethane Formation: Applications to Meet the Proposed THM Standard. *Jour. AWWA*, 72(10).

Koch, B., S.W. Krasner, M.J. Sclimenti, and W.K. Schimpff. 1991. Predicting the Formation of DBPs by the Simulated Distribution System. *Jour. AWWA*, 83(10):62.

Krasner, S., et al. 1989. The Occurrence of Disinfection By-Products in Drinking Water. *Jour. AWWA*, 81(8):41.

Lange, A.L., and E. Kawczynski. 1978. Controlling Organics: The Contra Costa County Water District Experience. *Jour. AWWA*, 70(11).

Miller, W.J., and P.C. Uden. 1983. Characterization of Nonvolatile Aqueous Products of Humic Substances. *Environ. Sci. & Tech.*, 17(3).

Moore, G.S., R.W. Tuthhill, and D.W. Polakoff. 1979. A Statistical Model for Predicting Chloroform Levels in Chlorinated Surface Water Supplies. *Jour. AWWA*, 71(9).

Morrow, N.M., and R.A. Minear. 1987. Use of Regression Models to Link Raw Water Characteristics to THM Concentration in Drinking Water. *Water Res.*, 21(41).

Oliver, B.G., and E.M. Thurman. 1983. Influence of Aquatic Humic Substances Properties on Trihalomethane Potential. In *Water Chlorination Environmental Impact and Health Effect*, Vol. 4. Jolley, R.L., ed.

Ozekin, K. 1994. Modeling Bromate Formation During Ozonation and Assisting Its Control. PhD thesis. University of Colorado.

Reckhow, D.A., and P.C. Singer. 1985. Mechanisms of Organic Halide Formation During Fulvic Acid Chlorination and Implications with Respect to Preozonation. In *Water Chlorination; Chemistry, Environmental Impact and Health Effect*, Vol. 5. Jolley, R.L., et al., eds. Chelsea, Mich.: Lewis Publishers.

Siddiqui, M., et al. (Unpublished). Ozone Enhanced Removal of Natural Organic Matter. Accepted for publication in *Water Res.*

Song, R., C. Donohoe, R.A. Minear, P. Westerhoff, K. Ozekin, and G.L. Amy. 1996. Empirical Modeling of Bromate Formation During Ozonation of Bromide-Containing Waters. *Water Res.*, 30(5).

Stevens, A.A., C.J. Slocum, D.R. Seeger, and G.G. Robeek. 1976. Chlorination of Organics in Drinking Water. *Jour. AWWA*, 68(11).

Trussell, R.R., and M.D. Umphres. 1978. The Formation of Trihalomethanes. *Jour. AWWA*, 70(11).

Urano, K., H. Wada, and T. Takemosa. 1983. Empirical Rate Equation for Trihalomethane Formation with Chlorination of Humic Substances in Water. *Water Res.*, 17(12).

CHAPTER • FOUR

Natural Organic Matter: Structural Characteristics and Reactive Properties

JEAN-PHILIPPE CROUÉ
JEAN-FRANÇOIS DEBROUX
GARY L. AMY
GEORGE R. AIKEN
JERRY A. LEENHEER

Natural organic matter (NOM) represents a heterogeneous mixture responsible for the formation of organic disinfection by-products (DBPs) during water treatment. A challenge facing water chemists has been to assign chemical identity to NOM, and to relate its chemical properties to its propensity to form DBPs. This chapter provides an overview of NOM, its isolation and fractionation, and its characterization, both in terms of general properties (structure and functionality) and reactive properties (disinfectant demand and DBP formation potential).

ORIGIN OF DISSOLVED ORGANIC CARBON IN AQUATIC SYSTEMS

Several processes including microbial degradation of organic matter, oxidative polymerization of phenolic compounds in plants and soils, and photolytic degradation of NOM result in the formation of many of the compounds that comprise dissolved organic carbon (DOC), especially nonvolatile organic acids that dominate the DOC content in most aquatic environments. Many of these organic acids are considered refractory because the rates of subsequent biodegradation are slower than for other fractions or classes of organic matter. It should be noted, however, that microbial processes continue to slowly alter the structure and chemical reactivity of these compounds. Once in the aquatic system, organic compounds, either anthropogenic or naturally derived, can be truly dissolved, associated with immobile phases, or associated with mobile particles (e.g., colloids).

In aquatic ecosystems, the sources of DOC can be categorized as (1) allochthonous, entering the system from the terrestrial watershed, or (2) autochthonous, being derived from biota (e.g., algae, bacteria, and macrophytes) growing in the water body. In temperate systems, the allochthonous sources are very important. Most of the DOC originates from the degradation and leaching of organic detritus within the watershed and is transported by streams and shallow groundwater flow.

The chemical characteristics of the DOC are not only influenced by the source materials, but also by the biogeochemical processes involved in carbon cycling within the terrestrial and aquatic systems. These processes include allochthonous flow of organic carbon to the aquatic system from the watershed, autochthonous carbon fixation by algae and aquatic plants, transformation and degradation of both autochthonous and allochthonous organic material by heterotrophic microbial activity, transport of particulate organic material to the sediments, remobilization of DOC from the sediments, and photodegradation by incident ultraviolet (UV) light. Each of these processes is likely to affect the DOC chemistry in a different way.

Organic matter derived from different source materials has distinctive chemical characteristics associated with those source materials. Aquatic humic substances derived from higher plants, for instance, are found to have relatively large amounts of aromatic carbon, are high in phenolic content, and are low in nitrogen content. Microbially derived humic substances (from algae and bacteria), on the other hand, have relatively large nitrogen contents, and low aromatic-carbon and phenolic content (Aiken et al. 1996). The relative contributions of allochthonous and autochthonous sources of DOC vary among different water bodies, but there is presently no way to quantify this variation based on chemical characterization of DOC. The problem of sorting out relative contributions to the DOC pool from different compartments of the environment is further complicated by the turnover and mixing that takes place in surface waters.

ISOLATION OF AQUATIC NOM

A large number of protocols have been developed to isolate aquatic NOM (Thurman 1985; Aiken and Leenheer 1993). These protocols are based on two different approaches: (1) concentration and fractionation, and (2) concentration only. Adsorption chromatography (e.g., using Amberlite XAD resins) and membrane filtration (reverse osmosis or nanofiltration), respectively, are the most commonly used isolation approaches at this time.

Adsorption Chromatography

Organic matter in water is a complex mixture of both high- and low-molecular-weight species, primarily organic acids that are relatively enriched in oxygen-containing functional groups. A common approach to the study of DOC is to isolate functionally distinct fractions of the organic matter to determine fundamental chemical information about the biogenesis and environmental roles of these materials. Because the purpose of isolating and fractionating DOC is to elucidate chemical properties, it is desirable to have an approach based on the chemical and reactive properties of the materials of interest. Adsorption chromatography on both ion-exchange resins (Leenheer 1981; Miles, Tuschall, and Brezonik 1983) and nonionic macroporous resins (Mantoura and Riley 1975; Thurman and Malcolm 1981; Aiken et al. 1992; Croué et al. 1993) has been successfully used for this purpose.

The XAD-8 resin method proposed by the United States Geological Survey (USGS) research group (Leenheer 1981; Thurman and Malcolm 1981) for the isolation of the so-called humic substances (i.e., humic and fulvic acids) is now considered by a large part of the scientific community as the reference method. Using this approach, the hydrophobic fraction of NOM (also called the *humic fraction*) that is adsorbed at acidic pH onto the XAD-8 resin column is separated from the hydrophilic fraction (traditionally referred to as the nonhumic fraction, although more recent work has demonstrated that a large part is "humic-like" NOM). Typically, 40 to 60 percent of the NOM can be recovered from a sample using the XAD-8 approach to isolate the hydrophobic acid fraction (Leenheer and Huffman 1976; Leenheer 1981; Thurman and Malcolm 1981; Martin-Mousset et al. 1997).

Another approach to the fractionation of the organic matter in a water sample is to use a two-column array of XAD-8 and XAD-4 resins (Aiken et al. 1992; Malcolm and McCarthy 1992; Croué et al. 1993; and Andrews and Huck 1994). This two-column setup (Figure 4-1) is an extension of the method of Thurman and Malcolm (1981) that allows for the isolation and separation of both the hydrophobic acid fraction (containing aquatic humic substances) and a portion of the organic matter that is more hydrophilic. The fraction obtained from the XAD-4 resin will be referred to in this chapter as *transphilic* because the compounds in this fraction are of intermediate polarity. In the literature, this fraction is also referred to as *hydrophilic acids* (Aiken et al. 1992; Croué et al. 1993), *XAD-4 acids* (Malcolm and McCarthy 1992), or *syn-fulvic acids* (Aiken et al. 1996). The efficiency of adsorption of organic matter on the XAD resins is a function of the aqueous solubility of the solute molecules and the resin used. The following five fractions of DOC can be obtained with this approach: hydrophobic acids (i.e., fulvic and humic acids) FA and HA respectively and neutrals (from XAD-8), transphilic acids and neutrals (from XAD-4), and a hydrophilic fraction (material that passes through both columns). The hydrophobic acid (HPOA) and transphilic acid (TPHA) fractions account for 50 to 90 percent of the DOC in most waters (Aiken et al. 1992).

Examples of fractionation data for select waters are given in Figure 4-2. In each case, the HPOA fraction (HA + FA) is the largest fraction of the DOC pool. These data can be used to better

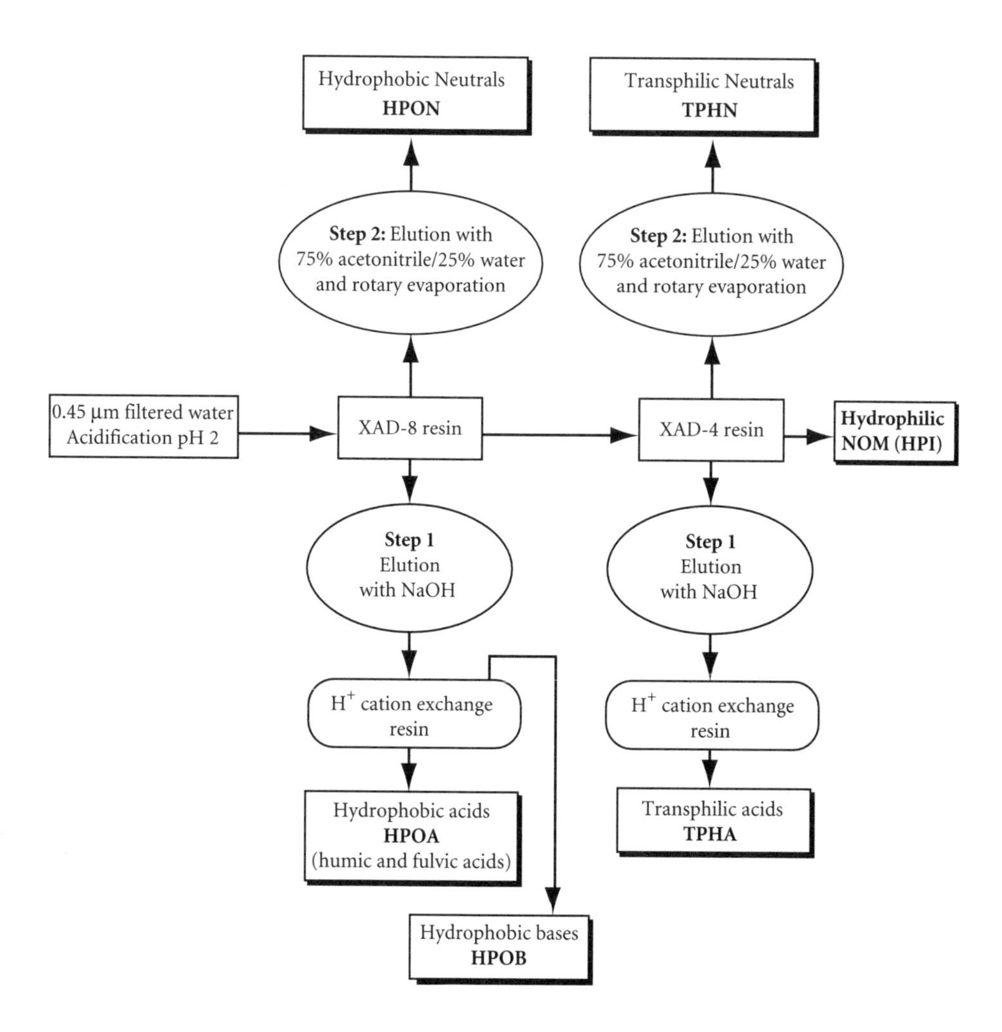

Figure 4-1 Schematic diagram of the XAD-8/XAD-4 isolation scheme

define the differences in DOC composition between different locations (Figure 4-2A) or at different sampling times (Figure 4-2B). In general, about 20 to 30 percent of dissolved organic matter in natural waters consists of hydrophilic organic compounds (HPI fraction) that do not adsorb onto XAD-8 or XAD-4 resins. The separation of inorganic salts from hydrophilic NOM is particularly challenging, and the ratio of DOC to salt concentration is frequently the determining factor in the recovery of hydrophilic

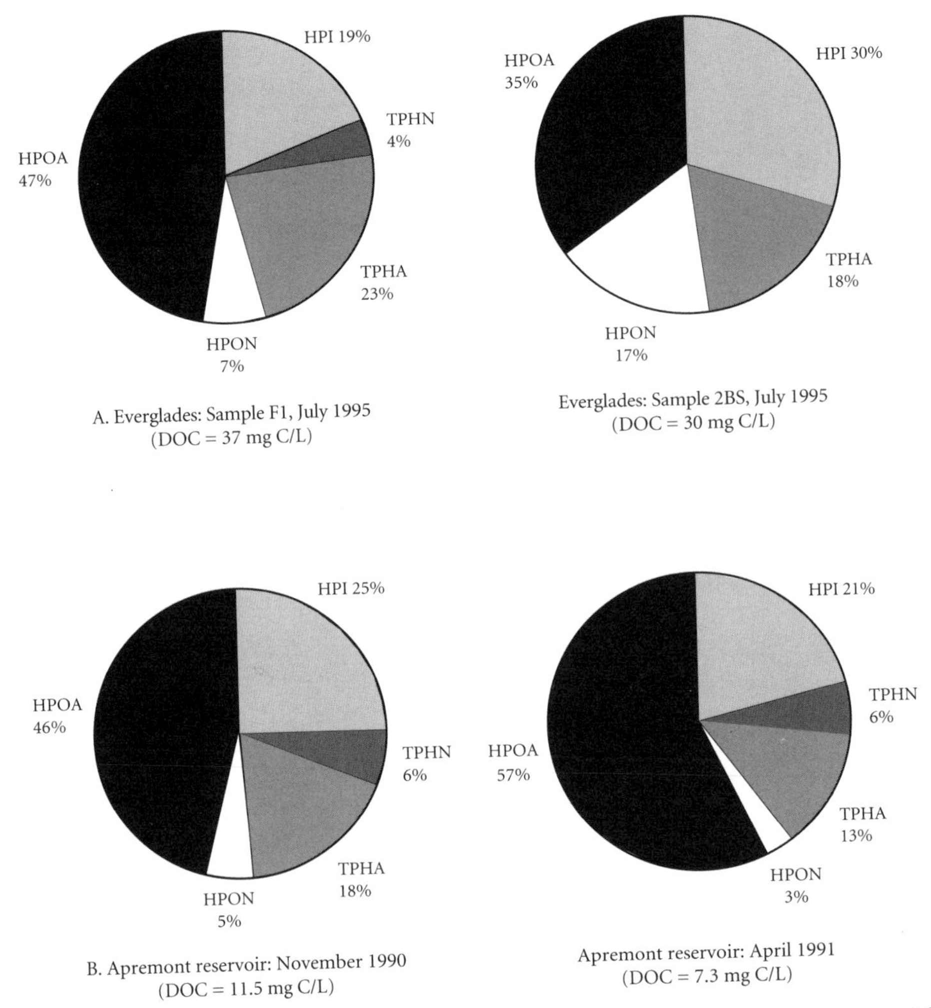

Figure 4-2 DOC distribution of the Everglades (Florida) (same time at two different locations) and the Apremont reservoir (France) (same location at two different times).

NOM. New approaches for isolating hydrophilic acids, bases, and neutrals were recently developed by Leenheer (1996). As a general trend, the acid fraction has been found to be the major part of the hydrophilic NOM. In addition, colloids can represent an important fraction of this group of NOM because of size exclusion phenomena.

Advantages and disadvantages of resin techniques. There are two primary advantages associated with the chromatographic methods of fractionation and isolation. First, it is possible to fractionate and isolate a total of approximately 55 to 90 percent of the DOC from a variety of aquatic environments. Second, the fractionation can be carried out on the original water sample without using a preconcentration step, such as reverse osmosis or vacuum evaporation, thereby maintaining fractionation consistency and comparability between samples.

Disadvantages of this approach include potential sample alteration associated with the use of various solvents and pH conditions necessary to bring about the sorption/desorption of the compounds of interest (Aiken 1988). Also, the fractions of organic matter obtained using these isolation methods are not sharply defined; there is some overlap between the different fractions.

Membrane Filtration: Reverse Osmosis

Only a few research studies have been devoted to the development of membrane processes for the isolation of aquatic NOM. Polyamide or polysulfone membranes have generally been preferred. Reverse osmosis is an efficient water concentration technique with a rejection coefficient for DOC above 90 percent. For most of the waters studied, the DOC recovery is generally above 85 percent, with few cases near 70 percent (Serkiz and Perdue 1990; Clair et al. 1991; Sun, Perdue, and McCarthy 1995; Croué et al. 1996).

The use of reverse osmosis for the isolation of NOM offers two major and important advantages. First, a large volume of water can be rapidly processed, which allows for the isolation of a large quantity of organic material in a short period of time. Second, NOM is never subjected to extreme pH values that could alter its structural properties (Serkiz and Perdue 1990).

Given the high desalting efficiency of reverse osmosis, the major drawback of this isolation procedure is the simultaneous concentration of salts that may represent a large proportion of the isolated material (Clair et al. 1991; Croué et al. 1996). Moreover, a portion of NOM can sorb to the membranes. Wiesner and Chellam (1992) reported that high-molecular-weight organics (i.e., humic substances) may represent a large part of the NOM adsorbed on membranes. Clark and Jucker (1993) observed that humic acids adsorb on the membrane studied to a greater extent than fulvic acids, finding that they were responsible for the lower apparent solubility (i.e., higher molecular weight) of humic acids. More recently, Nilson and DiGiano (1996) have shown hydrophobic NOM to be responsible for nearly all permeate decline and that it was more highly rejected by nanofiltration membranes than hydrophilic NOM.

Because only a few studies have focused on NOM isolation using reverse osmosis, and because of analytical interferences due to salts that are concentrated along with the NOM, few data have been published regarding the nature of NOM isolated with RO membranes. The most important finding is that RO isolates are more hydrophilic in character than XAD resin isolates. Serkiz and Perdue (1990) concluded that polysaccharides and polypeptides were more abundant in the organic materials isolated using reverse osmosis as compared to humic substances isolated from the same surface water (Suwannee River) using XAD-8 resin. Croué et al. (1996) also found that the RO isolate obtained from the Blavet River (France) was enriched in proteins and aminosugars as compared to the mixed XAD-8/XAD-4 isolate.

NATURAL ORGANIC MATTER CHARACTERISTICS

This section focuses mainly on the structural characteristics of XAD-8 (hydrophobic NOM) and XAD-4 (transphilic NOM) isolates from a large variety of waters.

Elemental Analysis

Among the various ways of characterizing NOM isolates, elemental analysis is generally the first approach that researchers use (Huffman and Stuber 1985). Elemental analysis commonly includes carbon (C), hydrogen (H), oxygen (O), nitrogen (N), sulfur (S), and ash contents of the isolates. Phosphorus and halogen analysis is more rarely performed. Results are given in percent by weight, and some specific ratios, such as C/H, C/O, and C/N are often used to describe the structural characteristics and origin of NOM.

Table 4-1 Elemental analysis of hydrophobic acids and transphilic acids isolated from surface waters

Source	Fraction	C %	H %	N %	O %	S %	Ash %
Yakima River	Hydrophobic acids*	56.1	4.95	2.2	35.5	0.97	1.1
(Washington) (Aiken et al. 1992)	Transphilic acids	50.5	4.4	3.0	40.6	1.2	3.9
Lake Skjervatjern	Humic acids	55.8	3.58	0.96	36.9	0.32	1.18
(Norway) (September 1990)	Fulvic acids	54.2	3.96	0.56	39.3	0.24	0.33
(Malcolm et al. 1993)	Transphilic acids	50.2	4.0	0.97	43.8	0.51	0.85
Apremont reservoir	Humic acids	48.1	4.9	3.04	36.1	2.58	4.7
(France) (April 1991)	Fulvic acids	49.7	4.9	2.14	39.5	1.88	1.5
(Martin 1995)	Transphilic acids	41.1	4.4	3.1	41.1	1.6	ND†

*Fulvic acids generally account for 90 percent of the hydrophobic acids.
†Not determined.

The elemental analysis database available in the literature mainly includes hydrophobic acid fractions (with or without fractionation into humic and fulvic acids), with some results on transphilic acid fractions. Information on the elemental composition of hydrophilic base and neutral fractions of NOM are restricted to recent isolation work conducted by USGS.

The most important findings to date are

- the acid fractions are characterized by lower C/O ratios compared to other fractions
- the base fraction shows the highest N content, which is consistent with its postulated nature, and the C/N ratios of the neutral fractions are lower than the C/N ratios calculated for the corresponding acid fractions
- the C/H ratio is generally lower for the neutral fractions compared to the acid fractions (more aliphatic in nature)

Table 4-1 compares the elemental composition of hydrophobic acids (i.e., humic and fulvic acids) and transphilic acids isolated from three different water sources. In this table, the hydrophobic acid fraction isolated by Aiken et al. (1992) from the Yakima River can be compared directly to the other fulvic acid fractions because humic acids generally represent a minor fraction of the overall hydrophobic acid fraction. The transphilic acids contain less carbon and more oxygen than the humic and/or fulvic acids from the same source. For all sources, fulvic acids show the lowest proportion of nitrogen, while the nitrogen content of transphilic and humic acids are similar.

Specific UV Absorbance

Specific UV absorbance (SUVA) is defined as the UV absorbance of a given sample determined at 254 nm divided by the DOC concentration of the solution. It is expressed in units of m^{-1} L/mg C. A strong correlation was noted by Chin, Aiken, and O'Loughlin (1994) between SUVA and the aromatic-carbon contents of a large number of fulvic acids as determined by nuclear magnetic resonance (^{13}C-NMR). Figures 4-3A and 4-3B, containing data from the literature, demonstrate the good relationship between these two parameters.

SUVA is, therefore, a good predictor of the aromatic-carbon content of NOM, and can be used to estimate the chemical nature of the DOC at a given location. It should be noted, however, that interference caused by the UV absorbance of nitrate may become problematic for low-DOC waters that have high nitrate concentrations.

Some features of SUVA include the following:

- The transphilic acid (TPHA) fraction contains organic matter that is of lower molecular weight (see below) and has a lower SUVA compared to the hydrophobic acid (HPOA) fraction. For a given water, the order of SUVA is HA > FA > TPHA.
- The SUVA of the whole water sample is largely dependent on the amount and SUVA of the HPOA fraction.
- For different water samples from a given aquatic system, variations in the SUVA of the whole water sample resulting from either temporal or spatial variability are reflected in differences in both the fractionation patterns and in the quality of the material contained in the fractions (Table 4-2).

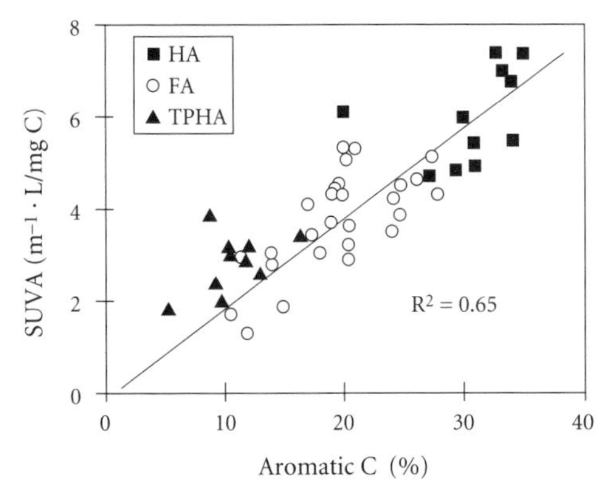

A. HA, FA, and TPHA isolates reported in the literature (Reckhow, Singer, and Malcolm 1990; Legube et al. 1990; Martin 1995; unpublished data from Aiken, Croué, Debroux, and Malcolm)

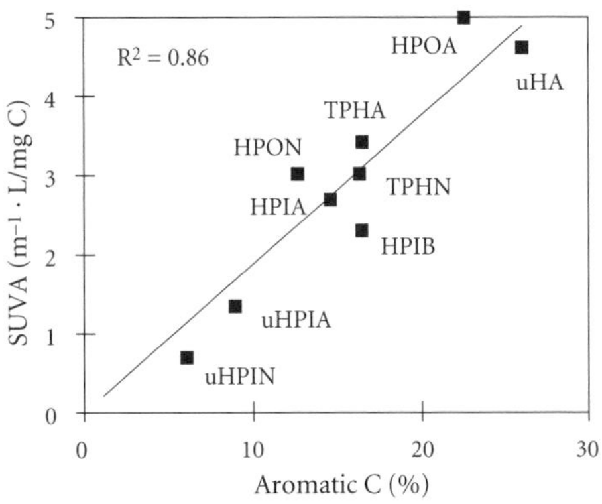

B. Suwannee River NOM isolates (Croué et al. 1998)

Figure 4-3 Relationship between SUVA and aromatic-carbon content determined by ^{13}C-NMR for various to NOM isolates

Table 4-2 Specific UV absorbance (SUVA) ($m^{-1}L/mg$ C) of the Apremont reservoir (France) and Florida Everglades isolates

	Apremont reservoir (France)		Florida Everglades	
	November 1990	April 1991	F1	2BS
Bulk water	2.6	3.7	3.8	2.2
Humic acids (HA)	4.6	5.1		
Fulvic acids (FA)	3.1	3.7	4.5*	3.6*
Hydrophobic neutrals (HPON)	2.0	2.3	3.6	ND†
Transphilic acids (TPHA)	2.0	2.3	2.3	2.7
Transphilic neutrals (TPHN)	ND	0.8	1.7	ND
Hydrophilics (HPI)	1.8	1.6	ND	1.2

*Values for the HPOA fraction.
†Not determined.

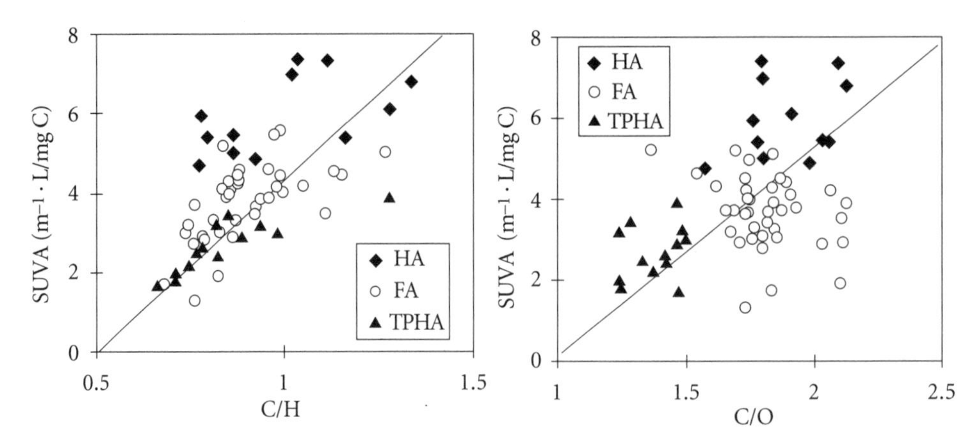

Figure 4-4 Relationship between elemental analysis and SUVA (after Reckhow, Singer, and Malcolm 1990; Legube et al. 1990; Martin 1995; and unpublished data from Aiken, Croué, Debroux, and Malcolm)

Elemental analysis and SUVA data from the literature are plotted in Figure 4-4. Because the C/H ratio is considered an indicator of the degree of unsaturation of organics, it is not surprising to observe that a general trend exists between this atomic ratio and SUVA. The correlation between SUVA and C/O is not as strong, however.

Molecular Weight

Problems associated with determining molecular weights of mixtures have long been recognized, particularly with regard to humic substances (Wershaw and Aiken 1985). Lansing and Kraemer (1935) pointed out that the usual methods for determining molecular weights of mixtures yield average molecular weights and, depending on the methods used, these averages are not directly comparable. Failure to consider this fact is one source of confusion in comparing molecular weight data for NOM.

In addition to methods that actually measure molecular weight, there are a number of methods commonly used in the study of NOM that determine molecular size (Wershaw and Aiken 1985). These methods include gel filtration, high-pressure size exclusion chromatography, ultrafiltration, and small-angle X-ray scattering. In these methods, model compounds of known molecular weight and composition are used to estimate the molecular weight of NOM. Problems can arise if the model compounds are not sufficiently similar to the NOM of interest (Beckett, Jue, and Giddings 1987). Choice of appropriate model compounds and instrumental operating conditions are hampered by a lack of detailed information about the chemical composition of NOM. Other problems arise due to methods of detection, such as UV absorbance, which are not universal detectors, but rely on a property of the organic matter that may not be shared equally by all molecules in the mixture (Huber and Frimmel 1991).

The majority of molecular weight data reported in the literature are for aquatic humic substances and the hydrophobic acid fraction. The reported values can vary widely depending on the methods used, and few studies have presented molecular weights determined by a variety of methods. Recently, Chin, Aiken, and O'Loughlin (1994) presented data for organic matter isolates and whole water samples that were determined by high-pressure size exclusion chromatography (HPSEC), vapor pressure osmometry ([VPO], a colligative property measurement), and field flow fractionation. While the values for different samples ranged from 490 to 14,500 daltons, there was good general agreement between the results obtained by the three methods for a given sample. The results of this work suggest that the dissolved NOM is smaller than previously believed, and the molecular size distribution is constrained within a narrow range. This study also showed that, in general, the molar absorptivity at 280 nm correlates well with both molecular weight and aromaticity of the aquatic humic substances used. Fewer data have been reported on the molecular weight and size of other fractions of NOM. Aiken et al. (1992) reported a molecular weight of 411 daltons for a sample of the transphilic organic acid fraction from the Yakima River at Kiona, Wash. The corresponding hydrophobic acid sample had a molecular weight of 650 daltons as determined by VPO.

^{13}C-NMR Spectrometry

^{13}C-NMR analysis provides both NOM organic structure information and carbon-containing functional group information. Because of the considerable overlap in the chemical shift for various types of

Table 4-3 Relative peak areas (in percent) of ^{13}C-NMR spectra of the Yakima River (WA) and Mayenne River (France) NOM isolates

		Integrated Values (percent) for Different Integration Ranges (ppm)					
		0–60	60–90	90–110	110–160	160–190	190–220
Source	Fractions	C-C/C-H	C-O	Anomeric C	Aromatic C	Carboxylic C	Carbonyl C
Yakima River (Aiken et al. 1992)	Fulvic acids	34.2	12.8	← 30.1 →		19.3	3.6
	Transphilic acids	32.0	18.0	← 23.4 →		22.8	3.8
Mayenne River (Martin 1995)	Humic acids	30.9	21.7	6.5	32.9	10.7	2.7
	Fulvic acids	39.1	17.6	4.2	20.4	15.6	3.1
	Transphilic acids	33.4	29.5	7.7	9.7	17.6	1.7

structures, quantitative determination of various structures in NOM depends on the judgment of the analyst. The structural characteristics of NOM depend on the origin and nature of the water source. However, some general observations can be made from this technique.

Two major structural differences exist between the two most abundant acid fractions (Table 4-3). First, the transphilic acids have greater carboxyl and heteroaliphatic (C-O of aliphatic alcohol, ester ether) carbon contents than the corresponding fulvic acids. This is in accordance with the higher C/O atomic ratio generally observed for fulvic acids as compared to transphilic acids. Second, fulvic acids have a greater amount of aromatic-carbon content than the corresponding transphilic acids.

The humic acid fraction is the least enriched in carboxyl carbon content and the most enriched in aromatic carbon, structural characteristics that partially explain its low aqueous solubility. A sharp and well-resolved peak in the 55 to 60 ppm chemical shift region is often seen in humic acid spectra; this is an indicator of methoxyl groups of lignin-type structures within the matrix. Finally, the spectra of humic acid fractions generally show peaks around 155 and 105 ppm, which are indicative of aromatic phenolic substituents that are known to be associated with chlorine reactivity (Hanna et al. 1991).

Isolation work conducted on two surface waters by Leenheer (1996) confirms that the hydrophilic acid fraction has the largest carboxylic acid content, while the hydrophilic neutral fraction has the largest carbohydrate content.

Pyrolysis Gas Chromatography/Mass Spectrometry (GC/MS)

During the last decade, a number of research projects have used conventional pyrolysis GC/MS (flash pyrolysis with final temperature around 700°C was more generally used) to trace the evolution of NOM as a function of season or across drinking water treatment processes, and also for the characterization of aquatic NOM fractions (Gadel and Bruchet 1987; Bruchet, Rousseau, and Mallevialle 1990; Krasner et al. 1996; Harrington et al. 1996; Gray, Simpson, and McAuliffe 1996). Data from this technique can be interpreted with a semiquantitative approach (Bruchet, Rousseau, and Mallevialle 1990).

The interpretation of NOM pyrochromatograms obtained from conventional pyrolysis GC/MS is generally based on the determination of the peak intensity of some specific pyrolysis fragments that are chosen as indicators of the presence and distribution of the four biopolymer classes: polysaccharides, proteins, aminosugars and polyhydroxyaromatics (Bruchet 1985; Gadel and Bruchet 1987; Peschel and Wildt 1988; Bruchet, Rousseau, and Mallevialle 1990; Biber, Gülacar, and Buffle 1996). In general, chromatograms of hydrophobic and transphilic acid fractions provide clear evidence of their aromatic character with the presence of a large peak of phenol (Martin 1995; Krasner et al. 1996; Harrington et al. 1996). These authors established good correlations between the intensity of the peak of phenol and the aromatic carbon content as determined by ^{13}C-NMR. Compared to fulvic acids, both transphilic and humic acids appear to contain a larger proportion of proteins (peaks of toluene, styrene, pyrrole, and benzonitrile) and aminosugars (peaks of acetamide and ethanamide) in their structures, which can be related to their higher nitrogen content (Martin 1995; Krasner et al. 1996). Bruchet (1985) also observed that humic acids are structurally more heterogeneous than fulvic acids showing carbohydrates as the most important class of constituents for some water sources.

REACTIVITY WITH CHLORINE

The quantity and speciation of DBPs in finished drinking water can vary not only by water quality (i.e., temperature, pH, bromide, and DOC concentrations; see chapter 2), but also by the distribution and properties of the organic molecules that comprise NOM.

Table 4-4 Adsorption of disinfection by-product precursors from the Apremont reservoir (France) (Nov. 21, 1990) on XAD-8 and XAD-4 resins used in series (Croué et al. 1993)

	DOC	UV Abs 254 nm	TOXFP		TTHMFP	
	(mg C/L)	*(cm^{-1})*	*(µg Cl/L)*	*(µg Cl/mg C)*	*(µg/L)*	*(µg/mg C)*
Raw water	11.3	0.29	1,819	161	658	58
XAD-8 filtrate	5.8	0.10	551	95	121	21
XAD-4 filtrate	3	0.05	168	56	21	7

NOTE: Conditions: 72-h contact time, Cl_2/C = 4 mg/mg, 20°C.

Adsorption of DBP Precursors Onto XAD Resins

The isolation of DBP precursors with XAD resins has been the most commonly used approach to study the reactivity of NOM with chlorine. Croué et al. (1993) have shown that 73 percent of the DOC and more than 90 percent of the DBP precursors of the Apremont reservoir (France) could be adsorbed onto the XAD-8 and XAD-4 resins when used in series (Table 4-4). Disinfection by-product precursors were preferentially adsorbed onto the XAD-8 resin (70 percent), but a significant fraction was also retained by the XAD-4 resin (21 percent). The most hydrophilic fraction of the NOM (i.e., the fraction of the NOM analyzed in the XAD-4 filtrate) contained a small portion of the DBP precursors of the water, but these precursors were found to be quite reactive with chlorine.

Distribution of DBP Precursors: Hydrophobic and Hydrophilic Fractions of NOM

Few studies have been devoted to comparing the reactivity of hydrophobic and hydrophilic NOM isolated from surface waters. More work has been done on the comparison of humic and fulvic acids, the major constituents of the hydrophobic NOM (Table 4-5).

The chlorination study conducted by Labouyrie-Rouillier (1997) on the NOM fractions isolated by Leenheer (1996) from the Suwannee River (Georgia) showed that all fractions of NOM are more or less reactive with chlorine (Table 4-6). The chlorine demand varied from 0.8 to 2.8 mg Cl_2/mg C and no direct correlation could

Table 4-5 Chlorine demand and DBP formation potentials of hydrophobic and transphilic acid fractions isolated from surface waters

Origin	Fraction	Chlorine Demand (*mg Cl_2/mg C*)	TOXFP (*μg Cl/mg C*)	THMFP (*μg $CHCl_3$/mg C*)	Reference
Apremont reservoir (France)	Humic acids	1.5	266	82	Martin (1995)*
	Fulvic acids	1.4	240	74	
	Transphilic acids	1.2	157	56	
Hellurude Lake (Norway)	Humic acids	2.9	354	93	"
	Fulvic acids	2.3	273	91	
	Transphilic acids	2.1	267	82	
Mayenne River (France)	Humic acids	2.9	231	79	"
	Fulvic acids	ND	174	41	
	Transphilic acids	ND	115	28	
Black Lake (North Carolina)	Humic acids	2.3	288	68	Reckhow, Singer, and Malcolm (1990)†
	Fulvic acids	1.5	208	48	
Coal Creek (Colorado)	Humic acids	2.0	268	59	"
	Fulvic acids	1.6	232	51	
Ogeechee River (Georgia)	Humic acids	2.1	262	59	"
	Fulvic acids	1.5	216	50	
Ohio River (Ohio)	Humic acids	2.1	232	47	"
	Fulvic acids	1.2	161	32	
Missouri River (Iowa)	Humic acids	1.1	230	50	"
	Fulvic acids	2.1	136	31	
Suwannee River (Georgia)	Humic acids	ND	ND	115	Oliver and Thurman (1983)‡
	Fulvic acids	ND	ND	81	
Biscayne Groundwater (Florida)	Humic acids	ND	ND	53	"
	Fulvic acids	ND	ND	45	

*pH 7.5, Cl_2/TOC = 4 mg/mg, 3-day contact time, 20°C
†pH 7, Cl_2/TOC = 4 mg/mg, 3-day contact time, 20°C
‡pH 7, Cl_2/TOC = 10 mg/mg, 7-day contact time, 20°C

be established between this parameter and DBP formation potentials (DBPFPs). The ultrahydrophilic humic acids (uHA [Leenheer 1996]) fraction that was found to be colloidal in nature was the most reactive with chlorine. With the exception of this particular fraction, the HPOA (i.e., humic substances) and TPHA, which were the most abundant fractions, were the highest total organic halide (TOX) and

Table 4-6 Chlorine demand and disinfection by-product formation potentials of Suwannee River (6A) NOM isolates (Labouyrie-Rouillier 1997)

Fraction	Chlorine Demand (*mg Cl_2/mg C*)	TOXFP (*μg Cl/mg C*)	THMFP (*μg $CHCl_3$/mg C*)	TCAAFP (*μg/mg C*)	DCAAFP (*μg/mg C*)
HPON	1.5	182	51	51	24
HPOA	1.4	268	55	59	25
TPHA	1.4	224	40	57	23
TPHN	1.2	204	40	44	22
HPIA	0.8	171	36	36	22
$HPIA_2$	1.1	150	41	34	17
uHPIA	0.78	121	28	30	21
HPIB	2.5	175	29	31	39
uHA	2.8	285	63	66	45
uHPIN	0.8	117	23	26	22

NOTES:
HPOA and HPON = hydrophobic acids and neutrals
TPHA and TPHN = transphilic acids and neutrals
HPIA and HPIB = hydrophilic acids and bases
$HPIA_2$ = a portion of HPIA fraction
uHA = ultrahydrophilic humic acids
uHPIA and uHPIN = ultrahydrophilic acids and neutrals (see Leenheer [1996] for details regarding the NOM fractionation)
TCAAFP = trichloroacetic acid formation potential
DCAAFP = dichloroacetic acid formation potential
Formation potential conditions: 72-h contact time, 20°C, pH 8, Cl_2/C = 4 mg/mg

trichloroacetic acid (TCAA) precursors. The hydrophobic NOM fractions (HPOA and HPON) gave the largest trihalomethane formation potentials (THMFPs), while no significant differences between fractions were observed regarding the dichloroacetic acid formation potential (DCAAFP) in general.

Martin (1995) compared the reactivity of humic, fulvic, and transphilic acids isolated from nine surface waters at pH 7.5 (three-day contact time). Table 4-5 shows the results obtained for three of them. As observed for all water sources, humic acids were more reactive with chlorine than fulvic acids, which were more reactive than transphilic acids. The average chlorine demands were 2.8, 2, and 1.7 mg Cl_2/mg C for humic, fulvic, and transphilic acids, respectively. For the same fractions, the average TOX formation

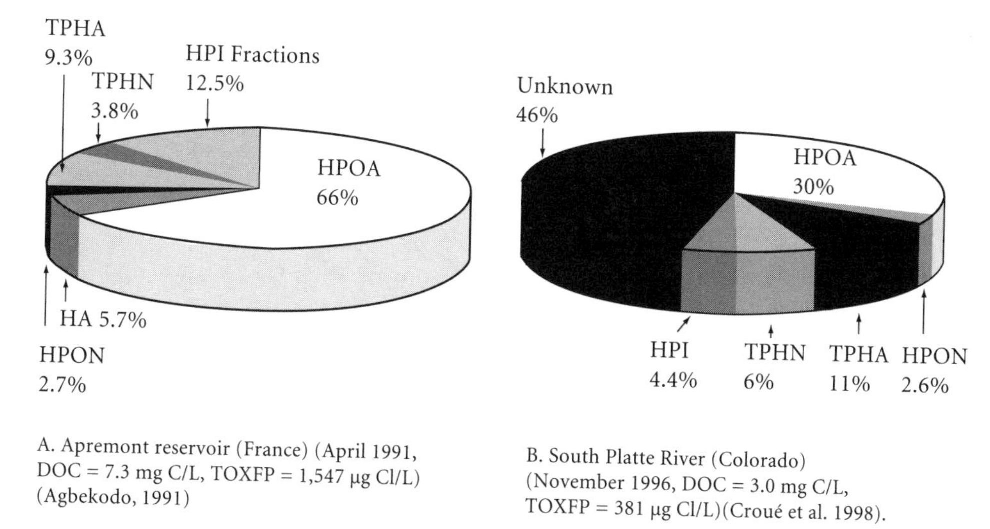

A. Apremont reservoir (France) (April 1991, DOC = 7.3 mg C/L, TOXFP = 1,547 µg Cl/L) (Agbekodo, 1991)

B. South Platte River (Colorado) (November 1996, DOC = 3.0 mg C/L, TOXFP = 381 µg Cl/L)(Croué et al. 1998).

Figure 4-5 Distribution of TOX precursors in two surface waters

potentials (TOXFPs) were 327, 247, and 188 µg Cl/mg C, and the average THMFPs were 97, 72, and 58 µg $CHCl_3$/mg C.

The work conducted by Reckhow, Singer, and Malcolm (1990) with five surface waters (pH 7; three-day contact time) led to similar conclusions (Table 4-5) showing humic acids to be greater chlorine consumers than the corresponding fulvic acids (2.1 as compared to 1.4 mg Cl_2/mg C as mean values). In all cases, DBPFPs were also greater for humic acids than for the corresponding fulvic acids from the same source (256 and 191 µg Cl/mg C, 56.8 and 42.6 µg $CHCl_3$/mg C, and 83 and 46.2 µg TCAA/mg C, as mean values). Using longer chlorine contact time (one week rather than three days), Oliver and Thurman (1983) reached similar conclusions (Table 4-5).

Because the HPOA fraction, and in particular fulvic acids, typically represent the most abundant fraction of the DOC of surface waters, and because they incorporate some of the most chlorine-reactive sites, hydrophobic acids (i.e., humic substances) can be considered as the major DBP precursors of humic-type waters. Figure 4-5A shows the contribution of the isolated hydrophobic and transphilic acid fractions of the Apremont reservoir (France) to the TOXFP of the raw water (Agbekodo 1991). In contrast, Figure 4-5B shows that for low DOC waters, such as the South Platte

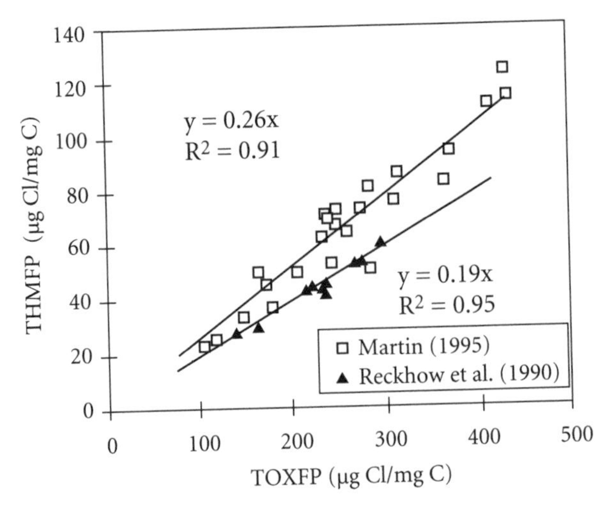

Figure 4-6 Relationship between THMFP and TOXFP for aquatic humic, fulvic, and transphilic acids (Reckhow, Singer, and Malcolm 1990: pH 7, 72 h, 20°C; Martin 1995: pH 7.5, 72 h, 20°C)

River (Colorado), the hydrophobic and transphilic fractions comprise only 50 percent of its TOX precursors, with the other half being comprised of hydrophilic colloids (Croué et al. 1998). For both waters, because the determination of the TOXFP of the isolated fractions was conducted in the absence of bromide, the assumption was made that the presence of bromide in the raw water would have no significant effect on the production of TOX.

Relative Proportion of the DBPs

Trihalomethanes (THMs), such as chloroform, represent an important fraction of the DBPs formed during the chlorination of humic materials. Figure 4-6 plots the THM (chloroform) yield (expressed as μg Cl/mg C) as a function of the TOX yield for a large variety of NOM isolates studied by Martin (1995) and Reckhow, Singer, and Malcolm (1990). A good relationship between the two parameters was obtained with both sets of data. For Martin (1995), who studied humic, fulvic, and transphilic acids, chloroform accounted for about 25 percent of the TOX, while for Reckhow,

Singer, and Malcolm (1990), who studied only humic and fulvic acids, the ratio was close to 20 percent. The difference can be attributed to the different pH conditions; the higher the pH, the higher the production of chloroform and the lower the formation of TOX (Reckhow and Singer 1985). It is quite interesting to observe that whatever the origin and nature of the acid fraction (i.e., hydrophobic or transphilic acids) of NOM, the proportion of chloroform into TOX remains almost constant.

Chlorination experiments conducted at pH 8 (72 hours, 20°C) by Labouyrie-Rouillier (1997) on the Suwannee River (6A) NOM fractions obtained from the quantitative isolation procedure developed by Leenheer (1996) showed that the THMFP/TOXFP ratio varied with the nature of the fraction, from 0.14 for the lowest (i.e., hydrophilic base fraction) to 0.24 for the highest (i.e., hydrophobic base fraction). Including the results obtained with the 10 isolated fractions (see Table 4-5), the linear regression gave the following equation: THMFP = 0.18 TOXFP with $R^2 = 0.75$.

Among the 10 humic and fulvic acid fractions studied by Reckhow, Singer, and Malcolm (1990) at neutral pH, only two fulvic acids produced more THM (chloroform) than TCAA (expressed as µg THM or TCAA/mg C), but for all 10 extracts, the sum of the yields of TCAA and dichloroacetic acid (DCAA) were greater than the THM yield. When expressed as a percentage of the TOX, the proportion of THM (average: 19.2 percent) was greater than the proportion of TCAA (average: 14.8 percent) for fulvic acids (with one exception). The opposite was found for humic acids (19.1 and 20.4 percent, respectively, as average values). Dichloroacetic acid accounted for 4.6 to 6.5 percent of the TOX, while the proportion of trichloroacetone (TCAC) and dichloroacetonitrile (DCAN) was less than 1 percent.

At pH 7.5, Croué (1987) observed that fulvic acids isolated from three surface waters produced slightly more chloroform than TCAA on chlorination (expressed as µg THM or TCAA/mg C). Again, the sum of the haloaceticacids (TCAA + DCAA) was, in all cases, greater than the THM concentration.

The production of TCAA was similar or higher than the production of chloroform at pH 8 for all fractions isolated from the Suwannee River (Table 4-5 [Labouyrie-Rouillier 1997]). As a general trend, THM and TCAA, accounted for 19 percent ($R^2 = 0.75$)

and 15 percent ($R^2 = 0.81$) of the TOX, respectively. The sum of THM, TCAA, and DCAA accounted for 37 to 52 percent (average: 42 percent) of the TOX depending on the fraction.

These results show that not only are the chlorination conditions important parameters that control the production and distribution of the DBPs (i.e., increasing pH above 6 enhances the production of chloroform but reduces the formation of TCAA [Reckhow and Singer 1985]), the origin and nature of the NOM plays an important role, too. In finished drinking waters, THMs are generally present as the largest class of DBPs, and haloacetic acids (HAAs) are the next most significant class. The opposite has been shown from the chlorination studies of humic substances. It is then reasonable to believe that the more hydrophilic fractions of NOM, which become predominant in finished drinking waters (i.e., humic materials are preferentially removed during conventional treatment; see chapter 12), are more significant precursors of THMs than HAAs. The presence of bromide, commonly present in natural waters but generally absent in NOM fraction studies, could also be an important factor.

Relation With Structure

Various organic compounds, or structures within these compounds, possess the propensity to form $CHCl_3$. De Laat, Merlet, and Doré (1982) and Reckhow and Singer (1985) found that various organic compounds can form chloroform on chlorination. These compounds possess varying efficiencies in the formation of chloroform, as electron-rich moieties are extremely vulnerable to electrophilic attack. This attack may be the first in a series of oxidation and substitution reactions that eventually produce smaller, commonly found DBPs (e.g., THMs and HAAs).

Specific UV absorbance was found to be a good predictor of the aromatic-carbon content of hydrophobic organics (see Figure 4-3). As expected, good relationships can also be established between this parameter and the TOXFP or THMFP of humic substances (Figure 4-7). However, pH conditions appear to be a more important parameter determining the linear regression

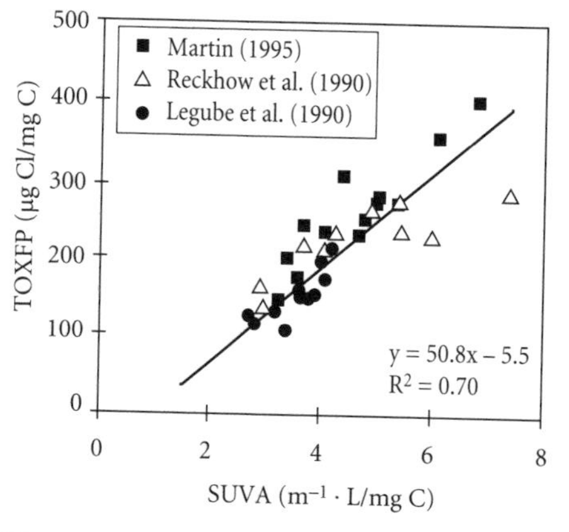

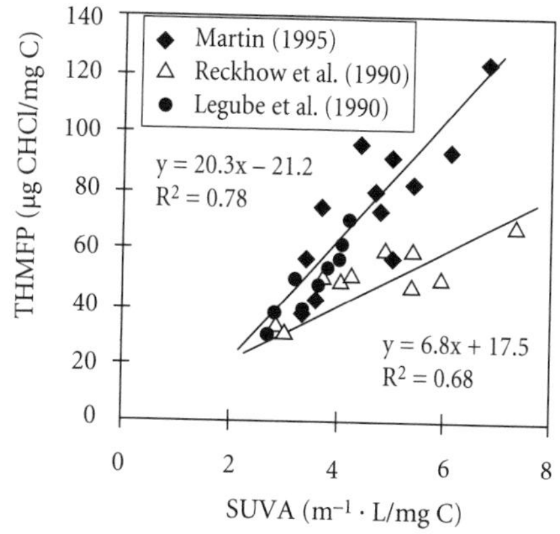

Figure 4-7 Relationship between TOXFP and THMFP and SUVA for chlorinated hydrophobic acids (experimental conditions: Reckhow, Singer, and Malcolm 1990: pH 7, 72 h, 20°C; Martin 1995 and Legube et al. 1990: pH 7.5, 72 h, 20°C)

relationship for the THMFP values than for the TOXFP values. Previous work by Oliver and Thurman (1983) showed that THMFP could be correlated with the specific absorbance of the NOM at 400 nm (color/mg C). Because SUVA was also found to be a good predictor of molecular weight (Chin, Aiken, and O'Loughlin 1994) and fluorescence efficiency (Belin et al. 1996) of hydrophobic NOM, correlation between these two parameters and THMFP are expected (Oliver and Thurman 1983; Debroux, unpublished data). Reckhow, Singer, and Malcolm (1990) and Debroux (unpublished data) noticed that correlations between the characteristics of humic materials and chlorine consumption were generally stronger than those for THMFP.

A comparison of the reactivity of NOM fractions (including hydrophobic and transphilic acids [Martin 1995]) with chlorine confirmed the relationship between SUVA and THMFP or TOXFP. However, as shown in Figure 4-8A, better relationships were obtained when each fraction was considered separately (similar observations were made for THMFP). This may indicate that the nature of the aromatic precursor sites differ with

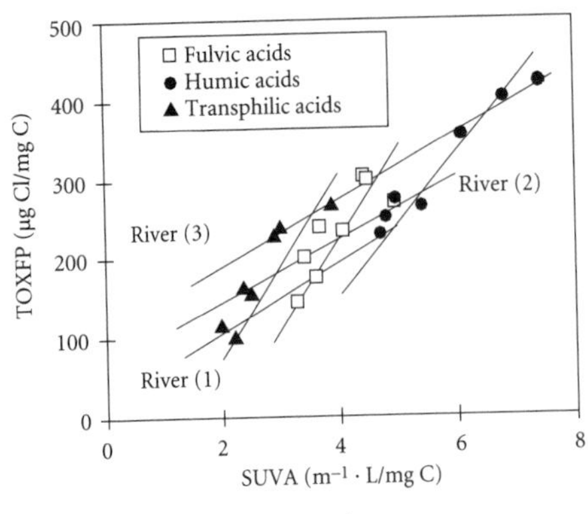

A. Comparison among several water sources (Martin 1995)

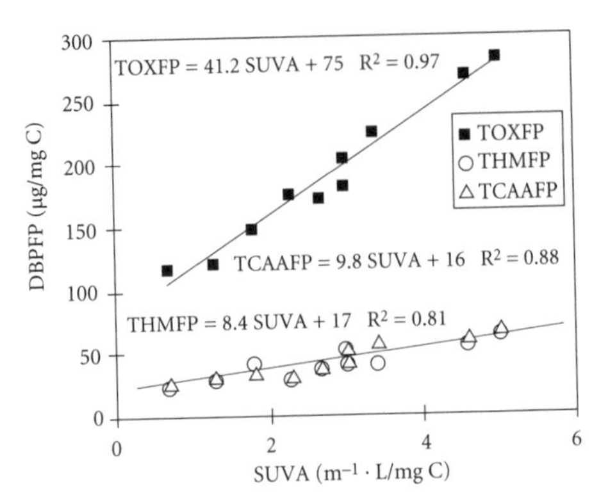

B. Comparison of the NOM fractions isolated from the Suwannee River (Labouyrie-Rouillier 1997)

Figure 4-8 Chlorination of hydrophobic and hydrophilic fractions of NOM: Relationship between SUVA and DBPFP

the characteristics of the NOM fraction. Reckhow, Singer, and Malcolm (1990) indicated that the nature of the aromatic sites, and in particular the relative abundance of activated and nonactivated rings, is one of the major parameters governing the reactivity of NOM with chlorine. An estimation of the concentration of activated aromatic structures on a probabilistic basis led to the determination of two distinct correlations for humic and fulvic acids.

Figure 4-8B shows that good correlations can be obtained between SUVA and TOXFP, THMFP, or TCAAFP for the isolates of a selected water. This supports the importance of aromaticity in the production of chlorinated DBPs regardless of the nature of the specific fraction. Nevertheless, the relationship between DBPFPs and SUVA for isolated fractions appears to differ between sites, indicating that the structural characteristics of the DBP precursors relate to the origin of the NOM.

CONCLUSIONS

The examples presented illustrate that NOM derived in different ecosystems can differ significantly in hydrophobicity. These differences are related to the varying behavior of NOM observed in water treatment and natural systems. Understanding these differences is paramount in refining treatment systems as health effects of DBPs become realized and stricter regulations are promulgated.

Powerful tools have been developed and applied to define the general properties (size, structure, and functionality) of NOM. Some (e.g., ^{13}C-NMR) require isolation and concentration of the NOM for analytical sensitivity. Others (e.g., SUVA) are applicable to bulk water samples. It has been demonstrated that SUVA is a good predictor of the aromatic-carbon content of NOM, regardless of the origin and nature of the NOM fractions.

Work on NOM fractions has helped identify problematic fractions that may be targeted for removal before application of a disinfectant. In general, hydrophobic acids (i.e., humic substances) and transphilic acids that account for 60 to 70 percent of the DOC of colored waters are the major DBP precursors. Good removal efficiency can be observed for these fractions of NOM using conventional drinking water processes. For low DOC, noncolored waters, the more hydrophilic fractions of NOM seem to comprise a large part of the DBP precursors. More work needs to be done to characterize these fractions of NOM.

There has been some success in linking general chemical properties to reactive properties of NOM (DBP formation and disinfectant demand), providing some mechanistic insight into NOM-disinfectant interactions. Although there is not currently a measurable physical-chemical parameter that solely identifies DBP precursors, SUVA appears to be a good surrogate parameter to predict DBP formation potentials, regardless of the nature of the NOM in a particular water source. However, it has been shown that NOM fractions isolated from various water sources did not exhibit the same chlorine reactivity (e.g., chlorine demand and DBP formation potential) even if they had similar SUVA values. As varying environments produce NOM of different structure and quantity, understanding these differences will improve the understanding of DBP formation.

REFERENCES

Agbekodo, K. 1991. Utilisation des Résines Macroporeuses XAD-8 et XAD-4 Pour l'extraction et la Caractérisation du Carbone Organique Dissous d'une eau de Barrage. Diplôme d'Etudes Approfondies Chimie et Microbiologie de l'Eau, Université de Poitiers, France.

Aiken, G.R. 1988. A Critical Evaluation of the Use of Macroporous Resins for the Isolation of Aquatic Humic Substances. *Humic Substances and Their Role in the Environment.* Frimmel, F.H., and R.F. Christman, eds. Dahlem Workshop reports. New York: John Wiley & Sons.

Aiken, G.R., D.M. McKnight, K.A. Thorn, and E.M. Thurman. 1992. Isolation of Hydrophilic Organic Acids From Water Using Nonionic Macroporous Resins. *Organic Geochemistry,* 18:567–573.

Aiken, G.R., and J. Leenheer. 1993. Isolation and Chemical Characterization of Dissolved and Colloidal Organic Matter. *Chemistry and Ecology,* 8:135–151.

Aiken, G.R., D.M. McKnight, R. Harnish, and R. Wershaw. 1996. Geochemistry of Aquatic Humic Substances in the Lake Fryxell Basin. Antarctica, *Biogeochemistry,* 34:157–188.

Andrews, S.A., and P.M. Huck. 1994. Using Fractionated Natural Organic Matter to Quantitate Organic By-products of Ozonation. *Ozone Science and Engineering,* 16:1–12.

Beckett, R., Z. Jue, and J.C. Giddings. 1987. Determination of Molecular Weight Distributions of Fulvic and Humic Acids Using Flow Field-Flow Fractionation. *Environ. Sci. & Tech.,* 21:289–295.

Belin, C., J-P Croué, M. Lamotte, A. Deguin, and B. Legube. 1996. Characterization of Natural Organic Matter Using Fluorescence Spectroscopy. In Proc. of the NOM workshop *Influence of Natural Organic Matter Characteristics on Drinking Water Treatment and Quality.* Poitiers, France.

Biber, M.V., F.O. Gülacar, and J. Buffle. 1996. Seasonal Variations in Principal Groups of Organic Matter in a Eutrophic Lake Using Pyrolysis GC/MS. *Environ. Sci. & Tech.,* 30:3501–3507.

Bruchet, A. 1985. Applications de la Technique de Pyrolyse-CG-SM à l'étude des Matières Organiques Non Volatiles des Eaux Naturelles et en Cours de Traitement. Thèse de 3ème cycle, Université de Poitiers, France.

Bruchet, A., C. Rousseau, and J. Mallevialle. 1990. Pyrolysis-GC-MS for Investigating High Molecular Weight THM Precursors and Other Refractory Organics. *Jour. AWWA,* 82:66–74.

Chin, Y., G. Aiken, and E. O'Loughlin. 1994. Molecular Weight, Polydispersity, and Spectroscopic Properties of Aquatic Humic Substances. *Environ. Sci. & Tech.,* 28:1853–1858.

Clair, T.H., J.R. Kramer, M. Sydor, and D. Eaton. 1991. Concentration of Aquatic Dissolved Organic Matter by Reverse Osmosis. *Water Res.,* 25:1033–1037.

Clark, M.M., and C. Jucker. 1993. Interactions Between Hydrophobic Ultrafiltration Membranes and Humic Substances. In *Proc. 1993 AWWA Annual Conference*. Denver, Colo.: American Water Works Association.

Croué, J-P 1987. Contribution à l'Étude de l'Oxydation par le Chlore et l'Ozone d'Acides Fulviques Naturels Extraits d'Eaux de Surface. Doctorat de l'Université de Poitiers, France.

Croué, J-P, B. Martin, P. Simon, and B. Legube. 1993. Les Matières Hyrophobes et Hydrophiles des Eaux de Retenue - Extraction, Caractérisation et Quantification. *Water Supply*, 11:51–62.

Croué, J-P, L. Labouyrie-Rouillier, D. Violleau, E. Lefèbvre, and B. Legube. 1996. Isolation and Characterization of Natural Organic Matter From Surface Water: Comparison of Resin Adsorption and Membrane Filtration Isolation Procedures. In Proc. of the NOM workshop *Influence of Natural Organic Matter Characteristics on Drinking Water Treatment and Quality*. Poitiers, France.

Croué, J-P, G.V. Korshin, J.A. Leenheer, and M.M. Benjamin. 1998. Isolation, Fractionation and Characterization of Natural Organic Matter in Drinking Water. AWWARF report, in preparation.

DeLaat, J., N. Merlet, and M. Doré. 1982. Chlorination of Organic Compounds: Chlorine Demand and Reactivity in Relationship to the Trihalomethane Formation. *Water Res.*, 16:1437–1450.

Gadel, F., and A. Bruchet. 1987. Application of Pyrolysis-Gas Chromatography-Mass Spectrometry to the Characterization of Humic Substances Resulting From Decay of Aquatic Plants in Sediments and Waters. *Water Res.*, 21:1195–1206.

Gray, K.A., A.H. Simpson, and K.S. McAuliffe. 1996. Use of Pyrolysis-Gas Chromatography-Mass Spectrometry to Study the Nature and Behavior of Natural Organic Matter in Water Treatment. *Water Disinfection and Natural Organic Matter: Characterization and Control.* Minear, R.A., and G.L. Amy, eds. ACS Symposium series 649, Amer. Chem. Soc., Washington, D.C.

Hanna, J.V., W.D. Johnson, R.A. Quezada, M.A. Wilson, and L. Xia-Qiao. 1991. Characterization of Aqueous Humic Substances Before and After Chlorination. *Environ. Sci. & Tech.*, 25:1160–1164.

Harrington, G.W., A. Bruchet, D. Rybacki, and P.C. Singer. 1996. Characterization of Natural Organic Matter and Its Reactivity With Chlorine. *Water Disinfection and Natural Organic Matter: Characterization and Control.* Minear, R.A., and G.L. Amy, eds. ACS Symposium series 649, Amer. Chem. Soc., Washington, D.C.

Huber, S.A., and F.H. Frimmel. 1991. Flow Injection Analysis of Organic and Inorganic Carbon in the Low-ppb-range. *Anal. Chem.*, 63:2122–2130.

Huffman, E.W.D., and H.A. Stuber. 1985. Analytical Methodology for Elemental Analysis of Humic Substances. *Humic Substances in Soil, Sediments and Water - Geochemistry, Isolation and Characterization.* Aiken, G.R., et al., eds. New York: Wiley-Interscience.

Krasner, S.W., J-P Croué, J. Buffle, and E.M. Perdue. 1996. Three Approaches for Characterizing NOM. *Jour. AWWA*, 88:66–79.

Labouyrie-Rouillier, L. 1997. Extraction et Caractérisation des Matières Organiques Naturelles Dissoutes d'Eaux de Surface: Etude Comparative des Techniques de Filtration Membranaire et d'Adsorption sur Résines Macroporeuses Non Ioniques. Doctorat de l'Université de Poitiers, France.

Lansing, W.D., and E.O. Kreamer. 1935. Molecular Weight Analysis of Mixtures by Sedimentation Equilibrium on the Svedberg Ultracentrifuge. *J. Am. Chem. Soc.*, 57:1369–1377.

Leenheer, J.A., and E.W.D. Huffman. 1976. Classification of Organic Solutes in Water by Using Macroreticular Resins. *J. Res. U.S. Geol. Surv.*, 4:737–751.

Leenheer, J.A. 1981. Comprehensive Approach to Preparative Isolation and Fractionation of Dissolved Organic Carbon From Natural Waters and Wastewaters. *Environ. Sci. & Tech.*, 15:578–587.

Leenheer, J.A. 1996. Fractionation, Isolation and Characterization of Hydrophilic Constituents of Dissolved Organic Matter in Water. In Proc. of the NOM workshop *Influence of Natural Organic Matter Characteristics on Drinking Water Treatment and Quality*. Poitiers, France.

Legube, B., F. Xiong, J-P Croué, and M. Doré. 1990. Etude Sur les Acides Fulviques Extraits d'Eaux Superficielles Françaises: Extraction, Caractérisation et Réactivité Avec le Chlore. *Rev. Fr. Sci. Eau*, 4:399–424.

Malcolm, R.L., and P. McCarthy. 1992. Quantitative Evaluation of XAD-8 and XAD-4 Resins Used in Tandem for Removing Organic Solutes From Water. *Environmental International*, 18:597–607.

Mantoura, R.F.C., and J.P. Riley. 1975. The Analytical Concentration of Humic Substances From Natural Waters. *Analyt. Chim. Acta*, 76:97–106.

Martin, B. 1995. La Matière Organique Naturelle Dissoute des Eaux de Surface: Fractionnement, Caractérisation et Réactivité. Doctorat de l'Université de Poitiers, France.

Martin-Mousset, B., J-P Croué, E. Lefèbvre, and B. Legube. 1997. Distribution and Characterization of Dissolved Organic Matter of Surface Waters. *Water Res.*, 31:541–553.

Miles, C.J., J.R. Tuschall, and P.L. Brezonik. 1983. Isolation of Aquatic Humus With Diethylaminocellulose. *Anal. Chem.*, 55:410–411.

Nilson, J.A., and F.A. DiGiano. 1996. Influence of NOM Composition on Nanofiltration. *Jour. AWWA*, 88:53–66.

Oliver, B.G., and E.M. Thurman. 1983. Influence of Aquatic Humic Substance Properties on Trihalomethane Potential. *Water Chlorination: Environmental Impact and Health Effects*, Vol. 4. Ann Arbor, Mich.: Ann Arbor Publishers., Inc.

Peschel, G., and T. Wildt. 1988. Humic Substances of Natural and Anthropogeneous Origin. *Water Res.*, 22:105–108.

Reckhow, D.A., and P.C. Singer. 1985. Mechanisms of Organic Halide Formation and Implications With Respect to Pre-ozonation. *Water Chlorination: Chemistry, Environmental Impact and Health Effects.* Chelsea, Mich.: Lewis Publishers.

Reckhow, D.A., P.C. Singer, and R.L. Malcolm. 1990. Chlorination of Humic Materials: Byproduct Formation and Chemical Interpretations. *Environ. Sci. & Tech.*, 24:1655–1664.

Serkiz, S.M., and E.M. Perdue. 1990. Isolation of Dissolved Organic Matter From the Suwannee River Using Reverse Osmosis. *Water Res.*, 24:911–916.

Sun, L., M. Perdue, and J.F. McCarthy. 1995. Using Reverse Osmosis to Obtain Organic Matter From Surface and Ground Waters. *Water Res.*, 29:1471–1477.

Thurman, E.M. 1985. Developments in Biochemistry: Organic Geochemistry of Natural Waters. Dordrecht, Holland: Nijhoff M. and Junk W. Publishers.

Thurman, E.M., and R.L. Malcolm. 1981. Preparative Isolation of Aquatic Humic Substances. *Environ. Sci. & Tech.*, 15:463–466.

Wershaw, R.L., and G.R. Aiken. 1985. Molecular Weight and Size Measurements of Humic. *Humic Substances in Soil, Sediment, and Water: Geochemistry, Isolation, and Characterization.* Aiken, G.R., et al., eds. New York: John Wiley.

Wiesner, M.R., and S. Chellam. 1992. Mass Transport Considerations for Pressure-Driven Membrane Processes. *Jour. AWWA*, 84:88–95.

CHAPTER • FIVE

*Epidemiology and Toxicology of Disinfection By-Products**

JENNIFER ORME ZAVALETA

FRED S. HAUCHMAN

MICHAEL W. COX

The discovery in 1976 that the disinfection by-product (DBP), chloroform, was an animal carcinogen (NCI 1976) caused questions to be raised concerning the risk trade-off of disinfection as a water treatment technology. Although there has been a concerted research effort over the past 20 years to ascertain health risks associated with various DBPs, some uncertainties concerning the health effects and risks of exposure to low levels of DBPs remain.

* This chapter has been reviewed within the US Environmental Protection Agency (USEPA) and represents the views of the authors; it does not necessarily reflect USEPA policy. Any mention of trade names or products does not constitute endorsement.

SUMMARY OF HEALTH RISKS

In 1995 USEPA released a draft research plan to address and reduce the uncertainties concerning the health effects and risks associated with DBP exposure (USEPA 1995). Implementation of and results from this research plan will support USEPA's efforts to regulate disinfectants, DBPs, and microbial pathogens. The key scientific research questions that shape the health effects research program are

- What are the health effects in communities receiving disinfected drinking water?
- What is the toxicity of individual by-products and mixtures of by-products?
- How do we characterize the risks?
- How do risks posed by DBPs compare with risks posed by microorganisms?

To address these questions, research is being conducted on a variety of health end points including cancer, reproductive and developmental effects, the nervous system, liver, kidney, and immune system. Cancer and adverse reproductive outcomes have been given the greatest attention from the public, research, and regulatory perspective.

Carcinogenicity

In developing drinking water regulations, the maximum contaminant level goal (MCLG), provides the health basis for the regulation. The MCLG is set at a level at which adverse health effects would not be expected, incorporating a margin of safety. Presently, the process for establishing MCLGs is based, as a matter of policy, on the weight of evidence for carcinogenicity for a particular contaminant following a drinking water exposure (USEPA 1994). The evidence for carcinogenicity for DBPs is provided from epidemiology studies evaluating communities served by disinfected drinking water and from laboratory studies in which animals were exposed to high concentrations of by-products over their lifetime. USEPA (1994) and the International Agency for

Research on Cancer (IARC 1992) have evaluated the combined weight-of-evidence for cancer associated with the consumption of chlorinated drinking water from human and animal studies, and have determined that there is insufficient evidence to associate a cancer risk with the consumption of chlorinated drinking water. However, results of the individual studies suggest that further study is needed.

Epidemiology studies. Epidemiology studies measure aggregate exposure to all DBPs as a mixture in drinking water. A number of epidemiology studies have been conducted which examine the effects associated with chlorination and chloramination. Results from these studies suggest there may be a slightly increased risk of bladder, colon, and rectal cancer following long-term exposure to chlorinated water. The types of studies that were conducted include both ecologic studies, the results of which tend to generate hypotheses for further study, and case-control studies, which provide more information concerning the magnitude of risk and information that can be used to determine a cause-and-effect relationship.

Animal studies are designed to focus on the health effects posed by individual DBPs. When chloroform was reported to produce tumors in animal studies (NCI 1976), several ecological epidemiology studies were initiated to compare effects in persons consuming chlorinated surface water with people consuming unchlorinated groundwater (Crump and Guess 1982). In this instance, characterization of exposure focused on the presumption that chlorinated surface water contained more trihalomethanes (THMs) or chloroform than groundwater on a broad geographical scale and did not include an analysis for other contaminants (Crump and Guess 1982; USEPA 1994). From the ecological studies, cancer associations were noted primarily for stomach, bladder, colon, and rectum. Although these associations were not consistent when viewed by sex, ethnicity, or geographic region (USEPA 1994), they provided hypotheses for further study.

Several case-control studies have been conducted to further evaluate the potential association of chlorinated water consumption with cancer. Some of these studies compared deceased cases involving specific cancer sites of interest with noncancer related deaths from the same geographic region. The limitation to this type of decedent study is that often there is a lack of information on residential history and risk factors for the individual cases. Incident case-control studies involve interviews with survivors or their families to gather this type of information. From these studies, odds ratios (ORs) can be determined that compare the odds of exposure among cases with controls. Odds ratios provide an estimate of relative risk. The OR is a measurement of the likelihood of developing some outcome (e.g., bladder cancer) in the presence of a particular factor, such as exposure to chlorinated water, as compared with the likelihood of the outcome developing without that factor present. Interpretation of ORs varies among the epidemiological community. Odds ratios from 1 to 1.5 or so could be considered in the range of background variability. Others have taken a stricter interpretation and view any increase greater than 1 to infer a positive causal association between exposure and risk. Another type of risk estimate is the relative risk (RR), which is the risk of disease in the exposed group divided by the risk of disease in the unexposed group. This type of risk estimate is more characteristic of cohort epidemiology studies. Monson (1980) concluded that RRs of 1.2 or less imply no association; 1.2–1.5 indicate a weak association; 1.5–3 is moderate; and greater than 3 would be considered strong.

Crump and Guess (1982) and Shy (1985) evaluated the associations of bladder, colon, and rectal cancers from decedent case-control studies. The exposures in these studies compared surface water with groundwater, and chlorinated groundwater with unchlorinated groundwater sources. Analyses were made comparing ORs by cancer site for males and females in several states (Table 5-1). In general, bladder cancer risk was increased for males in two of the five states studied and for females in one state. A slight increase in colon cancer was observed in males studied in two of the five states. Increased risk of rectal cancer was also reported for some of the study groups. Although several of the ORs are statistically significant, they could be attributed to incomplete information on exposures and confounding factors (factors inherent in the design or conduct of the study that affect a given result) (Crump and Guess 1982). In this case the ORs would not provide sufficient evidence for a causal association between consumption of chlorinated water and cancer incidence (USEPA 1994).

Table 5-1 Epidemiology cancer studies by tumor type, state in which study was conducted, and sex

Tumor Type	State	Males*	Females*
Bladder	North Carolina	Yes (OR = 1.54)†	Yes (OR = 1.54)
	New York	Yes (OR = 2.02)	No
	Louisiana, Wisconsin, Illinois	No	No
Colon	North Carolina	Yes (OR = 1.30)	No
	Wisconsin	Yes (OR = 1.35)	No
	Louisiana, Illinois	No	No
Rectal	North Carolina	Yes (OR = 1.54)†	Yes (OR = 1.54)†
	Louisiana	Yes (OR = 1.68)†	Yes (OR = 1.68)†
	Illinois	No	Yes (OR = 1.35)
	New York	Yes (OR = 2.33)	No

*Statistically significant increase.
†ORs determined for males and females combined.
Source: Adapted from Crump and Guess (1982); Shy (1985).

Perhaps the most noted epidemiology study purporting a link between chlorinated drinking water and cancer is the incident case-control study conducted by the National Cancer Institute, NCI (Cantor et al. 1985, 1987, 1990). The NCI study evaluated 10 areas to determine if there was a relationship between bladder cancer and artificial sweeteners (e.g., saccharin). Cantor and co-workers conducted an ancillary study that examined the consumption of chlorinated drinking water with bladder cancer incidence. Information on drinking water source and treatment over their lifetime was gathered for 1,244 cases and 2,550 controls. Smoking and beverage intake data were collected for some of these individuals. Cantor et al. (1985, 1987) found no association between bladder cancer incidence and duration of exposure. A two-fold increase in risk was reported for people consuming chlorinated surface water for 60 years or more who never smoked as compared to those individuals consuming unchlorinated groundwater (OR = 2.3).

Cantor et al. (1987, 1990) further analyzed these data considering the amount of water consumed, water supply source, and type of water treatment. An increased bladder cancer risk was found for males and females combined who mostly consumed tap-water-based beverages (OR = 1.43) compared to those who drank the least amount of these beverages. In general, the authors reported that the risk of bladder cancer increased in those individuals who consumed greater amounts of chlorinated surface water for 40 years or more. This study provides additional suggestive evidence for an association between chlorinated water consumption and cancer. In reviewing these data, however, Devesa et al. (1990) indicated a need for duplication of results and resolution of data inconsistencies before drawing a conclusion of causal association. Further refinement of exposure measurements to account for volatility of various DBPs is also needed (USEPA 1994).

A population-based, case-control study of bladder cancer and drinking water disinfection methods was conducted by McGeehin et al. (1993) from 1990 to 1993 in Colorado. The data were adjusted to minimize the potential for possible confounding effects, including years of tap-water intake, cigarette smoking, coffee consumption, and medical history. The authors reported an association between bladder cancer and consumption of chlorinated surface water for >30 years compared with no exposure (OR = 1.8).

Health Canada conducted a population-based, case-control study that evaluated the relationship between cancers of the bladder, colon, and rectum with exposures to chlorinated by-products in public water supplies (King and Marrett 1996). Levels of total trihalomethanes (TTHMs) were estimated based on the water source, chlorination dose, and treatment method. After controlling for potential confounders, the risk of bladder cancer increased with increasing duration of exposure to chlorinated surface water. A significant increase in risk was reported among those exposed for >35 years (OR = 1.41) compared with those exposed for <10 years. Exposure to TTHM levels of ≥ 50 μg/L for 35 or more years was associated with an increased risk compared to those exposed for <10 years (OR = 1.63). Colon cancer risk estimates associated with 35 or more years of consumption of chlorinated surface water was only suggestive of a slight excess risk (OR = 1.20). Exposure to THM levels of ≥ 50 or 75 μg/L for 35 or more years was associated with an increased risk of colon cancer (OR = 1.48 and 1.79, respectively) compared to exposure for <10 years. There was no increased risk of rectal cancer observed from consuming chlorinated drinking water.

In a prospective cohort investigation of postmenopausal women in Iowa, Doyle et al. (1997) found that women who resided in communities with surface water sources or drinking water with high levels of DBPs were at increased risk for cancer, particularly colon cancer (RR = 1.67). The authors stated that while their findings were similar to those found in some previous studies, interpretation of the results is complicated by a number of limitations in this study, such as potential misclassification of exposure, recall bias, and lack of residential history and water consumption in some cases.

In a statistical exercise that evaluated the results from 12 epidemiology studies and pooled the relative risks from 10 of those studies, Morris et al. (1992) reported a pooled relative risk of 1.21 for bladder cancer and 1.38 for rectal cancer. This metaanalysis, which is a process of statistically combining study results, would indicate that approximately 10,000 cancer cases each year are attributed to the consumption of chlorinated drinking water. Typically, metaanalysis is used to combine the results of small studies that have similar experimental design and exposure conditions. In this application, studies of varying experimental design and exposure information were combined. Thus, the results from this metaanalysis cannot be used to determine a causal association for chlorinated water and cancer (Murphy 1993). The USEPA has initiated an effort to conduct another metaanalysis. The intent of this analysis would be to build on the Morris study and account for differences in methodology, exposure definition and assessment, and bias among the studies that would be combined.

Animal studies. Results from animal studies in which controlled laboratory animals are exposed to individual DBPs over a lifetime indicate an increased incidence of liver, kidney, and intestinal cancers. These studies involved oral lifetime exposures in male and/or female rats and mice. By-products for which there is evidence for carcinogenicity potential include the four THMs, di- and trichloroacetic acid, chloral hydrate, bromate, and MX (Table 5-2). Estimates of cancer risks were developed for these compounds by applying mathematical models, such as the linearized multistage (LMS) model, to extrapolate from high-exposure doses in laboratory animal studies to predict the risk to humans exposed at lower concentrations that are encountered in the environment.

Table 5-2 Animal carcinogenicity studies for disinfection by-products

Disinfection By-Product	USEPA Classification and 10^{-6} Cancer Risk	Mutagenicity	Carcinogenicity
Chloroform	B2: Probable human carcinogen 10^{-6} = 6 μg/L	Inconclusive	Liver (mice); kidney (rats) (NCI 1976; Jorgenson et al. 1985)
Bromodichlo-romethane	B2 10^{-6} = 0.6 μg/L	Positive and negative studies	Kidney (male mice, rats); liver (female mice); large intestine (rats) (NTP 1987)
Dibromochlo-romethane	C: Possible human car-cinogen Not determined (ND)	Inconclusive	Liver (mice) (NTP 1985)
Bromoform	B2 10^{-6} = 4 μg/L	Positive and negative studies	Large intestine (rats) (NTP 1989)
Dichloroacetic acid	B2 ND	Positive and negative studies	Liver (mice, male rat) (DeAngelo et al. 1991, 1996)
Trichloroacetic acid	C ND	Positive and negative studies	Liver (male mice) (Bull et al. 1990)
Chloral hydrate	C ND	Positive	Liver (mice) (Rijhsing-hani et al. 1986; Daniel et al. 1992)
Bromate	B2 10^{-6} = 0.05 μg/L	Mostly positive	Kidney (rats); other sites (Kurokawa et al. 1986a, b)
MX	Not classified ND	Very positive	Liver, thyroid, other sites (rats) (Komulainen et al. 1997)

The LMS model assumes a linear relationship between exposure and effect such that exposure to even one molecule could result in a cancer effect. The resulting risk estimates are expressed as a probability of increasing one's risk per number of people exposed, such as one million.

As noted above, chloroform was found to produce liver tumors in male and female mice and kidney tumors in male rats when administered by corn oil gavage (NCI 1976). Jorgenson et al. (1985) further evaluated the carcinogenic potential of chloroform administered to male rats and female mice in drinking water for two years. They reported a dose-related increase in kidney tumors in male rats, but no significant increase in the incidence of liver tumors in female mice. Drinking water concentrations associated with a one in one million (10^{-6}) excess cancer risk for chloroform are listed in Table 5-2. EPA also considered a non-zero MCLG for chloroform after considering the scientific data related to the way chloroform produces cancer in animals. However, given the remaining uncertainty, EPA promulgated a final MCLG of zero for chloroform (USEPA 1998).

Bromodichloromethane (NTP 1987), bromoform (NTP 1989), and to a lesser extent dibromochloromethane (NTP 1985), also show evidence for carcinogenicity. As with the NCI chloroform study, each of these compounds was given to the animals in a corn oil vehicle. Liver tumors were observed in male and female mice exposed to dibromochloromethane and in female mice given bromodichloromethane. Bromodichloromethane exposure also resulted in increased kidney tumors in male mice and in male and female rats, and tumors of the large intestine in male and high-dose female rats. Intestinal tumors were evident in bromoform-treated animals. Maximum contaminant level goals of zero were also promulgated for bromodichloromethane and bromoform (USEPA 1998).

Although there is some evidence of carcinogenicity for dibromochloromethane, there are sufficient uncertainties concerning the conduct of the study, mode of chemical administration, and resulting data that preclude the development of an estimate of cancer risk until additional data are gathered. As a result of this uncertainty, USEPA set an MCLG of 0.06 mg/L based on noncancer effects for dibromochloromethane (USEPA 1998).

The USEPA and NTP are evaluating the carcinogenicity of bromodichloromethane given to animals in drinking water. The results from these studies may provide further information on the influence of corn oil gavage on the results previously reported by NTP and the scientific defensibility of the risk estimates derived from the earlier studies.

Much of the concern over the possible health effects of DBPs has focused on the presence of THMs. However, interest in the significance of the potential cancer risk posed by haloacetic acids, chloral hydrate, bromate, and MX is increasing.

Research conducted or sponsored by USEPA has shown that di- and trichloroacetic acid exposure at high doses in drinking water produces liver tumors in mice and rats. Dichloroacetic acid produced liver tumors in male and female mice and male rats (Herren-Freund and Pereira 1986; DeAngelo et al. 1991, 1996). Bull et al. (1990) reported that trichloroacetic acid exposure increased liver tumors in male mice only. Chloral hydrate, a metabolite of the chlorinated acids, also produces liver tumors in mice (Rijhsinghani et al. 1986; Daniel et al. 1992). The National Center for Toxicological Research is further evaluating the carcinogenic potential of chloral hydrate. These DBPs are also formed as metabolites of the carcinogenic or potentially carcinogenic drinking water contaminants vinyl chloride and trichloroethylene. The role of the chlorinated acids and chloral hydrate in the etiology of the cancer responses observed with vinyl chloride and trichloroethylene has yet to be determined. The USEPA set MCLGs of zero for dichloroacetic acid based on carcinogenicity. Maximum contaminant level goals of 0.3 and 0.04 mg/L were set for trichloroacetic acid and chloral hydrate, respectively, based on noncancer health effects with limited evidence for carcinogenicity (USEPA 1998). Quantitative cancer risks for di- and trichloroacetic acid and chloral hydrate have not been estimated. Available data suggest that a more biologically based approach for estimating cancer risk, rather than the linear, no threshold methodology, is warranted. Although EPS set an MCLG of zero for bromate (1998), further study of the disposition of these chemicals once they are ingested, and of the mechanism of action for carcinogenicity, is needed to develop a biologically based risk estimate. Parallel studies on the brominated acetic acids (i.e., dibromoacetic acid and bromodichloroacetic acid) are also being initiated.

Bromate, a by-product most commonly associated with the use of ozone as a disinfectant, increased the incidence of kidney and other tumors in male and female rats (Kurokawa et al. 1986a, b). Although USEPA set an MCLG of zero for bromate (1998), further study is under way to confirm the findings from the Kurokawa study and discern the potential mechanism of action for bromate-induced carcinogenicity. Available data suggest that bromate may interact with the DNA in the kidney, which may lead to tumor formation. Thus, a more biologically based method for quantifying risk may be appropriate for bromate.

MX, a chlorinated hydroxyfuranone (Kronberg et al. 1988), is another by-product formed from the oxidation of organic material in water by chlorine, and to a lesser extent by other disinfectants. MX is a potent mutagen producing positive mutagenic responses in a variety of cell cultures and whole-animal assays (WHO 1996). Although highly reactive, the available data indicate that MX can be absorbed and induce tumor formation in animals. In a recently published study by Komulainen et al. (1997), MX was found to be carcinogenic in both male and female rats at multiple sites following exposure in drinking water for two years. The strongest tumorigenic response was observed in the thyroid glands and liver. Tumors (adenomas) were also observed in the adrenal glands, lungs, pancreas, and mammary glands among rats given the higher doses of MX in their drinking water. Although MX was found to be a more potent carcinogen than other by-products, the limited monitoring conducted to date suggests that it is present in much lower concentrations than the other by-products of potential concern. To better characterize the shape of the dose–response curve for MX, particularly in the low-dose region, the National Toxicology Program will soon initiate a new chronic bioassay of MX in rats and mice.

The USEPA recently proposed revisions to the cancer risk assessment guidelines (USEPA 1996). The proposed guidelines place greater emphasis on information related to the mechanism and circumstances, such as route of exposure, by which chemicals induce cancer in humans or animals. This information may reduce the uncertainty associated with characterizing and estimating cancer risk depending on whether the response was considered linear (no threshold) or nonlinear (threshold). The agency recently sponsored a workshop with the International Life Science Institute (ILSI) to apply the proposed guidelines to the data for chloroform, dichloroacetic acid, and other chemicals. The ILSI report suggested that the available data could support a non-linear assessment of carcinogenicity. This report prompted USEPA to consider a non-zero MCLG for chloroform. However, as mentioned above, remaining uncertainties led the agency to set a zero goal for chloroform.

Uncertainties in cancer studies. There is a body of evidence that suggests that specific DBPs induce cancer in animals if consumed at certain concentrations for a significant period of time. In assessing risk to protect public health, the conservative presumption has been that if a response is observed in animals, then a response is likely in humans. However, extrapolating the results observed in animal studies to predict human risk introduces uncertainties in the resulting risk assessment. Some of the uncertainties specific to the DBP risk assessments include the mode of chemical administration, specifically corn oil gavage, and whether or not that mode affected the response. The mechanism of carcinogenic action also seems to differ among the by-products, with some chemicals interacting with DNA as part of the carcinogenic response and others apparently not. This is also evident from the mutagenicity data presented in Table 5-2. In addition to the uncertainties associated with applying a no-threshold linear quantitative model (LMS) to estimate risk, the types of tumors found in the animal studies (e.g., liver and kidney) are not entirely consistent with results from the epidemiology studies (e.g., bladder, colon, and rectal). Further research comparing DBP disposition and mechanism(s) of carcinogenicity between humans and animals may help to clarify the significance of this difference in response.

The epidemiology studies, while suggestive of a cancer risk, tend to raise additional questions, such as the toxicity of a by-product mixture compared with individual chemical exposure and the role of source water quality on chemical or mixture toxicity. The shortcomings of exposure assessment approaches compound the problem. Although a conclusion regarding the carcinogenic potential for chlorinated water cannot be drawn at this time, it is important to note that while the risk appears to be small, the exposed population is large (more than 200 million).

Conducting a well-designed epidemiology study that addresses water quality, exposure (including volume of water consumed), and effects issues noted from the existing database may help to reduce some of the uncertainty surrounding the carcinogenic potential of chlorinated water. However, it is likely that significant uncertainties will still remain, due to the difficulties associated with trying to refine risk estimates involving weak associations (i.e., OR <1.5), and the methodologic problems in reconstructing historical exposures to by-products over several decades.

Reproductive and Developmental Toxicity

Much of the health concern attributed to disinfected water has focused on carcinogenicity, which may be related to the policy for developing drinking water MCLGs based on the weight of evidence for a chemical's carcinogenic potential in drinking water. This policy, along with USEPA's practice of developing risk assessments for lifetime exposures, places greater emphasis on studies involving chronic, long-term exposures. However, recent public concern about reproductive and developmental effects associated with DBPs in drinking water indicate a need to compare cancer or other chronic effects with those effects resulting from shorter durations of exposure. In many cases, reproductive or developmental effects may result from higher exposure concentrations than those that result in other chronic effects. Thus, a careful evaluation of all potential health effects resulting from contaminant exposure should be conducted before assessing a risk that serves as a basis for a regulation. Regarding exposure to disinfected water and DBPs, the limited epidemiological and animal studies conducted to date indicate the need for additional research to determine if adverse reproductive outcomes are a public health concern. For most by-products, it does not appear at this time that these end points are more sensitive than cancer or other health effects.

Epidemiology studies. A few studies have been conducted to examine the relationship between adverse reproductive or developmental outcomes with exposure to chlorinated drinking water. Kramer et al. (1992) conducted a case-control study in Iowa that focused on effects related to chloroform or other THM levels in drinking water. They reported slightly positive associations for infant growth retardation (OR = 1.8) and low birth weight (OR = 1.3) in communities where chloroform levels exceeded 10 µg/L. The authors considered these findings to be preliminary due to limitations in determining individual exposures resulting from residential mobility and fluctuations in THM levels.

Aschengrau, Zierler, and Cohen (1993) conducted a case-control study in Massachusetts comparing community drinking water quality with events such as stillbirths, congenital deformities, and infant deaths. The investigators linked 2,348 women to drinking water samples from 155 towns as part of the case-control study. A positive association was reported between chlorinated water and incidence of stillbirth (OR = 2.6). The authors indicated that these findings were preliminary and that further research was needed that would address a specific hypothesis and improve the exposure assessment.

Another study conducted by Bove and co-workers (Bove et al. 1995) in New Jersey compared exposures to drinking water contaminants with birth weights and defects. Public drinking water supplies from both groundwater and surface water sources were used in the study. For waters containing THM levels greater than 80 μg/L, positive associations were found for low birth weight (OR = 1.34), central nervous system defects (OR = 2.6), neural tube defects (OR = 2.98), and cardiac defects (OR = 2.0). Limitations to this study include possible investigator bias, exposure misclassification, and incomplete accounting of confounding risk factors (Bove et al. 1995).

Savitz, Andrews, and Pastore (1995) conducted a population-based, case-control study in which information on water source, amount of water consumed, and TTHM concentration in water was compared with data on various adverse reproductive outcomes of possible concern, including miscarriage, preterm delivery, and low birth weight. The authors concluded that a strong association between chlorinated by-products and these outcomes could not be demonstrated. However, the limitations of the study design and the observation of an increased miscarriage risk in the high-exposure group suggested that additional analysis of this issue was warranted.

Kanitz et al. (1996) conducted an epidemiology study in Italy on the association between various adverse reproductive outcomes and exposure to drinking water treated with either chlorine dioxide or sodium hypochlorite. The authors reported an increased risk of smaller body length and smaller cranial circumference in infants born to mothers who were exposed to water treated with either disinfectant. Neonatal jaundice was associated with maternal exposure to water treated with chlorine dioxide. Questions about the experimental design of this study, particularly with respect to the exposure assessment, raise the need for a more rigorous investigation to verify the results that were reported.

An expert panel that convened to review the reproductive effects of drinking water DBPs concluded that the evaluation of adverse reproductive outcomes associated with chlorinated water and by-products was too limited to draw any conclusions (Reif et al. 1996). The panel suggested a course for further epidemiological study that would address the limitations in exposure analysis, confounders, data gaps, and small sample size observed from the available studies.

Recently, a new study provided additional information suggesting an association between chlorinated drinking water consumption and adverse reproductive outcome. In this study, Waller et al. (1998) evaluated THM levels in tap water with pregnancy outcome. With the full cooperation of utilities, the authors obtained records of THM measurements taken during the time study participants were pregnant. The TTHM level in a participant's home tap water was estimated by averaging water distribution system TTHM measurements taken during the participant's first three months of pregnancy. This first trimester TTHM level was combined with self-reported tap-water consumption to create a TTHM exposure level.

Exposure levels of the individual THM species (chloroform, bromoform, etc.) were estimated in the same manner. Actual THM concentrations in the home tap water were not measured.

Women with high TTHM exposure in home tap water (drinking five or more glasses per day of cold home tap water containing at least 75 μg/L TTHM) had an early-term miscarriage rate of 15.7 percent, compared with a rate of 9.5 percent among women with low TTHM exposure (drinking less than five glasses per day of cold home tap water or drinking any amount of tap water containing less than 75 μg/L TTHM). An adjusted odds ratio for early-term miscarriage of 1.8 (95 percent confidence interval 1.1–3.0) was determined.

When the four individual THMs were studied, only high bromodichloromethane (BDCM) exposure, defined as drinking five or more glasses per day of cold home tap water containing >18 μg/L bromodichloromethane, was associated with early-term miscarriage. An adjusted odds ratio for early-term miscarriage of 3.0 (95 percent confidence interval 1.4–6.6) was determined. The authors concluded that their study shows an association between ingestion of large amounts of tap water containing high levels of TTHMs and/or bromodichloromethane and early-term miscarriage.

As with other studies of this nature, the Waller et al. (1998) study provides additional information, but does not necessarily demonstrate a cause-and-effect relationship. More research is needed in other parts of the country to repeat the findings of this study and to perhaps establish a dose–response relationship.

Animal studies. A number of laboratory animal studies have been conducted on DBPs to determine their potential for affecting reproductive or developmental processes. Limited data on reproductive processes have focused on impairments to sperm mobility, morphology, and maturation. By-products affecting sperm at experimental dose levels include bromodichloromethane, di- and trichloroacetic acid, bromoacetic acid, dibromoacetic acid, chloral hydrate, and bromate (Klinefelter and Linder 1996). Linder et al. (1995) reported that fertility was decreased in male rats receiving dibromoacetic acid at dose levels as low as 10 mg/kg (which is still several orders of magnitude higher than ambient exposure levels). Other by-products appear to produce sperm effects at higher dose levels (Reif et al. 1996). While these dose levels correspond to by-product concentrations greater than those found in drinking water, many by-products have not been tested for these or other effects, including effects to female reproductive processes.

Effects observed in the offspring of animals exposed to experimental levels of DBPs include decreased pup weight and viability, and heart malformations. In developmental studies for the THMs, reduced body weights were observed in rat fetuses whose mothers received high doses of chloroform in corn oil by gavage during days 6–15 of gestation. Although body weights were decreased in the high-dose group, no other developmental or teratogenic effects were noted in the fetuses (Thompson, Warner, and Robinson 1974). Decreased pup viability has been reported in pregnant rats given bromoform or bromodichloromethane in corn oil by gavage (Ruddick, Villeneuve, and Chu 1983; Kavlock and Narotsky 1992).

Exposure to the chlorinated and brominated acetic acids appears to produce more overt developmental effects in animals exposed *in utero*. Pups whose mothers were given either chlorinated or brominated acetic acid in drinking water during days 6–15 of pregnancy exhibited heart defects. These effects have been reported in studies of chloroacetic acid, di- and trichloroacetic acid, and bromoacetic acid (Smith et al. 1989, 1990, 1993; Randall et al. 1991). The individual haloacids have been shown to exhibit a range of potencies in inducing developmental effects in whole embryo cultures (Richard and Hunter 1996; Hunter et al. 1996). Further study with these halogenated by-products is needed to discern the mechanism for this adverse effect and to determine dose levels of concern so that they may be compared with other end points of toxicity.

Uncertainties for reproductive and developmental effects. Many of the uncertainties noted with the carcinogenicity epidemiology database also apply to the limited database for adverse reproductive effects. The available epidemiology studies are useful for generating hypotheses for further study, but are insufficient (due to study design and incomplete exposure evaluations) to draw any conclusions regarding a cause-and-effect relationship between consumption of chlorinated drinking water and adverse reproductive outcomes. Future studies will need to improve exposure assessment and control for important risk factors, such as smoking, and exposure from other drinking water contaminants (Reif et al. 1996). In addition, these studies will need to be conducted in such a way that a dose–response relationship can be determined, thus reducing uncertainty in the resulting quantitative risk assessment.

The design for epidemiology studies should include a consideration of the results of reproductive and developmental toxicology studies as they become available. Many of the by-products have gone through screening tests to determine their potential for causing reproductive and developmental effects. Further toxicity studies should focus on the mechanisms by which DBPs cause these effects to gain a better understanding of the significance of this risk.

RECENT FINDINGS

The health effects of DBPs have been the subject of several scientific meetings. A workshop held in October 1995 discussed the emerging health effects data and addressed critical issues in health effects research for DBPs (Fawell et al. 1997). Some of the major conclusions from the workshop include the following: (1) efforts to reduce the risk from by-products should not compromise the microbial quality of the water; (2) although the risks from by-products appear low (given the inconclusiveness of the human and animal data), they are still important because of the large number of people potentially exposed; (3) there is a need to identify additional by-products because only about half of the halogenated compounds in chlorinated drinking water have been identified; (4) because of limited resources it is critical for future research to target those by-products that pose the greatest public health concerns and to develop a prioritization process for selecting those by-products for future testing; (5) studies on interactions and mixtures should be hypothesis-driven and should focus on by-products for which the mode or mechanism of action has been established; (6) there is a need to improve exposure assessments for epidemiology studies; and (7) there is sufficient screening-level toxicity data for chlorination by-products.

The workshop conclusions were based on a ranking of by-products considering health effects and levels of occurrence in drinking water (Fawell et al. 1997). The health end point used in this analysis was usually cancer. The results of the ranking suggested that research should be prioritized for dichloroacetic acid and bromate to expand the existing health effects database and that no further chronic bioassay research on chloroform was necessary because extensive research indicates that chloroform is unlikely to pose a risk at concentrations found in drinking water.

CONCLUSION

Disinfection is essential for preventing waterborne disease. The process of disinfection, however, can introduce chemical by-products that may pose a risk into the water. The experimental data describing the health effects of various by-products suggest a potential for risk. When compared with the available epidemiological studies, the risk is difficult to characterize. The cancer epidemiology studies provide weak associations for some cancers but are not sufficient to determine a cause-and-effect relationship between the consumption of chlorinated drinking water and cancer. Thus, if there is a cancer risk posed by drinking chlorinated water, it could be inferred that the risk would be small. However, given the uncertainty of a causal association and because of the exposure to large populations from chlorinated water, potentially conservative estimates of carcinogenic risk have been derived for DBPs. Further information concerning chemical disposition and mode or mechanism of carcinogenicity in humans would reduce the uncertainty regarding carcinogenic potential and possibly reconcile the differences observed between the human and animal studies.

For reproductive and developmental effects linked to drinking water disinfection, the conclusions are similar to those drawn for cancer risk. That is, the data are considered insufficient to causally link drinking water chlorination with an adverse reproductive outcome. The animal and epidemiological data provide limited evidence for a potential risk. However, further research is needed to show any causal relationships.

Management actions to reduce the potential risk posed by DBPs must not compromise the microbiological quality of the drinking water (ILSI 1993). There is much less uncertainty concerning microbial risk and the benefit of drinking water disinfection as a means to control and prevent this risk. Further epidemiological and toxicological research is needed to determine the important by-products of concern, the important health effects posed by those by-products in drinking water, the mechanism(s) by which the various by-products exert their toxic or carcinogenic effect(s), and how the estimate of by-product risk compares with microbial risk. Reducing the uncertainty in the health effects of DBPs will improve the confidence in risk management decisions concerning the presence and control of these by-products in drinking water.

REFERENCES

Aschengrau, A., S. Zierler, and A. Cohen. 1993. Quality of Community Drinking Water and the Occurrence of Late Adverse Pregnancy Outcomes. *Arch. Env. Health*, 48:105–113.

Bove, F., M.C. Fulcomer, J.B. Klotz, J. Esmart, E.M. Duffy, and J.E. Savrin. 1995. *Am. Jour. of Epidemiology*, 141:850.

Bull, R.J., I.M. Sanchez, M.A. Larsen, and A.J. Lansing. 1990. Liver Tumor Induction in B6C3F1 Mice by Dichloroacetate and Trichloroacetate. *Toxicol.*, 63:341–359.

Cantor, K.P., R. Hoover, P. Hartge, T.J. Mason, D.T. Silverman, and L.I. Levin. 1985. Drinking Water Source and Bladder Cancer: A Case Control Study. Chapter 12. In *Water Chlorination: Chemistry, Environmental Impact, and Health Effects*, Vol. 5. Chelsea, Mich.: Lewis Publishers.

Cantor, K.P., R. Hoover, P. Hartge, T.J. Mason, D.T. Silverman, R. Altman, D.F. Austin, M.A. Child, C.R. Key, L.D. Marrett, M.H. Myers, A.S. Narayana, L.I. Levin, J.W. Sullivan, G.M. Swanson, D.B. Thomas, and D.W. West. 1987. Bladder Cancer, Drinking Water Source, and Tap Water Consumption: A Case Control Study. *J. NCI*, 79:1269–1279.

Cantor, K.P., R. Hoover, P. Hartge, T.J. Mason, and D. Silverman. 1990. Bladder Cancer, Tap Water Consumption and Drinking Water Source. In *Water Chlorination: Chemistry, Environmental Impacts and Health Effects*, Vol. 6. Chelsea, Mich.: Lewis Publishers.

Crump, K.S., and H.A. Guess. 1982. Drinking Water and Cancer: Review of Recent Epidemiological Findings and Assessment of Risks. *Ann. Rev. of Pub. Health*, 3:339–357.

Daniel, F.B., A.B. DeAngelo, J.A. Stober, G.R. Olson, and N.P. Page. 1992. Hepatocarcinogenicity of Chloral Hydrate, 2-Chloroacetaldehyde, and Dichloroacetic Acid in the B6C3F1 Mouse. *Fund. Appl. Tox.*, 19:159–168.

DeAngelo, A.B., F.B. Daniel, J.A. Stober, and G.R. Olson. 1991. The Carcinogenicity of Dichloroacetic Acid in the Male B6C3F1 Mouse. *Fund. Appl. Tox.*, 16:337–347.

DeAngelo, A.B., F.B. Daniel, B.M. Most, and G.R. Olson. 1996. The Carcinogenicity of Dichloroacetic Acid in the Male Fischer 344 Rat. *Toxicol.*, 114:207–221.

Devesa, S.S., D.T. Silverman, J.K. McLaughlin, C.C. Brown, R.R. Connelly, and J.T. Fraumeni, Jr. 1990. Comparison of the Descriptive Epidemiology of Urinary Tract Cancers. *Can. Causes Cont.*, 1:133–141.

Doyle, T.J., W. Zheng, J.R. Cerhan, C.P. Hong, T.A. Sellers, L.H. Kushi, and A.R. Folsom. 1997. The Association of Drinking Water Source and Chlorination By-products With Cancer Incidence Among Postmenopausal Women in Iowa: A Prospective Cohort Study. *Am. J. Pub. Health*, 87:1168–1176.

Fawell, J., D. Robinson, R. Bull, L. Birnbaum, G. Boorman, B. Butterworth, P. Daniel, H.G. Gorchev, F. Hauchman, P. Julkunen, C. Klaassen, S. Krasner, J.O. Zavaleta, J. Reif, and R. Tardiff. 1997. Disinfection By-Products in Drinking Water: Critical Issues in Health Effects Research. *Envir. Health Perspectives*, 105:108–109.

Herren-Freund, S., and M. Pereira. 1986. Carcinogenicity of By-products of Disinfection in Mouse and Rat Liver. *Envir. Health Perspectives*, 69:59–65.

Hunter, S., E.H. Rogers, J.E. Schmid, and A. Richard. 1996. Comparative Effects of Haloacetic Acids in Whole Embryo Culture. *Teratology*, 54:57–64.

ILSI. 1993. *Safety of Water Disinfection: Balancing Chemical and Microbial Risks.* Craun, G., ed. Washington, D.C.: International Life Sciences Institute Press.

Jorgenson, T.A., E.F. Meierhenry, C.J. Rushbrook, R.J. Bull, and M. Robinson. 1985. Carcinogenicity of Chloroform in Drinking Water to Male Osborne-Mendel Rats and Female B6C3F1 Mice. *Fund. Appl. Tox.*, 5:760–769.

Kanitz, S., Y. Franco, V. Patrone, M. Caltabellotta, E. Raffo, C. Riggi, D. Timitilli, and G. Ravera. 1996. Association Between Drinking Water Disinfection and Somatic Parameters at Birth. *Envir. Health Perspectives*, 104(5):516–520.

Kavlock, R.J., and M. Narotsky. 1992. Effects of Bromoform and Bromodichloromethane in an *in vivo* Developmental Toxicity Test. HERL/EPA Internal Report to the Office of Water.

King, W.D., and L.D. Marrett. 1996. Case-Control Study of Bladder Cancer and Chlorination By-products in Treated Water (Ontario, Canada). *Can. Causes Cont.*, 7(6):596–604.

Klinefelter, G., and R. Linder. 1996. Recent Reproductive Effects Associated With Disinfection By-products. In *Disinfection By-Products in Drinking Water: Critical Issues in Health Effects Research - Workshop Report.* Washington, D.C.: International Life Sciences Institute Press.

Komulainen, H., V.-M. Kosma, S.-L. Vaittinen, T. Vartianinen, E. Kaliste-Korhonen, S. Lotjonen, R.K. Tuominen, and J. Toumisto. 1997. Carcinogenicity of the Drinking Water Mutagen 3-chloro-4-(dichloromethyl)-5-hydroxy-2(5H)-furanone in the Rat. *Jour. Nat. Can. Inst.*, 89(12):848–856.

Kramer, M.D., C.F. Lynch, P. Issacson, and J.W. Hanson. 1992. The Association of Waterborne Chloroform With Intrauterine Growth Retardation. *Epidemiology*, 3:407–413.

Kronberg, L., B. Holmbom, M. Reunanen, and L. Tikkanen. 1988. Identification and Quantification of the Ames Mutagenic Compound 3-chloro-4-(dichloromethyl)-5-hydroxy-2(5H)-furanone and its Geometric Isomer (E)-2-chloro-3-(dichloromethyl)-4-oxybutenoic acid in Chlorine Treated Humic Water and Drinking Water Extracts. *Environ. Sci. & Tech.*, 22:1097.

Kurokawa, Y., S. Takayama, Y. Konishi, Y. Hiasa, S. Asahina, M. Takahashi, A. Maekawa, and Y. Hayashi. 1986. Long-term *in vivo* Carcinogenicity Tests of Potassium Bromate, Sodium Hypochlorite, and Sodium Chlorite Conducted in Japan. *Envir. Health Perspectives*, 69:221–236.

Kurokawa, Y., S. Aoki, Y. Matsushima et al. 1986. Dose Response Studies on the Carcinogenicity of Potassium Bromate in F344 Rats After Long-Term Oral Administration. *J. NCI*, 77:977–982.

Linder, R.E., G.R. Klinefelter, L.F. Strader, M.G. Narotsky, J.D. Suarez, N.L. Roberts, and S.D. Perreault. 1995. Dibromoacetic Acid Affects Reproductive Competence and Sperm Quality in the Male Rat. *Fund. Appl. Tox.*, 28:9–17.

McGeehin, M.A., J.S. Reif, J.C. Becher, and E.J. Mangione. 1993. Case-Control Study of Bladder Cancer and Water Disinfection Methods in Colorado. *Am. J. Epidemiol.*, 138:492–501.

Monson, R. 1980. *Occupational Epidemiology*. Boca Raton, Fla.: CRC Press, Inc.

Morris, R.D., A. Audet, I.O. Angelillo et al. 1992. Chlorination, Chlorination By-products and Cancer: A Meta Analysis. *Am. J. Pub. Health*, 82:955.

Murphy, P.A. 1993. Quantifying Chemical Risk From Epidemiology Studies: Application to the Disinfectant By-product Issue. In *Proc. Safety of Water Disinfection: Balancing Chemical and Microbial Risks*. Washington, D.C.: International Life Sciences Institute Press.

NCI. 1976. Report on Cancer Carcinogenesis Bioassay of Chloroform. NTIS PB-264018. National Cancer Institute.

NTP. 1985. Toxicology and Carcinogenesis Studies of Chlorodibromomethane in F344/N Rats and B6C3F1 Mice (Gavage Studies). Tech. Rpt. Series No. 282. National Toxicology Program.

———. 1987. Toxicity and Carcinogenesis Studies of Bromodichloromethane in F344/N Rats and B6C3F1 Mice (Gavage Studies). Tech. Rpt. Series No. 321. National Toxicology Program.

———. 1989. Toxicity and Carcinogenesis Studies of Bromoform in F344/N Rats and B6C3F1 Mice (Gavage Studies). Tech. Rpt. Series No. 350. National Toxicology Program.

Randall, J.K., S.A. Christ, P. Horton-Perez, G. Nolan, E.J. Read, and M.K. Smith. Developmental Effects of 2-Bromoacetic Acid in the Long-Evans Rat. *Teratology*, 43:454 (abstract).

Reif, J.S., M.C. Hatch, M. Bracken, L.B. Holmes, B.A. Schwetz, and P.C. Singer. 1996. Reproductive and Developmental Effects of Disinfection By-Products in Drinking Water. *Envir. Health Perspectives*, 104:1056.

Richard, A.M., and S. Hunter. 1996. Quantitative Structure-Activity Relationships for the Developmental Toxicity of Haloacetic Acids in Mammalian Whole Embryo Culture. *Teratology*, 53:352–360.

Rijhsinghani, K.S., M.A. Swerdlow, T. Ghose, C. Abrahams, and K.V.N. Rao. 1986. Inductions of Neoplastic Lesions in the Liver of C57BL x C3HF1 Male Mice by Chloral Hydrate. *Can. Detec. Prev.*, 9:279–288.

Ruddick, J.A., D.C. Villeneuve, and I. Chu. 1983. A Teratology Assessment of Four Trihalomethanes in the Rat. *J. Env. Sci. Health*, 1318:333–349.

Savitz, D.A., K.W. Andrews, and L.M. Pastore. 1995. Drinking Water and Pregnancy Outcome in Central North Carolina: Source, Amount, and Trihalomethane Levels. *Envir. Health Perspectives*, 103(6):592–596.

Shy, C. 1985. Chemical Contamination of Water Supplies. *Envir. Health Perspectives*, 62:399–406.

Smith, M.K., J.L. Randall, E.J. Read, and J.A. Stober. 1989. Teratogenic Activity of Trichloroacetic Acid in the Rat. *Teratology*, 40:445–451.

———. 1990. Developmental Effects of Chloroacetic Acid in the Long-Evans Rat. *Teratology*, 41:452 (abstract).

———. 1993. Developmental Toxicity of Dichloroacetate in the Rat. *Teratology*, 46:217–223.

Thompson, D.J., S.D. Warner, and V.B. Robinson. 1974. Teratology Studies in Orally Administered Chloroform in the Rat and Rabbit. *Tox. and Appl. Pharm.*, 29:348–357.

USEPA. 1994. National Primary Drinking Water Regulations; Disinfectants and Disinfection By-Products; Proposed Rule. Part II. *Fed. Reg.*, 59:38668–38829.

———. 1995. Draft Research Plan for Microbial Pathogens and Disinfection By-Products in Drinking Water. Office of Research and Development and Office of Water.

———. 1996. Proposed Guidelines for Carcinogen Risk Assessment; Notice. *Fed. Reg.*, 61:17960–18011.

———. 1998. National Primary Drinking Water Regulations; Disinfectants and Disinfection By-Products; Final Rule. *Fed. Reg.*, 69390–69476.

Waller, K., S.H. Swan, G. DeLorenze, and B. Hopkins. 1998. Trihalomethanes in Drinking Water and Spontaneous Abortion. *Epidemiology*, 9(2):134–140.

WHO. 1996. *Guidelines for Drinking Water Quality. Health Criteria and Other Supporting Information.* Vol. 2, 2nd ed. Geneva, Switzerland: World Health Organization.

CHAPTER • SIX

Balancing Microbial and Chemical Risks

PATRICIA A. MURPHY
GUNTHER F. CRAUN

In the early part of the 20th century, improvements in sewage collection and treatment, protection of drinking water sources, and the filtration and disinfection of drinking water helped to end the large-scale waterborne epidemics of cholera and typhoid in the United States. Unfortunately, illness and death from waterborne cholera, diarrhea, and other diseases are still a fact of life for many people in the developing countries of Africa, Asia, and Latin America where these sanitary engineering practices are not yet commonplace (Craun 1996; Swerdlow et al. 1992; Craun et al. 1991). Even in highly industrialized countries, waterborne outbreaks continue to occur when treatment is inadequate, reminding us that waterborne pathogens still pose a formidable risk (Kramer et al. 1996; Badenoch 1990).

The benefits of water disinfection are well recognized, but the discovery in 1974 that chlorine reacts with organic matter to form chloroform (Rook 1974; Bellar, Litchtenberg, and Kroner 1974) raised questions about its possible health risks. In the past 25 years, an increasing number of disinfection by-products (DBPs) have been identified, and concerns have intensified about health risks, such as cancer and adverse outcomes of pregnancy (Craun 1993; Galal-Gorchev 1996b).

INFECTIOUS WATERBORNE DISEASE RISKS

Infectious diseases have far-reaching, socioeconomic implications in both industrialized and developing countries, especially for the very young, the aged, the nutritionally deficient, and those whose immune systems may be compromised by disease or therapeutic agents. In AIDS patients, *Cryptosporidium* infection may have a prolonged clinical course, contributing to death, and no drug therapy has been found effective for this infection (Soave 1990). The World Health Organization estimates that in developing countries about half the population suffers from health problems associated with insufficient or contaminated water, 80 percent of all disease and over one third of deaths are water-related, and as much as one tenth of a person's productive time is sacrificed to water-related diseases (Galal-Gorchev 1993, 1996a, b). In the poorest countries of the world, the median mortality for children under 5 years of age is 193 per 1,000 live births, 20-fold greater than in the healthiest countries (Okun 1993). Unsafe, inadequate water supplies play a major role in the transmission of diarrheal diseases in countries with poor sanitation, and 1.6 billion cases of diarrhea and over 3 million associated deaths occur annually among children under 5 years of age (Galal-Gorchev 1996a). Mortality and diarrheal episodes can be dramatically reduced (22 to 26 percent) in infants and children by improvements in water supply and sanitation (Galal-Gorchev 1993). This staggering burden of disease cannot be overlooked when governments develop DBP regulations and policies for the use of chlorine and other water disinfectants.

Although infectious diseases are largely under control in the United States, waterborne outbreaks continue to occur. Endemic waterborne risks have also been identified. Since 1981, 58 percent of the 345 reported waterborne outbreaks occurred in water systems with inadequate or no disinfection (Craun 1996). Just six years ago, *Cryptosporidium* contamination caused an estimated

403,000 cases of illness, 1,000 hospitalizations, and 100 deaths in Milwaukee, Wisc. (MacKenzie 1994; Blair 1994; Hoxie, Davis, and Vergeront 1996). Bacteria, such as *Shigella* and enterohemorrhagic *E. coli*, have also caused deaths and severe illness during waterborne outbreaks in the United States (Craun 1992). Waterborne outbreaks of Norwalk virus have also been reported, and increased concern about the possible waterborne transmission of other pathogens has caused the US Environmental Protection Agency (USEPA) to consider regulations for a number of microbial contaminants, including *Cyclospora*, microsporidia, *Legionella*, *Aeromonas*, *Mycobacterium*, and adenoviruses (USEPA 1997).

In a community where drinking water met or exceeded current regulations for quality and no outbreaks were reported, randomized prospective epidemiological studies found that 14 to 35 percent of gastrointestinal illness may be waterborne (Payment et al. 1991a, b, 1997). The first study observed illness over a 15-month period among members of 307 eligible households consuming tap water and 299 households that had been provided reverse-osmosis filters; 35 percent of self-reported, mild gastroenteritis was attributed to consumption of municipal tap water. A second intervention study compared self-reported gastroenteritis over a 16-month period among 1,400 families randomly allocated to one of the following four groups: tap water, tap water from a continuously purged water service line, bottled water directly from the municipal treatment plant, and purified bottled water. Using the purified bottled water group as the baseline, 14 percent excess gastroenteritis was found associated with tap water; children 2 to 5 years old had a 17 percent excess gastroenteritis in the tap-water group and 40 percent in the purged-tap group. No increased illness was found among users of bottled plant water, suggesting that contamination of the distribution system may be partly responsible for endemic gastroenteritis. Studies have also suggested that 40 to 70 percent of endemic giardiasis may be associated with unfiltered surface water supplies (Chute, Smith, Baron 1987; Birkhead and Vogt 1989; Fraser and Cooke 1991).

Because waterborne disease often goes unrecognized and unreported in the United States, the risk may be even greater than indicated by current statistics (Craun 1986, 1990). Hauschild and Bryan (1980) estimated that 1,400,000 to 3,400,000 cases of food- and waterborne disease occur annually in the United States. Berger (1987) estimated that 350,000 to 875,000 cases of epidemic waterborne disease may occur each year, and Carlsten (1993) estimated 700,000 cases. Morris and Levin (1995) believe that 1.8 million cases of epidemic and endemic waterborne disease and 1,800 associated deaths occur annually. Bennett et al. (1987) estimated the annual incidence to be more than 900,000 cases, with 900 associated deaths. Although there is a great uncertainty associated with these estimates, they are a reminder that waterborne infectious disease risks cannot be ignored even in highly industrialized countries.

MANAGING MICROBIAL AND DBP RISKS

The reaction of chemical disinfectants with naturally occurring organic matter (NOM) and bromide in water sources will produce a number of DBPs, including chloroform and the other trihalomethanes (THMs), haloacetic acids (HAAs), haloacetonitriles (HANs), haloketones (HKs), chloral hydrate, cyanogen chloride, and aldehydes (Stevens et al. 1990; Singer 1993; Krasner et al. 1996). The concentration and mix of DBPs will depend on the disinfectant; its point of application, concentration, and contact time; the water pH and temperature; and concentration of bromide and organic precursor material in source waters. The following three primary strategies are used to reduce the formation of DBPs: moving the point of disinfection, using an alternative disinfectant, and removing DBP precursors from the water. Some DBPs can also be removed after their formation. Since no feasible means are available to reduce the formation of all DBPs, it is important to identify those DBPs that may pose the greatest risk.

Changing water treatment practices for the management of possible DBP risks must be approached with caution because these changes may increase the microbial risk and also affect the mix and concentration of DBPs. For example, decreasing the water pH may reduce the formation of total trihalomethanes (TTHMs) but may increase the formation of haloacetic acids (Singer 1993). Removal of total organic carbon (TOC) from source waters will reduce the formation of chlorinated by-products, but brominated by-products, which toxicologists believe may present a greater health risk than chlorinated by-products (Bull and Kopfler 1991), will be reduced to a lesser extent or their formation may even be increased. Ozone reacts with bromide to form bromate, which toxicological studies suggest may have a high carcinogenic potency (Bull and Kopfler 1991).

Chlorine, the most commonly used water disinfectant throughout the world, and the frequently used alternatives, such as chloramines, ozone, and chlorine dioxide, are all reactive, and all will produce DBPs. Use of an alternative chemical disinfectant can produce other DBPs that may also pose health risks. Selecting a water disinfectant requires consideration not only of its ability to form potentially harmful DBPs but also its ability to effectively inactivate waterborne pathogens. Water disinfected with chloramines, for example, generally contains low levels of THMs, but chloramines are less effective disinfectants than free chlorine. Chloramines can also cause acute effects, such as methemoglobinemia, hemolysis, and anemia in kidney dialysis patients. Chlorine dioxide does not produce THMs, but breaks down to yield chlorite and chlorate, which may cause acute health effects at low concentrations. Trihalomethanes are not formed when ozone is used; however, other DBPs are produced, and little is known about their possible health effects. While ozone is an effective disinfectant, no residual ozone disinfectant can be maintained in the water distribution system.

Costs of Controlling Microbial and DBP Risks

One of the most cost-effective investments that society can make is a central water system with either chlorination or conventional water treatment (filtration and chlorination) to remove and control pathogens (Clark, Hurst, and Regli 1993; Clark, Goodrich, and Ireland 1984). Conservative estimates suggest that the benefits associated with water treatment systems will far exceed their costs. Costs saved by the prevention of infectious disease may be as much as eight to twelve times greater than the cost of the water system (Clark, Hurst, and Regli 1993; Craun 1996). Water treatment systems that employ properly designed and operated filtration and disinfection processes will virtually eliminate the risk of waterborne disease (Okun 1993). For developing countries, however, a focus on drinking water protection and treatment alone may not be sufficient, and an investment in improved sanitation may also be required to achieve significant reductions in waterborne disease risks (Esrey et al. 1990). In many developing countries, the availability of water itself is also important (Okun 1993).

Coagulation and clarification can be optimized to help remove both pathogens and precursor material for formation of DBPs, but other available treatment options for removal of precursors or DBPs may be expensive and technologically complex. Granular activated carbon (GAC) is effective for removing organic precursors and DBPs, but will be expensive, especially for small systems (Clark and Adams 1993). Some membrane technologies are able to remove both microbial contaminants and DBPs, but because of high costs, and operational and maintenance considerations, may be appropriate only for water systems in more technologically advanced and wealthy communities (Clark and Adams 1993). Bromide, for example, can only be removed from source water by reverse osmosis, an expensive technology.

CONSIDERATION OF COMPETING RISKS

The risk assessment issue is essentially that of a risk trade-off between exposure to DBPs and exposure to waterborne pathogens under various water treatment scenarios (Graham and Weiner 1995). It is important to identify (1) the DBPs that pose the greatest health risk, (2) how a proposed treatment strategy or technology will affect the removal (and risk) of these DBPs and microbial contaminants, and (3) the costs associated with these water technologies. Key issues surrounding the health risks of DBPs include regional differences in source water quality, which affect the levels and types of DBPs; the amount of precursor organic material and bromide present in water sources; water treatment technologies available to reduce the formation of DBPs; and the community's financial resources to commit to water treatment.

Before attempting to balance microbial and chemical risks, a quantitative assessment is required of the competing risks posed by disinfection and other water treatment technologies used to reduce the formation of DBPs. But how can birth defects and other adverse outcomes of pregnancy, cancer, and diarrheal disease be compared? In many developing countries life expectancy is low and the death toll among infants and children suffering from diarrhea is so large, microbial risks easily outweigh cancer and other risks. In many industrialized countries, cancer morbidity and mortality are important, and the risk trade-offs are not obvious even though deaths are also associated with infectious disease and disinfected water is likely to make only a small contribution to total cancer deaths (Pike, Fawell, and Miller 1993). Cancer will occur late in life with costs focused on medical treatment rather than lost productivity and wages, important consequences of infectious disease and important costs for developing countries.

BALANCING RISKS—THE PROBLEM STATEMENT

The basic question that must be answered for populations in both developing and developed countries is "What specific water treatment option(s) will provide the optimum reduction in adverse health consequences from both DBPs and waterborne microbial contaminants (or provide the maximum health benefit at an acceptable cost) for any particular water source?" To answer this question, we must be able to compare, in a formal decision analysis framework, health risks of DBPs and pathogens that will differ in severity, duration of illness, costs of medical treatment, economic consequences, and perceived importance by the public. There is a notable imbalance between the real and perceived risks associated with disinfection of drinking water (Otterstetter 1996). Diarrhea or gastroenteritis is usually perceived to be of lesser importance and there is the perceived lack of a common measure that would allow the comparison of these health outcomes. Any approach must accommodate both microbial and chemical risk information in the same analysis and be able to provide information about the health and economic consequences of choosing specific water treatment options for controlling DBPs.

Waterborne pathogens can cause a wide range of effects, from relatively mild illness to death, and the assessment of microbial risk requires information about infectious or toxic dose, viability and virulence of the organism, exposure–response relationships (probability of infection, disease, or death), occurrence in the water, organism–host interaction, susceptibility and immune-status of different populations, secondary person-to-person transmission, and chronic effects or associated sequelae of infectious disease (e.g., liver damage, myocarditis, and aseptic meningitis). Traditional risk assessment models for chemicals do not fully accommodate microbial contaminants, and several risk assessment approaches for pathogens have recently been proposed (ILSI 1996; Hurst, Clark, and Regli 1996; Hurst and Murphy 1996).

The risk assessment must also consider ways to evaluate complex mixtures of contaminants (bacteria, viruses, protozoa, inorganic chemicals, DBPs, and other organic chemicals) that vary with the water source, amount of bromide and precursor organic material in the water, sources of contamination, season, water treatment technologies employed, residence time in storage reservoirs and the water delivery system, and other factors. Conceptually, an infinite number of such mixtures can exist, but it has not yet been feasible to toxicologically test even a small number of chemical mixtures likely present. Other concerns include identifying all of the health risks that may be associated with microbial agents and disinfectant residuals and DBPs (D/DBPs) and estimating the probability of the health risk under various exposure conditions.

A public-health-focused approach, the prevention–effectiveness analysis, can be used to simultaneously compare risks of dissimilar health end points and from differing exposures (NRC 1990; Teutsch 1992; Gorsky and Teutsch 1995; Haddix et al. 1996). Prevention–effectiveness analyses have been successfully used to evaluate medical therapies and public health prevention programs, and this approach offers a structured, quantitative way to compare health risks of exposure to D/DBPs and waterborne pathogens under various water quality conditions and water treatment scenarios.

Briefly discussed in the remainder of this chapter are the reasons for proposing this approach (i.e., why the approach is considered applicable in decision-making about water treatment strategies to balance microbial and DBPs risks) and ways to quantitatively measure DBP and microbial risks so they can be compared.

PREVENTION–EFFECTIVENESS ANALYSIS

The prevention–effectiveness approach is outcome-oriented and has been used to evaluate the effectiveness of medical therapies and immunization, both of which can have benefits and risks. The approach seeks a direct link between the medical or public health intervention and the health outcome(s) of interest using both decision and economic analysis to compare the cost-effectiveness of the intervention. An analogy can be made between the treatment of drinking water and a public health intervention. Since the primary purpose of water treatment is to prevent the waterborne transmission of pathogens, it should be evaluated like other public health intervention programs (CDC 1992a, b, c, d, 1993a, b, c, d). Similar to an immunization program, there may be undesirable side effects when a disinfectant is used. Microbial risks are targeted for prevention, and the health risks that may be associated with D/DBPs are the undesired side effects or competing risks. By viewing these disparate health outcomes as interdependent instead of as unrelated entities, they can be evaluated simultaneously. Duration and severity of illness and the public's perception of the degree of risk (or dread) are also considered in the analysis.

Communities and regulatory agencies should consider these risk trade-offs in a formal way when choosing or approving water treatment technologies, and the proposed framework will help them to make informed choices. The framework is sufficiently flexible to accommodate such variables as type of source water, water quality, type of disinfection and other water treatment processes, chemicals and microbes likely to be present, and type of risk anticipated. The framework allows a lack of data to be handled in a consistent fashion.

PROPOSED APPROACH TO ASSESS RISKS FROM D/DBPs AND MICROBES

The prevention–effectiveness framework is proposed as a logical way to compare the chemical and microbial risks that may be associated with D/DBPs. The analysis should include both primary effects (e.g., cancer, adverse pregnancy outcomes, and gastroenteritis) and secondary effects (e.g., long-term sequelae of acute disease, side effects of treatment of acute disease, and costs of energy use and disposal of water treatment residuals). None should be excluded initially because of the lack of data or low probability of occurrence, as these can be adequately considered in the decision tree and by sensitivity analysis. Both the occurrence and severity of disease should be taken into account, and the analysis should reflect the financial costs associated with each water treatment option. It should be emphasized that sufficient data are not currently available to use the approach for choosing among the various water disinfectants and treatment technologies with any certainty, but the approach can be used to provide a direction for assessing research priorities.

The mixture of DBPs and pathogens to which humans are exposed can vary with time and location within a water distribution system. Consideration of the health effects from exposure to DBPs and pathogens combined is not realistic at present, but consideration of dose–response data for a mixture(s) of DBPs may be appropriate. If toxicity data are available for the chemical mixture, the quantitative risk assessment can be done directly from these data. If not, data for a "sufficiently similar" mixture (USEPA 1986) can be used, or the mixture can be evaluated by an analysis of its components. Most of the available toxicity data are for single DBPs, but some reports are available on DBP mixtures (Kool, van Kreigl, and Zoeteman 1982; Yang et al. 1989; Chapin et al. 1989; Germolec et al. 1989; Hong, Yang, and Borman 1992). These risk estimates should be compared with results of epidemiologic studies in which exposures are to the entire complex mixture.

Water Treatment Scenarios to Define Exposure

Many factors are involved in defining DBP mixtures that may be in drinking water, but a limited number of representative water treatment scenarios should be considered for comparison of risks. These include

- type and characteristics of source water
- type and amount of disinfectant used and contact time with precursor material
- ranges of water pH and temperature

- organic material and bromide ion concentration in source water
- water treatments to remove precursor material or DBPs (e.g., enhanced coagulation, GAC, packed-tower air stripping, and membrane filtration)

Once these characteristics have been considered, existing data sources (e.g., USEPA's Information Collection Rule) should be examined to determine if additional data collection and analysis are needed to identify the DBPs and pathogens of greatest concern and the concentrations that may be expected. Although water distribution system hydraulic models (e.g., EPANET) are not yet able to predict levels of DBPs or pathogens, they should be evaluated for their possible use in improving exposure estimates. These models, along with water quality information, will be important to help develop exposure scenarios.

Risk Estimates

Although highly uncertain at the present time because of a lack of methodologically sound studies, estimates of the population attributable risk for cancer and adverse outcomes of pregnancy can be obtained from clinical, epidemiological, and toxicological studies. Epidemiological studies, clinical dose–response data for specific pathogens, and modeling of infectious disease and waterborne outbreaks can also provide estimates of microbial risks, but additional data are required. The possible role of the water distribution system in increasing microbial risks should also be considered (Payment et al. 1997). Biofilms or layers of bacterial flora may colonize the interior of water distribution pipes and shield pathogens from inactivation by disinfectant residuals (White 1995; Arrage and White 1996).

Framing the Prevention–Effectiveness Analysis for Drinking Water

Farnham, Ackerman, and Haddix (1996) outline 13 key steps that must be addressed before either conducting or evaluating a prevention–effectiveness analysis. The prevention strategies being evaluated must be clearly indicated, including the baseline water treatment technology that represents the current practice (i.e., chlorine as the primary disinfectant). The analysis can accommodate any number of comparisons between technologies. The societal perspective is considered appropriate, analyzing all benefits of a program no matter who receives them, and all costs of a program, no matter who pays them (Farnham, Ackerman, and Haddix 1996) because the goal is to analyze the allocation of societal resources among competing activities.

A major impediment to successfully undertaking a comparative risk analysis has been the need to express all the health outcomes with a common metric. Graham and Weiner (1995) recommend focusing on the following characteristics to facilitate a credible weighing of one risk against another: (1) the magnitude or probability of the risk; (2) the size of the population affected; (3) the certainty in the risk estimates; (4) the type of adverse outcome and its duration and severity; (5) the population distribution of the risk; and (6) the timing of the risk. Measures of the lengthening or shortening of life, usually expressed as either lives or life-years lost or gained, have been commonly used in medical decision making, public health, and economics (Putnam and Graham 1993; Gardner and Sandborn 1990) and should be considered as a common outcome measure for comparison of D/DBP and microbial risks. These measures can also be weighted by an economic value or subjective assumptions about quality of life in measures such as years of potential life lost, disability-adjusted life years, quality-adjusted life years, or years of healthy life (Murray 1994; CDC 1982; Gardner and Sandborn 1990; Erickson, Wilson, and Shannon 1995; Murphy 1993). The weighting must reflect some social valuation of the outcome characteristics.

Analytic Method

The prevention–effectiveness approach is essentially an economic analysis, and three analytic methods (Haddix et al. 1996) are commonly applied, all incorporating some form of subjective social valuations (e.g., dollars or quality of life). In cost–benefit analysis (CBA), all consequences and benefits are converted into monetary values and results are expressed as either the net present value or the dollar amount of benefit per dollars expended (Clemmer and

Haddix 1996). In cost-effectiveness analysis (CEA), monetary values are assigned to the cost of care for persons with the conditions of interest and the costs associated with lost productivity relating to morbidity and mortality (Haddix and Shaffer 1996). Results are usually presented in the form of cost per health outcome (e.g., cost per case prevented or cost per life saved).

Cost–utility analysis (CUA) is a type of CEA but differs in the way the health outcomes are measured. Quality-of-life adjustments are specifically built into the calculations. In CUA, benefits are expressed as the number of life years saved, adjusted to account for loss of quality from morbidity of the health outcome or side effects from the intervention. The most common outcome measure for this type of analysis is the quality-adjusted life year.

Using a Decision-Tree Model

The underlying theory of decision analysis is that of the expected utility. This assumes that, if given a choice among different options, the option with the outcome that has the highest expected utility should be selected (Snider, Holtgrave, and Dunet 1996). The expected utility of any particular option is calculated from estimates of the probability of the different possible outcomes and the strength of preference for each outcome. There are several different ways to structure a decision analysis. The decision tree is most commonly used (Haddix et al. 1996) and should be appropriate for use in a prevention–effectiveness study of the D/DBP–microbe problem. In any decision analysis, the first step is to separate the specific problem into components that are small enough to be understood and analyzed and then used to model the essential elements of the problem. Using this methodology, the possible chains of events, i.e., decisions, chance events, and final outcomes, are identified in a structured tree that clearly specifies the sequence of events.

Uncertainty

Major sources of uncertainty in decision analysis (Morgan and Henrion 1990) include the health effects data, effectiveness of a specific treatment option in reducing DBPs and pathogens, achieving specific health outcomes, and the role of the distribution system in increasing or decreasing exposures. Additional uncertainties will be associated with various parameters in the decision-tree model. It is important to distinguish between parameter uncertainty, which implies that more research is needed, and parameter variability, which implies the need to modify the structure of the model.

Sensitivity and influence analyses will help identify the factors that are most important to the final outcome. The type of sensitivity analysis that will be conducted depends on the types and amount of information available. For example, if reliable data on population distributions of certain exposure variables can be assembled, then a formal Monte Carlo type analysis might be undertaken. When very sparse information exists on the variables of concern, expert judgment might be used to select extreme high and low values to evaluate their effect on the outcome.

CONCLUSIONS

Preventing infectious waterborne disease must remain a priority for both highly industrialized and developing countries. The disinfection of water is a primary public health intervention that results in a greatly reduced incidence of waterborne infection when effectively applied. Chlorine is the most widely used and cost-effective disinfectant for drinking water. In most parts of the world, chlorine cannot be readily replaced in the foreseeable future as a reliable and cost-effective water disinfectant.

Disinfection by-product regulations and control strategies must not increase microbial risks. Where risks for DBPs are likely to be very small relative to risks from waterborne pathogens, abandoning chlorination or drastically reducing its use is simply not appropriate. For example, concerns in the United States about DBPs were believed to have deterred officials in some areas of Peru from assiduously applying chlorine at sufficient levels to prevent the waterborne transmission of cholera (Salazar-Lindo 1993).

Any public policy for drinking water disinfection must consider not only the magnitude of competing risks from acute microbial and long-term chemical exposures but also a country's resources to deal with these risks. In developing countries with inadequate water supplies, poor sanitation, insufficient economic resources, and low life expectancy, the high risk of enteric disease from inadequately treated water would be of much greater concern than the relatively low carcinogenic health risks from the long-term consumption of disinfected water. As a country's economic and sanitation standards improve and waterborne disease is no longer a major threat, smaller risks from DBPs, which were once tolerated, may assume greater importance.

For countries considering regulations and strategies to reduce DBPs, the prevention–effectiveness approach can be used to compare microbial and chemical risks. This approach has been used to evaluate medical therapies and public health intervention programs, such as immunization against infectious diseases, where there may be both benefits and risks associated with the medical treatment or immunization. The prevention–effectiveness framework simultaneously considers disparate health outcomes in a cost–benefit and cost–utility analysis, and it allows direct analytical use of expert judgment, microbial risk models, DBP and microbial exposures, and human and animal toxicity data to identify water treatment options that can provide the maximum health benefits at an acceptable cost. The utility of this approach should be thoroughly examined by an appropriate peer-review panel, and case studies should be developed for several drinking water exposures of interest. Although exposure and health risk data are not yet sufficient to allow the use of this approach for regulatory decisions, use of expert judgment and sensitivity analyses can help identify areas of uncertainty that have the greatest effect on the outcome of the decision analysis. Thus, at the present time, this approach should be used primarily to provide a direction for additional research and development of appropriate decision-tree models.

REFERENCES

Arrage, A.A., and D.C. White. 1996. Biofilm Induced Decrease in Chlorine Effectiveness Against *Escherichia coli* and *Legionella bosmanii*. Manuscript in preparation.

Badenoch, J. 1990. Cryptosporidium *in Water Supplies*. London: HMSO.

Bellar, T.A., J. Litchtenberg, and R.C. Kroner. 1974. The Occurrence of Organohalides in Chlorinated Drinking Water. *Jour. AWWA*, 66:703.

Bennett, J.V., S.D. Holmberg, M.F. Rogers, and S.L. Solomon. 1987. Infectious and Parasitic Diseases. In *Closing the Gap: The Burden of Unnecessary Illness.* Amler, R.W., and H.B. Dull, eds. New York: Oxford Univ. Press.

Berger, P.S. 1987. Unpublished data.

Birkhead, G., and R.L. Vogt. 1989. Epidemiological Surveillance for Endemic *Giardia lamblia* Infection in Vermont. *Am. Jour. of Epidemiology,* 129(4):762–68.

Blair, K. 1994. *Cryptosporidium* and Public Health. *Health and Envir. Digest,* 8(8):61–3.

Bull, R.J., and F.C. Kopfler. 1991. *Health Effects of Disinfectants and Disinfection By-Products.* Denver, Colo.: American Water Works Association.

Carlsten, C. 1993. Water Disinfection By-products: The Regulatory Negotiation as a Means to Grapple With Uncertainty. Unpublished manuscript.

Centers for Disease Control. 1982. Introduction to Table V: Premature Deaths, Monthly Mortality, and Monthly Physician Contacts - United States. *Morbidity Mortality Weekly Report,* 31:109.

———. 1992a. Public Health Focus: Fluoridation of Community Water Systems. *Morbidity Mortality Weekly Report,* 41:372–75, 381.

———. 1992b. Public Health Focus: Mammography. *Morbidity Mortality Weekly Report,* 41:454–59.

———. 1992c. Public Health Focus: Surveillance, Prevention, and Control of Nosocomial Infections. *Morbidity Mortality Weekly Report,* 41:783–87.

———. 1992d. Public Health Focus: Effectiveness of Smoking-Control Strategies - United States. *Morbidity Mortality Weekly Report,* 41:645–47, 653.

———. 1993a. Public Health Focus: Effectiveness of Rollover Protective Structures for Preventing Injuries Associated With Agricultural Tractors. *Morbidity Mortality Weekly Report,* 42:57–59.

———. 1993b. Public Health Focus: Prevention of Blindness Associated With Diabetic Retinopathy. *Morbidity Mortality Weekly Report,* 42:191–95.

———. 1993c. Mandatory Bicycle Helmet Use - Victoria, Australia. *Morbidity Mortality Weekly Report,* 42:359–63.

———. 1993d. Public Health Focus: Physical Activity and the Prevention of Coronary Heart Disease. *Morbidity Mortality Weekly Report,* 42:669–72.

Chapin, R.E., J.L. Phelps, B.A. Schwetz, and R.S.H. Yang. 1989. Toxicology Studies of a Chemical Mixture of 25 Groundwater Contaminants. III. Male Reproduction Study in B6C3F$_1$ Mice. *Fund. Appl. Tox.,* 13:388–98.

Chute, C.G., R.P. Smith, and J.A. Baron. 1987. Risk Factors for Endemic Giardiasis. *Am. J. Pub. Health,* 77(5):585–87.

Clark, R.M., and J.Q. Adams. 1993. Control of Disinfection By-products: Economic and Technological Considerations. In *Safety of Water Disinfection: Balancing Chemical and Microbial Risks.* Craun, G.F., ed. Washington, D.C.: ILSI Press.

Clark, R.M., C.J. Hurst, and S. Regli. 1993. Costs and Benefits of Pathogen Control in Drinking Water. In *Safety of Water Disinfection: Balancing Chemical and Microbial Risks*. Craun, G.F., ed. Washington, D.C.: ILSI Press.

Clark, R.M., J.A. Goodrich, and J.C. Ireland. 1984. Costs and Benefits of Drinking Water Treatment. *Jour. Envir. Sys.*, 4(1):30.

Clemmer, B., and A.C. Haddix. 1996. Cost-Benefit Analysis. In *Prevention Effectiveness: A Guide to Decision Analysis and Economic Evaluation*. Haddix, A.C., et al., eds. New York: Oxford University Press.

Craun, G.F. 1986. Statistics of Waterborne Outbreaks in the U.S. (1920–1980). In *Waterborne Diseases in the United States*. Craun, G.F., ed. Boca Raton, Fla.: CRC Press.

———, ed. 1990. *Methods for the Investigation and Prevention of Waterborne Disease Outbreaks*. Cincinnati, Ohio: US Environmental Protection Agency.

———. 1992. Waterborne Disease Outbreaks in the United States of America: Causes and Prevention. *World Health Statistics Quarterly*, 45:192.

———, ed. 1993. *Safety of Water Disinfection: Balancing Chemical and Microbial Risks*. Washington, D.C.: ILSI Press.

———. 1996. *Water Quality in Latin America: Balancing the Microbial and Chemical Risks in Drinking Water Disinfection*. Washington, D.C.: ILSI Press.

Craun, G.F., D. Swerdlow, R. Tauxe, et al. 1991. Prevention of Cholera in the United States. *Jour. AWWA*, 83:40.

Erickson, P., R. Wilson, and I. Shannon. 1995. Years of Healthy Life. Statistical Notes No. 7 (April 1995). US Dept. of Health and Human Services, National Center for Health Statistics.

Esrey, S.A., J.B. Potash, L. Roberts, and C. Shiff. 1990. *Health Benefits from Improvements in Water Supply and Sanitation: Survey and Analysis of the Literature on Selected Diseases*. Technical Report No. 66, Water and Sanitation for Health Project. Washington, D.C.: US Agency for International Development.

Farnham, P.G., S.P. Ackerman, and A.C. Haddix. 1996. Study Design. In *Prevention Effectiveness: A Guide to Decision Analysis and Economic Evaluation*. Haddix, A.C., et al., eds. New York: Oxford University Press.

Fraser, G.G., and K.R. Cooke. 1991. Endemic Giardiasis and Municipal Water Supply. *Am. J. Pub. Health*, 81:760–63.

Galal-Gorchev, H. 1993. WHO Guidelines for Drinking Water Quality. *Safety of Water Disinfection: Balancing Chemical and Microbial Risks*. Craun, G.F., ed. Washington, D.C.: ILSI Press.

———. 1996a. Disinfection of Drinking Water and By-products of Health Concern. In *Water Quality in Latin America: Balancing the Microbial and Chemical Risks in Drinking Water Disinfection*. Craun, G.F., ed. Washington, D.C.: ILSI Press.

———. 1996b. WHO Guidelines for Drinking Water Quality and Health Risk Assessment for Disinfectants and Disinfection By-products. In *Water Quality in Latin America: Balancing the Microbial and Chemical Risks in Drinking Water Disinfection.* Craun, G.F., ed. Washington, D.C.: ILSI Press.

Gardner, J.W., and J.S. Sandborn. 1990. Years of Potential Life Lost (YPLL) - What Does It Measure? *Epidemiology*, 1:322.

Germolec, D.R., R.S.H. Yang, M.F. Ackermann, G.J. Rosenthal, G.A. Boorman, P. Blair, and M.I. Luster. 1989. Toxicology Studies of a Chemical Mixture of 25 Groundwater Contaminants. II. Immunosuppression in $B6C3F_1$ Mice. *Fund. Appl. Tox.*, 13:377–87.

Gorsky, R.D., and S.M. Teutsch. 1995. Assessing the Effectiveness of Disease and Injury Prevention Programs: Costs and Consequences. *Morbidity Mortality Weekly Report*, 44 (No. RR-10):1–10.

Graham, J.D., and J.B. Weiner. 1995. *Risk vs. Risk: Trade-offs in Protecting Health and the Environment.* Cambridge, Mass.: Harvard University Press.

Haddix, A.C., and P.A. Shaffer. 1996. Cost-effectiveness Analysis. In *Prevention Effectiveness: A Guide to Decision Analysis and Economic Evaluation.* Haddix, A.C., et al., eds. New York: Oxford University Press.

Haddix, A.C., S.M. Teutsch, P.A. Shaffer, and D.O. Dunet. 1996. *Prevention Effectiveness: A Guide to Decision Analysis and Economic Evaluation.* New York: Oxford University Press.

Hauschild, A., and F. Bryan. 1980. Estimate of Cases of Foodborne and Waterborne Illness in Canada and the United States. *Jour. Food Protection*, 43:435.

Hong, H.L., R.S.H. Yang, and G.A. Boorman. 1992. Alterations in Hemotopoietic Responses in B6C3Fa Mice Caused by Drinking a Mixture of 25 Groundwater Contaminants. *Jour. Envir. Path. Toxicol. Oncol.*, 11:65–74.

Hoxie, N.J., J.P. Davis, and J.M. Vergeront. 1996. Cryptosporidiosis-associated Mortality Following a Massive Waterborne Outbreak in Milwaukee, Wisconsin. Report of Wisconsin Department of Health and Social Services, Madison, Wisc.

Hurst, C.J., and P. Murphy. 1996. The Transmission and Prevention of Infectious Disease. In *Modeling Disease Transmission and Its Prevention by Disinfection.* Hurst, C.J., ed. Cambridge, Great Britain: Cambridge University Press.

Hurst, C.J., R. Clark, and S.E. Regli. 1996. Estimating the Risk of Acquiring Infectious Disease from Ingestion of Water. In *Modeling Disease Transmission and Its Prevention by Disinfection.* Hurst, C.J., ed. Cambridge, Great Britain: Cambridge University Press.

International Life Sciences Institute Pathogen Risk Assessment Working Group. 1996. A Conception Framework to Assess the Risks of Human Disease Following Exposure to Pathogens. *Risk Analysis*, 16(6):841.

Kool, H.J., C.F. van Kreijl, and B.C. Zoeteman. 1982. Toxicology Assessment of Organic Compounds in Drinking Water. *CRC Crit. Rev. Environ. Control*, 12:307–57.

Kramer, M.H., B.L. Herwaldt, G.F. Craun, et al. 1996. Waterborne Disease: 1993 and 1994. *Jour. AWWA*, 88(3):66–80.

Krasner, S.W., M.J. McGuire, J.G. Jacangelo, N.L. Patania, K.M. Reagan, and E.M. Aieta. 1989. The Occurrence of Disinfection By-products in U.S. Drinking Water. *Jour. AWWA*, 81:41–53.

MacKenzie, W.R., N.J. Hoxie, M.E. Proctor, et al. 1994. A Massive Outbreak in Milwaukee of *Cryptosporidium* Infection Transmitted Through the Public Drinking Water Supply. *New England Journal of Medicine*, 331(3):161–7.

Morgan, M.G., and M. Henrion. 1990. *Uncertainty - A Guide to Dealing With Uncertainty in Quantitative Risk and Policy Analysis.* Cambridge, Great Britain: Cambridge University Press.

Morris, R.D., and R. Levin. 1995. Estimating the Incidence of Waterborne Infectious Disease Related to Drinking Water in the United States. In *Assessing and Managing Health Risks from Drinking Water Contamination: Approaches and Applications.* Reivhard, E., and G. Zapponi, eds. IAHS Publication 233, p. 75–88.

Murphy, P.A. 1993. Quantifying Chemical Risk from Epidemiologic Studies: Application to the Disinfectant By-product Issue. In *Safety of Water Disinfection: Balancing Chemical and Microbial Risks.* Craun, G.F., ed. Washington, D.C.: ILSI Press.

Murray, C.J.L. 1994. Quantifying the Burden of Disease: The Technical Basis for Disability-Adjusted Life Years. *Bull. WHO*, 72:429–45.

National Research Council. 1990. *Valuing Health Risks, Costs, and Benefits for Environmental Decision Making: Report of a Conference.* Washington, D.C.: National Academy Press.

Okun, D.A. 1993. Global Water Supply Issues From a Public Health Perspective. *Safety of Water Disinfection: Balancing Chemical and Microbial Risks.* Craun, G.F., ed. Washington, D.C.: ILSI Press.

Otterstetter, H. 1996. Preface. In *Water Quality in Latin America: Balancing the Microbial and Chemical Risks in Drinking Water Disinfection.* Craun, G.F., ed. Washington, D.C.: ILSI Press.

Payment, P., L. Richardson, J. Siemiatycki, R. Dewar, M. Edwardes, and E. Franco. 1991. A Randomized Trial to Evaluate the Risk of Gastrointestinal Disease Due to Consumption of Drinking Water Meeting Current Microbiological Standards. *Am. J. Pub. Health*, 81:703.

Payment, P., E. Franco, L. Richardson, and J. Siemiatycki. 1991. Gastrointestinal Health Effects Associated With the Consumption of Drinking Water Produced by Point of Use Domestic Reverse Osmosis Filtration Units. *Applied Environ. Micro.*, 57:945.

Payment, P., J. Siemiatycki, L. Richardson, G. Renaud, E. Franco, and M. Prevost. 1997. A Prospective Epidemiological Study of Gastrointestinal Health Effects Due to the Consumption of Drinking Water. *Int. J. Environ. Health Res.*, 7:5–31.

Pike, E.B., J.K. Fawell, and D.G. Miller. 1993. Economic Factors in Balancing Chemical and Microbial Risks. *Safety of Water Disinfection: Balancing Chemical and Microbial Risks.* Craun, G.F., ed. Washington, D.C.: ILSI Press.

Putnam, S.W., and J.D. Graham. 1993. Chemicals Versus Microbials in Drinking Water: A Decision Sciences Perspective. *Jour. AWWA,* 85(3):57–61.

Rook, J.J. 1974. Formation of Haloforms During Chlorination of Natural Waters. *Water Treatment and Examination,* 23:34–43.

Salazar-Lindo, E. 1993. The Peruvian Cholera Epidemic and the Role of Chlorination in Its Control and Prevention. *Safety of Water Disinfection: Balancing Chemical and Microbial Risks.* Craun, G.F., ed. Washington, D.C.: ILSI Press.

Singer, P.C. 1993. Formation and Characterization of Disinfection By-products. In *Safety of Water Disinfection: Balancing Chemical and Microbial Risks.* Craun, G.F., ed. Washington, D.C.: ILSI Press.

Snider, D.E., D.R. Holtgrave, and D.O. Dunet. 1996. Decision Analysis. In *Prevention Effectiveness: A Guide to Decision Analysis and Economic Evaluation.* Haddix, A.C., et al., eds. New York: Oxford University Press.

Soave, R. 1990. Treatment Strategies for Cryptosporidiosis. *Annals New York Academy of Sciences,* 616:442–51.

Stevens, A.A., L.A. Moore, and C.J. Slocum, et al. 1990. By-products of Chlorination at Ten Operating Utilities. In *Water Chlorination: Chemistry, Environmental Impact and Health Effects.* Jolley, R.L, et al., eds. Chelsea, Mich.: Lewis Publishers.

Swerdlow, D.L., et al. 1992. Waterborne Transmission of Epidemic Cholera in Trujillo, Peru: Lessons for a Continent at Risk. *Lancet,* 340:28–32.

Teutsch, S.M. 1992. A Framework for Assessing the Effectiveness of Disease and Injury Prevention Programs. *Morbidity Mortality Weekly Report,* 41(No. RR-3):1–12.

USEPA. 1986. Guidelines for the Health Risk Assessment of Chemical Mixtures. *Fed. Reg.,* 51(185):34014–34025.

———. 1997a. Announcement of the Draft Drinking Water Contaminant Candidate List: Notice. *Fed. Reg.,* 62(193):52194–52219.

White, D.C. 1995. Chemical Ecology: Possible Linkage Between Macro- and Microbial Ecology. *Oikos,* 74:174–181.

Yang, R.S.H., T.J. Goehl, R.D. Brown, A.T. Charham, D.W. Arneson, R.C. Buchanan, and R.K. Harris. 1989. Toxicology Studies of a Chemical Mixture of 25 Groundwater Contaminants. I. Chemistry Development. *Fund. Appl. Tox.,* 13:366–376.

CHAPTER • SEVEN

Regulation of Disinfection By-Products

FREDERICK W. PONTIUS

Regulation of disinfection by-products (DBPs) has been the subject of regulatory activity by the US Environmental Protection Agency (USEPA) for over 20 years. Key milestones and deadlines for regulating microbials and DBPs are summarized in Table 7-1. At the time of this writing, USEPA had just completed a multiyear rulemaking process resulting in issuance of a final Stage 1 DBP Rule in 1998. The background and related technical considerations in the development of this rule are reviewed in this chapter.

EARLY HISTORY OF DBP REGULATION

The first drinking-water-related regulation, established in 1912, targeted the prevention of waterborne disease by prohibiting the use of the common cup on carriers of interstate commerce, such

Table 7-1 Key milestones and deadlines for regulation of microbials and DBPs

Dec. 24, 1975—National Interim Primary Drinking Water Regulations (NIPDWRs) promulgated; coliform bacteria and turbidity maximum contaminant levels (MCLs) were set to prevent bacterial contamination, based on the 1962 US Public Health Service Standards.

July 14, 1976—USEPA published an Advance Notice of Proposed Rulemaking (ANPRM) regarding control of organic contaminants in drinking water, including disinfection by-products (DBPs) formed as a result of using chlorine for disinfection.

Feb. 9, 1979—USEPA proposed to amend the NIPDWRs to regulate total trihalomethanes (TTHMs) in drinking water.

Nov. 29, 1979—Rule promulgated to amend the NIPDWRs to regulate TTHMs.

Nov. 29, 1981—TTHM regulation became effective for water systems serving 75,000 persons or more.

Mar. 5, 1982—USEPA proposed a rule to amend the NIPDWRs to clarify application of best available technology (BAT) for TTHMs.

Feb. 28, 1983—Rule promulgated amending the NIPDWRs to clarify application of BAT for TTHMs.

Oct. 5, 1983—USEPA published an ANPRM discussing development of revised National Primary Drinking Water Regulations (NPDWRs) in four phases, including DBPs.

Nov. 29, 1983—TTHM regulation became effective for water systems serving 10,000 persons or more but fewer than 75,000 persons.

Nov. 13, 1985—USEPA proposed recommended maximum contaminant levels (RMCLs) for microbiological contaminants.

June 16, 1986—Safe Drinking Water Act (SDWA) amendments for 1986 were enacted and new requirements were imposed on the USEPA to regulate contaminants.

Nov. 3, 1987—USEPA proposed the Surface Water Treatment Rule (SWTR) and Total Coliform Rule (TCR) to satisfy 1986 SDWA amendment requirements.

Jan. 22, 1988—USEPA published the first Drinking Water Priority List (DWPL) of contaminants for possible regulation under the SDWA, including several disinfectants and DBPs.

June 29, 1989—USEPA promulgated the SWTR and TCR.

Sept. 22, 1989—USEPA issued a strawman proposal for regulation of DBPs.

December 1990—USEPA floated a conceptual framework for DBP regulations.

Dec. 31, 1990—Effective date of the SWTR and TCR.

Jan. 15, 1991—USEPA promulgated stay of certain TCR provisions regarding variances.

Table 7-1 Key milestones and deadlines for regulation of microbials and DBPs (continued)

June 20, 1991—USEPA released a status report on development of regulations for disinfectants and DBPs (D/DBPs).

June 1992—USEPA released an updated status report on development of regulations for D/DBPs.

Sept. 15, 1992—Because of the complex issues involved, USEPA published a notice of intent to form an advisory committee under the Federal Advisory Committee Act (FACA) and to conduct a regulatory negotiation (or reg neg) with stakeholders to develop a proposed rule for D/DBPs.

June 29, 1993—New SWTR requirements for filtration performance and disinfection criteria took effect for systems using filtration and disinfection.

Spring 1994—Reg-neg committee completed work, concluding that a three-rule package consisting of the Information Collection Rule (ICR), the D/DBP Rule, and the enhanced SWTR (ESWTR) is needed to regulate DBPs to ensure microbial quality.

Feb. 10, 1994—Proposed ICR was published.

July 29, 1994—Proposed D/DBP Rule and ESWTR were published.

1995—USEPA conducted an extensive reassessment of its drinking water program, including a series of stakeholder meetings to gain input on regulatory issues and priorities.

May 14, 1996—Final ICR was promulgated.

May 1996—USEPA initiated a series of meetings to inform stakeholders about ICR implementation and related D/DBP and ESWTR issues.

June 1996—USEPA released the National Drinking Water Program Redirection Strategy; high priority was given to development of the total microbials/disinfection by-products (M/DBP) cluster consisting of six rules—the ICR, interim and final ESWTRs, Stage 1 and Stage 2 D/DBP Rules, and the Ground Water Rule (GWR).

Aug. 6, 1996—The SDWA amendments of 1996 were enacted, requiring USEPA to promulgate the D/DBP rules and ESWTR. Delay of the ICR means that the Stage 1 D/DBP Rule and IESWTR must be promulgated by November 1998.

Feb. 21, 1997—USEPA announced formation of an advisory committee under the FACA to assist the agency in expediting regulations, guidance, and policies to address microorganisms and D/DBPs in drinking water.

Mar. 13–14, 1997—First formal meeting of the M/DBP advisory committee. USEPA outlined its intention to publish a notice of data availability (NODA) of new data that have become available since the D/DBP and ESWTR were proposed. This information was used in developing an expedited rule to meet the November 1998 deadline.

(table continues on next page)

Table 7-1 Key milestones and deadlines for regulation of microbials and DBPs (continued)

Apr. 15, 1997—A series of Technical Work Group meetings began to provide analytical and technical support to the M/DBP advisory committee.

July 1997—ICR monitoring began.

July 15, 1997—M/DBP advisory committee completed work; agreement in principle regarding the expedited rules was signed.

Nov. 3, 1997—USEPA published NODAs regarding new information available since the proposed D/DBP and ESWTR were published.

Dec. 16, 1998—USEPA promulgated the expedited Stage 1 D/DBP Rule and IESWTR.

September 1999—USEPA expected to propose the GWR.

November 2000—USEPA expected to promulgate a long-term ESWTR to apply to systems serving <10,000 people (LT1ESWTR).

November 2000—USEPA expected to promulgate the final GWR.

May 2002—USEPA expected to promulgate the Stage 2 D/DBP Rule.

as trains (Pontius 1997). The first microbiological standards were set in 1914 and included a $100/cm^3$ limit for total bacterial count and not more than one of five 10 cm^3 portions of sample containing *E. coli* (now called *Escherichia coli*). These standards were only binding on water supplies used by interstate carriers, but many state and local governments adopted them as guidelines (Borchardt and Walton 1971).

By 1925, large cities applying either filtration, chlorination, or both, encountered little difficulty in complying with the 2 coliforms per 100 mL limit in force at that time. The US Public Health Service (USPHS) set revised standards to reflect the experience of water systems with excellent records of safety against waterborne disease. The limit was changed to 1 coliform per 100 mL, and the principle of attainability (i.e., standards were adopted that reflected what water systems with good safety records could attain) was established (USPHS 1925). Inherent in the development of the 1925 standards was the concept of relative risk. The availability of adequate treatment methods and the risk of contracting disease from contaminated drinking water relative to other sources influenced their development. As noted in the preamble to these standards (USPHS 1925):

> The first step toward the establishment of standards which will insure safety of water supplies conforming to

> them is to agree upon some criterion of safety. This is necessary because "safety" in water supplies, as they are actually produced, is relative and quantitative, not absolute. Thus, to state that a water supply is "safe" does not necessarily signify that absolutely no risk is ever incurred in drinking it. What is usually meant, and all that can be asserted from any evidence at hand is that the danger, if any, is so small that it cannot be discovered by available means of observation.

The 1925 USPHS standards and the revisions that followed in 1942, 1946, and 1962 did not include any DBPs because none had yet been discovered. Priority was given to ensuring that drinking water was microbially safe.

This early history progression is important because three basic public water supply principles emerged that continue to shape DBP regulation today. First is the high priority given to ensuring microbiological safety of drinking water so that predictable fatalities caused by consumption of microbially contaminated water are prevented. Second is recognition that "safe" drinking water is not necessarily risk-free, but that risks associated with drinking water consumption should be as small as feasible, especially when compared with other offsetting risks. Third, drinking water standards adopted must consider the ability of well-designed and operated water systems to meet the standard, otherwise a regulation might be set that was technologically impossible to meet, causing more harm than good, or costing more than customers are willing to pay (and, hence, would not be implementable).

REGULATION OF TRIHALOMETHANES (THMs)

The 1974 Safe Drinking Water Act (SDWA) gave USEPA the general authority to regulate drinking water contaminants, including DBPs. Toxicological studies in the early 1970s suggested that chloroform caused cancer in animals (NAS 1977). A discussion of prevalent health effects, available control technology, and costs of regulating organic contaminants was published July 14, 1976 (USEPA 1976). A two-part rule to control organic contaminants was proposed Feb. 9, 1978 (USEPA 1978). The first part of the proposal involved setting a maximum contaminant level (MCL) for total trihalomethanes (TTHMs), defined as the sum of the concentrations of chloroform, bromodichloromethane, dibromochloromethane, and bromoform. The second part of the proposal required the use of granular activated carbon (GAC) or equivalent technology by water systems subject to significant contamination of source water by synthetic organic chemicals.

The two parts of the proposal, although published together, were separate, but many of the 598 written public comments submitted showed confusion between the two. Some people assumed that GAC was proposed to control TTHMs, but this was not the case. The GAC treatment technique portion of the proposal was very controversial and was eventually withdrawn (USEPA 1981). Symons (1984) has reviewed the history of the attempted regulation of GAC treatment.

Regulations for TTHMs were promulgated Nov. 29, 1979 (USEPA 1979), based on knowledge of the health effects of the THM compounds available in the late 1970s. At that time, high exposure to chloroform was found to cause cancer in laboratory animals (rats and mice). Human evidence was inconclusive, although several epidemiological studies found positive correlations with some cancers (see chapter 5). The proposed rule reviewed 18 retrospective epidemiological studies that investigated some aspect of cancer mortality or morbidity and drinking water.

The USEPA concluded that THMs pose a carcinogenic risk at concentrations found in drinking water. In the final rule, the agency noted that no safe level can be deemed to exist for human exposure to carcinogens and concentrations of these contaminants should be reduced to the extent feasible. This general approach towards regulation of carcinogens in drinking water is still used today. The 1979 MCL established for TTHMs was 0.10 mg/L. Measures taken to meet the MCL for TTHMs were also expected to provide the additional benefit of reducing exposure to other undefined DBPs. Selection of the 1979 TTHM MCL was based on a balancing of public health considerations and the feasibility of public water systems achieving the MCL. The final rule notes (USEPA 1979):

> Thus, the interim MCL should not be construed as an absolutely "safe" level, but rather a feasible level achievable with water treatment technology available since

> 1974. The preponderance of the current scientific thought on human exposure to substances that have been demonstrated to be carcinogens in animals in appropriate tests is that they be considered potential carcinogenic risks to humans. The presumptions are that human health risk related to the extent of exposure and that no threshold level without risk can be experimentally demonstrated for a genetically diverse population. Translated into regulatory policy, exposure should be minimized so as to minimize unnecessary risks. Therefore, public water systems should strive to reduce TTHMs and related contaminant concentrations to levels as low as is economically and technologically feasible without compromising protection against the transmission of pathogenic microorganisms via drinking water.

In this rulemaking, USEPA discussed various treatment techniques that it believed public water systems might use to achieve compliance with the TTHM MCL. The wording of the rule, however, was ambiguous. A proposal amending the 1979 TTHM rule was published Mar. 5, 1982 (USEPA 1982), and promulgated Feb. 28, 1983 (USEPA 1983). This rule identified the best technologies, treatment techniques, or other means USEPA has determined to be "generally available," taking costs into consideration, and additional methods not determined to be "generally available." A rule published Aug. 3, 1993, later amended the analytical methods approved for compliance monitoring (USEPA 1993).

The 1979 TTHM regulation applied only to community water systems that served 10,000 people or more and that added a disinfectant to the water in any part of the treatment process. Requirements for water systems serving less than 10,000 people were at the discretion of the state primacy agency or USEPA if the state did not have primacy. USEPA believed at that time that exempting small systems (below 10,000 persons served) would not result in reduced health protection for these systems for several reasons summarized in Table 7-2 (USEPA 1979).

To meet the 1979 TTHM rule, many water utilities had to reconsider their disinfectant practice. A variety of strategies were developed, some involving significant capital improvements. To allow the necessary time for systems serving 75,000 or more people

Table 7-2 Considerations in regulating TTHMs and other DBPs in small systems (<10,000 persons)

- The majority of small systems (about 80 percent) use groundwater that is low in TTHM precursors.
- The proportion of small community water systems that use chlorine is generally less than that of large systems.
- The transport time within the distribution system is generally shorter in small systems.
- Small systems incur a greater risk of adversely affecting the microbiological quality of their drinking water when action is taken to lower TTHMs.
- The majority of waterborne disease outbreaks attributed to inadequate treatment occurred in small systems, highlighting the need to maintain strong disinfection practices in these systems.
- Small systems have limited or no access to the resources and professional expertise needed to control TTHMs.
- Small systems serve only about 20 percent of the population, but constitute the vast majority of the water systems, thus making careful regulatory oversight effectively impossible.

Source: USEPA (1979).

to comply, the effective date of the TTHM MCL was set for Nov. 29, 1981, two years after promulgation. For systems serving 10,000 or more people but less than 75,000 people, the TTHM MCL became effective Nov. 29, 1983, four years after promulgation.

REGULATION OF DBPs UNDER THE 1986 SDWA AMENDMENTS

The 1986 SDWA amendments required USEPA to develop a list of contaminants, known or anticipated to occur in public water systems, that may require regulation under the SDWA. At least 25 contaminants on each list were to be regulated within 24 months after the list was first published. The list was to be updated every 3 years, with a minimum of at least 25 additional contaminants regulated from each updated list. Disinfectants (Ds) and DBPs were included on the list for possible regulation. USEPA grouped several of them together in a D/DBP rulemaking covering disinfectants and DBPs to partially satisfy the requirement for regulating at least 25 contaminants every 3 years.

The 1986 SDWA amendments also required USEPA to issue regulations requiring filtration and disinfection for water systems using surface waters and to revise regulations for total coliforms. To comply with the 1979 TTHM MCL, many water utilities modified their disinfection practices (AWWA 1992). These two rules, the Surface Water Treatment Rule (SWTR) (USEPA 1989a) and the Total Coliform Rule (TCR) (USEPA 1989b), are the principal rules governing the microbial quality of drinking water and limited any compromises in disinfection as a result of treatment changes to control TTHMs.

1989 Strawman

Activity by USEPA focusing on potential D/DBP regulatory scenarios began in the spring of 1989. The USEPA released an initial strawman proposal for regulating D/DBPs on Sept. 22, 1989 (USEPA 1989c). This proposal was developed by USEPA to stimulate dialogue on this rule and to help focus research, data gathering, and analysis. The strawman, summarized in Table 7-3, was reviewed by the Science Advisory Board and discussed at a public meeting on Dec. 4, 1989.

The agency began to develop regulations for D/DBPs and released a conceptual framework in December 1990. A status report on development of regulations for D/DBPs was released for public information by USEPA on June 20, 1991 (USEPA 1991; Pontius 1991). An updated status report was distributed in June 1992.

Regulatory Negotiation

In the fall of 1992, USEPA determined that a regulation for DBPs would best be developed using the negotiated rulemaking process under the authority of the Negotiated Rulemaking Act of 1990 (P.L. 101-648), also called *regulatory negotiation* (or reg neg), because of the complex issues faced in this rulemaking. Alternative disinfectants to chlorine each have limitations, and no clear solution to the problem of minimizing DBPs while maintaining microbial quality was evident. USEPA viewed a negotiated rulemaking process as being beneficial in helping stakeholders understand the complexities of the risk–risk trade-off issues, mobilizing additional resources outside the agency to address the issues at

Table 7-3 Key elements of USEPA's 1989 D/DBP Rule Strawman

- Set MCL goals (MCLGs) and MCLs for total THMs (TTHMs), haloacetic acids, chlorine dioxide, chlorite, chlorate, chlorine, and chloramines; and potentially for chloropicrin, cyanogen chloride, hydrogen peroxide, bromate, iodate, and formaldehyde.
- Preferred option was to set the MCL for TTHMs at 50 μg/L or 25 μg/L. Other MCLs based on analyses of feasibility similar to TTHMs.
- Set treatment technique requirements or provide guidance for MX* (as a surrogate for mutagenicity), total oxidizing substances (as a surrogate for organic peroxides and epoxides), and assimilable organic carbon (AOC) (as a surrogate for microbiological quality of oxidized waters).
- List best available technologies (BAT) as precursor removal (50 percent removal of TTHM-formation potential), alternate oxidants, and by-product removal.
- Precursor removal as BAT to include conventional treatment modifications; granular activated carbon† (GAC) up to 30 minutes empty-bed contact time and 3 months regeneration; membranes may not be BAT because of lack of full-scale experience.
- Alternate oxidants allowed as BAT, assuming MCLG values are met for disinfectants, included chlorine dioxide with chlorite residual removal and chloramines, and ozone plus chloramines. TTHM MCL of 25 μg/L was lowest considered that would allow continued use of free chlorine.
- By-product removal BAT included aeration (for some systems); GAC adsorption (not for most chlorination by-products; effectiveness for ozonation by-products was unknown); reducing agents for MX, total oxidizing substances, possibly chloropicrin and cyanogen chloride; reducing agents or free chlorine for hydrogen peroxide; BAT for bromate and iodate were uncertain; and caveat ozone use with possible future need for post-GAC treatment for controlling AOC (bioassimilation) or removal of other by-products by adsorption.

*3-chloro-4-(dichloromethyl)-5-hydroxy-2(5H)-furanone.
†GAC not universally feasible because of water quality conditions.

hand, and in reaching a consensus on a sensible regulation until results of additional research became available.

A notice of intent to form an advisory committee for the D/DBP Rule was published Sept. 15, 1992 (USEPA 1992a). USEPA formed the advisory committee and its first meeting was held Nov. 23–24, 1992 (USEPA 1992b). The 18-member advisory committee discussed the key issues summarized in Table 7-4 during the reg-neg process. A series of technical documents were prepared and a Technologies Working Group (TWG) was formed to support the advisory committee.

Table 7-4 Key questions addressed during regulatory negotiation of the proposed D/DBP Rule

1. Which disinfectants and disinfection by-products present the greatest risks, and how should they be grouped for regulation?
2. Which categories of public water suppliers should be regulated?
3. Should the regulation establish MCLs or be technology driven?
4. How effective are advanced technologies and alternative disinfectants for the removal of disinfectants, disinfection by-products, and microbial risks?
5. How should disinfectants, disinfection by-products, and microbial risks be compared, given differences in the type and certainty of their effects?
6. What levels of disinfectants, disinfection by-products, and microbial risk are acceptable, and at what cost?
7. How should the achievement of acceptable levels of risk be defined?
8. How might risk–risk models be used, if at all, in the development of MCLs within the current regulatory schedule?
9. How should the needs of sensitive populations be taken into account in the rule?
10. How should BAT be defined for the removal of disinfectants and disinfection by-products?
11. Should a comprehensive D/DBP regulation be issued in 1995, or should the control of certain D/DBPs be deferred until research confirms the safety of alternative treatment methods?
12. How should affordability be factored into judgments regarding feasibility of treatment techniques?
13. How should monitoring requirements be defined?
14. How can the rule be drafted to be easily understood by both state regulators and small-system operators?

The reg-neg committee drew several key conclusions summarized in Table 7-5 and developed draft language for the following three rules:

- a two-stage DBP rule, with Stage 1 based on current knowledge of health risks, treatment costs, and competing microbial risks, and Stage 2 to be developed after new information is collected
- an interim enhanced Surface Water Treatment Rule (IESWTR) to expand the SWTR to protect against *Cryptosporidium* and to provide assurance that microbial

Table 7-5 Key DBP regulatory negotiating committee conclusions

- Individual DBPs cannot be predicted well.
- Control of one DBP may cause problems with control of another DBP.
- Aggregate measurements, such as TTHMs and HAA5,* allow assessment of treatment technologies.
- How BAT is defined determines feasible MCLs, because lower MCLs can be achieved if the most expensive technologies are considered BAT.
- Data does not exist for determining whether or not a substantial lowering of the TTHM MCL or new DBP MCLs will increase microbial risks.
- The DBP rule must be set with corresponding revisions of the SWTR to address potential microbial risks.
- The DBP rule should be set in two stages; Stage 1 based on current knowledge and Stage 2 when new information is available.

*HAA5 is the sum of monochloroacetic acid, dichloroacetic acid, trichloroacetic acid, monobromoacetic acid, and dibromoacetic acid. Only five of the nine haloacetic acids (HAAs) are included because occurrence data were not available for the remaining four.

risks will be controlled as water systems comply with the Stage 1 D/DBP rule

- an Information Collection Rule (ICR) to collect data for use in developing the Stage 2 D/DBP Rule and final Enhanced Surface Water Treatment Rule (ESWTR)

The D/DBP Rule and the companion ESWTR were proposed July 29, 1994 (USEPA 1994b, c). The ICR was proposed Feb. 10, 1994, and finalized May 14, 1996 (USEPA 1994a, 1996). In developing these proposals, the agency sought to adhere to two key intentions of the reg-neg committee. First, avoid premature shifts in disinfectant use, i.e.,

> ...The committee does not want to force large numbers of additional systems to switch disinfectants before more information is available.... (59 FR 38742)

Second, ensure that the microbial safety of drinking water is not compromised as a result of water treatment changes to meet a D/DBP regulation, i.e.,

> One of the major goals addressed by the Committee was to develop an approach that would reduce the level of exposure from disinfectants and DBPs without undermining the control of pathogens. (59 FR 38833)

The ICR, D/DBP Rule, and ESWTR are integrally connected and detailed reviews of their development are available (Pontius 1993a, b, c, 1996, 1997). Information Collection Rule monitoring began in July 1997 and continued through December 1998.

The proposed D/DBP Rule will apply to all community and nontransient, noncommunity water systems that treat their water with a chemical disinfectant regardless of the population served. The Stage 1 D/DBP Rule is to be promulgated and implemented concurrently with the IESWTR. The major provisions of the proposed Stage 1 D/DBP Rule are summarized in Table 7-6.

REGULATION OF DBPs UNDER THE 1996 SDWA AMENDMENTS

The 1996 SDWA amendments require USEPA to promulgate an IESWTR, a final (long-term) ESWTR, a Stage 1 D/DBP Rule, and a Stage 2 D/DBP Rule in accordance with a specified schedule (USEPA 1994a). If a delay occurs with respect to the promulgation of any rule in the schedule, all subsequent rules are to be completed as expeditiously as practicable but no later than a revised date that reflects the interval or intervals for the rules in the schedule. This schedule adjustment only applies to promulgation date, and is not tied to delays in implementation. The 1996 amendments also require USEPA to issue a Ground Water Rule (GWR) requiring disinfection, as necessary, for water systems using groundwater.

The final ICR (USEPA 1996) was promulgated May 14, 1996, after a 23-month delay caused by technical and administrative problems. As a result, USEPA was required to promulgate the Stage 1 D/DBP Rule and IESWTR no later than November 1998. Information Collection Rule data is not expected to be available until after the year 2000. Hence, if initial ICR data were to be considered, the final Stage 1 D/DBP Rule and IESWTR would not be expected until well after the year 2002. Slippage of ICR implementation and the deadlines imposed by the 1996 SDWA amendments caused the agency to rethink its approach for promulgation of the Stage 1 D/DBP Rule and IESWTR.

Table 7-6 Major provisions of the proposed Stage 1 D/DBP Rule

A. Proposed MCLGs, MCLs, and Monitoring

By-product	MCLG* (*mg/L*)	MCL (*mg/L*)	Compliance Monitoring
Chloroform	zero	NA†	NA†
Bromodichloromethane	zero	NA†	NA†
Bromoform	zero	NA†	NA†
Dibromochloromethane	0.06	NA†	NA†
Dichloroacetic acid	zero	NA†	NA†
Trichloroacetic acid	0.3	NA†	NA†
Chloral hydrate	0.04	NA†	NA†
TTHMs	NA‡	0.080	Tailored to system size and source water; compliance based on running annual average
HAA5	NA‡	0.060	Tailored to system size and source water; compliance based on running annual average
Bromate	zero	0.010	Monthly monitoring; compliance based on annual average; only for systems using ozone
Chlorite	0.08	1.0	Monthly monitoring; compliance based on annual average; only for systems using chlorine dioxide

*MCLGs were not negotiated, but determined by USEPA following the agency's standard risk assessment approach for drinking water contaminants.
†Not applicable. MCLs and compliance monitoring not proposed for individual DBPs. TTHMs and HAA5 are aggregate measurements that allow assessment of treatment technologies to control both known and unknown DBPs.
‡Not applicable. MCLGs are not proposed for TTHMs and HAA5 because they are aggregate measurements. MCLGs proposed for individual DBP species when health effects data available.
Source: USEPA (1994b).

B. Proposed Best Available Technology (BAT) for Meeting MCLs

By-Product	BAT
TTHMs HAA5	Enhanced coagulation (or enhanced softening) or GAC, with a 10-minute empty-bed contact time (180-day reactivation cycle), referred to as GAC10.
Bromate	Controlling the ozonation process
Chlorite	Controlling chlorine dioxide application

Table 7-6 Major provisions of the proposed Stage 1 D/DBP Rule (continued)

C. Proposed Maximum Residual Disinfectant Level Goals (MRDLGs), MRDLs, and BAT

Disinfectant	MRDLG	MRDL*	BAT
Chlorine†	4.0 (as Cl_2)	4.0 (as Cl_2)	Control of disinfection process
Chloramine‡	4.0 (as Cl_2)	4.0 (as Cl_2)	Control of disinfection process
Chlorine dioxide§	0.3 (as ClO_2)	0.8 (as ClO_2)	Control of disinfection process

*Compliance with MRDLs would be based on running annual averages, computed quarterly. USEPA recognizes that increasing the residual disinfectant above the MRDL for short periods of time may be necessary to address specific water quality problems, such as microbial regrowth.

†Measured as free chlorine.

‡Measured as total chlorine.

§MRDL for chlorine dioxide also applies to transient, noncommunity water systems because of the short-term health effects from high chlorine dioxide levels.

D. Precursor Removal

- Stage 1 will include a treatment technique requirement for removing DBP precursors (DBPPs). DBPPs are organic compounds in the source water that can react with the disinfectant to form DBPs. By removing the precursors, the formation of both known and unknown DBPs can be lowered.
- Because direct measurement of DBPPs is not feasible, total organic carbon (TOC) is used as a substitute, or surrogate, measurement to indicate the effectiveness of precursor removal. TOC measures the total amount of carbon associated with organic compounds in a sample. The lower the TOC, expressed in milligrams per litre, the lower the amount of organic compounds present.
- Stage 1 precursor-removal requirements will apply only to conventional water treatment plants (coagulation, flocculation, clarification, and filtration) and to softening plants. It will not apply to systems using direct filtration, slow sand filters, or diatomaceous earth filters. Also, it will not apply to systems using groundwater not under direct influence of surface water.
- Systems required to remove precursors will have to practice enhanced coagulation or enhanced softening. A specified percentage of the TOC in the source water will need to be removed before addition of the disinfectant.
- Systems meeting any one of the following requirements will not have to practice enhanced coagulation or enhanced softening:
  - — treated water TOC less than 2.0 mg/L
  - — source water TOC less than 4.0 mg/L, alkalinity greater than 60 mg/L, TTHMs less than 0.040 mg/L, and HAA5 less than 0.030 mg/L (using any disinfectant)
  - — TTHMs less than 0.040 mg/L and HAA5 less than 0.030 mg/L when chlorine is the disinfectant
- System makes a clear and irrevocable financial commitment to install technologies to limit TTHMs to less than 0.040 mg/L and HAA5 to less than 0.030 mg/L
- Systems practicing softening and removing at least 10 mg/L of magnesium hardness (as $CaCO_3$)

(table continues on next page)

Table 7-6 Major provisions of the proposed Stage 1 D/DBP Rule (continued)

E. Proposed Stage 2 D/DBP Rule
• The D/DBP Stage 2 rule, proposed at the same time as Stage 1, will serve as a placeholder. USEPA intends to revisit the proposed Stage 2 provisions after the results of ICR monitoring and supplemental research are available.
• Regulation of additional DBPs might be considered in Stage 2, and Stage 1 requirements will be revised if necessary. The proposed TTHM and HAA5 MCLs in Stage 2 are 0.040 mg/L and 0.030 mg/L, respectively. These limits were intended to apply only to systems using surface water or groundwater under direct influence of surface water, and conventional water treatment (coagulation, flocculation, clarification, and filtration).

The agency initiated stakeholder and technical meetings in the fall of 1996 to explore whether any or all of the elements of the Stage 1 D/DBP Rule could be promulgated to meet the November 1998 deadline. Sensing that agreement of stakeholders on an expedited rule was possible, the agency initiated the formation of an advisory committee under the Federal Advisory Committee Act (FACA). Using the 1994 proposed rules (USEPA 1994b, c) as starting points, the committee discussed, evaluated, and provided advice on data, analysis, and approaches to be included in a notice of data availability (NODA). Technical and policy issues involved in developing a Stage 1 D/DBP Rule and an IESWTR were examined to discuss which portions of the Stage 1 D/DBP Rule and IESWTR could be included in an expedited rulemaking.

This advisory committee met four times from March through June 1997, with the initial objective to reach consensus, where possible, on the elements to be contained in the Stage 1 D/DBP and IESWTR NODA. Where consensus was not reached, the committee sought to develop options and/or to clarify key issues and areas of agreement and disagreement. The committee recommended that the USEPA base the applicable sections of its anticipated NODAs on the elements of an agreement in principle signed July 15, 1997 (USEPA 1997a). The NODAs regarding the proposed D/DBP Rule (USEPA 1997b) and the IESWTR (USEPA 1997c) were published Nov. 3, 1997, and March 31, 1998 (1998a). Key provisions of the D/DBP NODA are summarized in Table 7-7.

The final Stage 1 D/DBP Rule (USEPA 1998b) and final IESWTR (USEPA 1998c) were published Dec. 16, 1998. To assist

Table 7-7 Major provisions of the notice of data availability

- The expedited DBP rule will include the proposed MCLs for TTHMs, HAA5, and bromate at the concentrations proposed in July 1994 (see Table 7-6).
- The proposed enhanced coagulation requirements should be modified based on a substantial body of new information available since the 1994 proposed rule. A revised TOC removal table will be adopted.
- Specific UV absorbance (SUVA) will be used for determining whether or not systems would be required to use enhanced coagulation. The use of a raw water SUVA < 2.0 L/mg-m as a criterion for not requiring a system to practice enhanced coagulation should be added to those criteria proposed for avoiding enhanced coagulation.
- For a system required to practice enhanced coagulation or enhanced softening, the use of a finished water SUVA < 2.0 L/mg-m should be added as a step 2 procedure. This criterion would be in addition to the proposed step 2 procedure, not in lieu of it.
- A microbial benchmarking procedure should be established in the IESWTR to provide a methodology and process by which public water systems most likely to make a change in disinfection practice and the state, working together, will ensure that there will be no significant reduction in microbial protection as the result of modifying disinfection practices in order to meet MCLs for TTHM and HAA5.
- Consistent with the existing provisions of the SWTR, credit for compliance with applicable disinfection requirements should continue to be allowed in the DBP rule for disinfection applied at any point prior to the first customer. USEPA will develop guidance on the use and costs of oxidants that control water quality problems (e.g., zebra mussels, Asiatic clams, iron, manganese, and algae) and whose use will reduce or eliminate the formation of DBPs of public health concern.
- Because filtration is an important barrier against passage of *Cryptosporidium* and other pathogens into treated water, and turbidity is used in the SWTR to indicate filtration performance, SWTR turbidity performance requirements should be tightened and new requirements imposed for individual filter monitoring in the IESWTR.
- An MCLG should be established for *Cryptosporidium* to protect public health.
- All surface water systems that serve more than 10,000 people and are required to filter must achieve at least a 2-log removal of *Cryptosporidium*.
- USEPA will issue a risk-based proposal of the final ESWTR for *Cryptosporidium* embodying the multiple-barrier approach (e.g., source water protection, physical removal, inactivation, etc.), including, where risks suggest appropriate, inactivation requirements.
- Sanitary surveys should be required in accordance with the USEPA/Association of State Drinking Water Administrators (ASDWA) joint guidance document (USEPA 1995).

Source: USEPA (1997b).

states and water utilities in complying with the Stage 1 D/DBP Rule and the IESWTR, USEPA has issued guidance on a number of topics that were raised during discussions of the advisory committee. These topics include enhanced coagulation, disinfection benchmarking and profiling, turbidity, alternative disinfectants and oxidants, simultaneous compliance with other regulations, sanitary surveys, unfiltered systems, and uncovered finished water reservoirs.

MICROBIALS/DISINFECTION BY-PRODUCTS CLUSTER REGULATORY SCHEDULE AND FUTURE RULES

The 1996 SDWA amendments require that USEPA develop the Stage 1 D/DBP Rule and IESWTR in accordance with the schedule published in the proposed Information Collection Rule (USEPA 1994a). That schedule was based on the SDWA statutory requirement in effect at that time, i.e., that national primary drinking water regulations become effective 18 months after promulgation. In recognition that this compliance time frame is unreasonably short, especially if major capital improvements are needed, Congress lengthened this time frame in the 1996 SDWA amendments to three years, with up to two additional years possible if needed for capital improvements. This issue was considered by USEPA and discussed in the D/DBP NODA (USEPA 1997b). The agency concluded that a five-year compliance period for the Stage 1 D/DBP Rule is necessary for small systems to provide for capital improvements. Table 7-8 summarizes the anticipated compliance time frames for the Stage 1 D/DBP Rule, IESWTR, the long-term 1ESWTR (LT1ESWTR), Stage 2 rules, and the GWR, together known as the microbial/disinfection by-products (M/DBPs) cluster.

Regulation of DBPs is perhaps the most difficult SDWA mandate USEPA faces. The expedited Stage 1 D/DBP Rule and interim ESWTR will place new requirements on water suppliers, state regulators, and USEPA. These rules serve as a starting point whereby the risks posed by DBPs can be addressed while ensuring the microbial safety of drinking water. Regulatory agencies and water suppliers alike must develop new strategies for disinfection, precursor removal, and DBP control. As these rules are implemented, the experience and knowledge gained will lay the groundwork for future regulation of DBPs.

Table 7-8 Anticipated compliance time frame for the Stage 1 D/DBP Rule and IESWTR

Rule	Surface Water Systems		Groundwater Systems	
(Promulgation)	**≥10,000**	**<10,000**	**≥10,000**	**<10,000**
Stage 1 D/DBP (December 1998)	December 2001	December 2003	December 2003	December 2003
IESWTR (December 1998)	December 2001	NA*	NA	NA
LTESWTR (December 2000)	December 2003 (if required)	December 2003	NA	NA
GWDR (December 2000)	NA	NA	December 2003	December 2003

*NA = not applicable.
Ref: USEPA (1997b, 1998b, 1998c).

REFERENCES

AWWA. 1992. AWWA Water Quality Division Disinfection Committee. Survey of Water Utility Disinfection Practices. *Jour. AWWA*, 84(9):121–128.

Borchardt and Walton. 1971. *Water Quality and Treatment.* Denver, Colo.: American Water Works Association.

NAS. 1977. National Academy of Sciences. *Drinking Water and Health.* Washington, D.C.: National Academy Press.

Pontius, F.W. 1991. Disinfectants–Disinfection By-Products Rule Update. *Jour. AWWA*, 83(12):24.

———. 1993a. Reg-Neg Process Draws to a Close. *Jour. AWWA*, 85(9):18.

———. 1993b. D–DBP Rule to Set Tight Standards. *Jour. AWWA*, 85(11):20.

———. 1993c. Strengthening the Surface Water Treatment Rule. *Jour. AWWA*, 85(12):20.

———. 1996. Inside the Information Collection Rule. *Jour. AWWA*, 88(8):16.

———. 1997. Chapter 1; History of the Safe Drinking Water Act. *SDWA Advisor on CD-ROM.* Denver, Colo.: American Water Works Association.

Symons, J.M. 1984. A History of Attempted Regulation Requiring GAC Adsorption for Water Treatment. *Jour. AWWA*, 76(8):34.

USEPA. 1976. Organic Chemical Contaminants. Control Options in Drinking Water. *Fed. Reg.*, 41(136):28991.

———. 1978. Control of Organic Chemical Contaminants in Drinking Water. *Fed. Reg.*, 43(28):5756.

———. 1979. Control of Trihalomethanes in Drinking Water. Final Rule. *Fed. Reg.*, 44(231):68624.

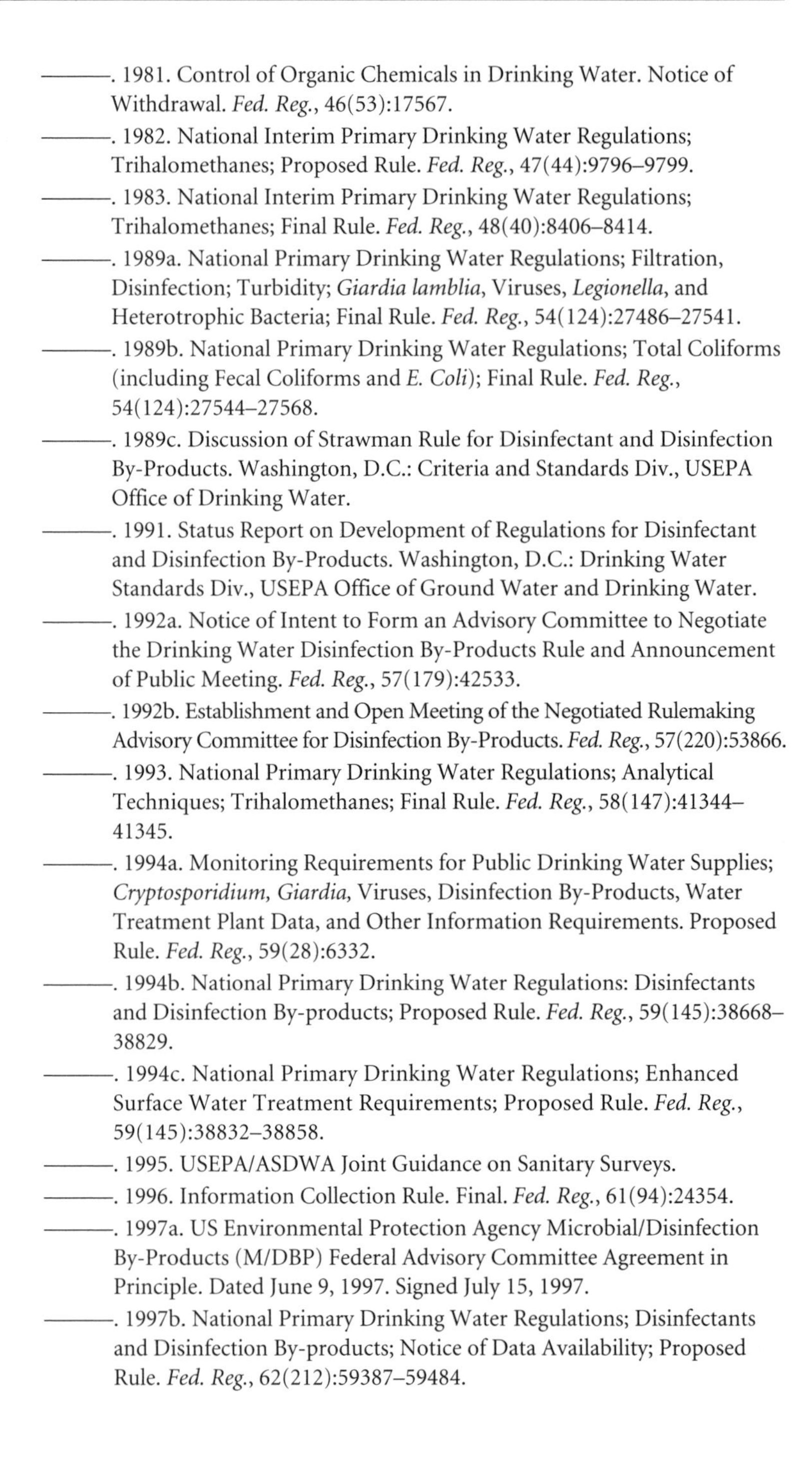

———. 1981. Control of Organic Chemicals in Drinking Water. Notice of Withdrawal. *Fed. Reg.*, 46(53):17567.

———. 1982. National Interim Primary Drinking Water Regulations; Trihalomethanes; Proposed Rule. *Fed. Reg.*, 47(44):9796–9799.

———. 1983. National Interim Primary Drinking Water Regulations; Trihalomethanes; Final Rule. *Fed. Reg.*, 48(40):8406–8414.

———. 1989a. National Primary Drinking Water Regulations; Filtration, Disinfection; Turbidity; *Giardia lamblia*, Viruses, *Legionella*, and Heterotrophic Bacteria; Final Rule. *Fed. Reg.*, 54(124):27486–27541.

———. 1989b. National Primary Drinking Water Regulations; Total Coliforms (including Fecal Coliforms and *E. Coli*); Final Rule. *Fed. Reg.*, 54(124):27544–27568.

———. 1989c. Discussion of Strawman Rule for Disinfectant and Disinfection By-Products. Washington, D.C.: Criteria and Standards Div., USEPA Office of Drinking Water.

———. 1991. Status Report on Development of Regulations for Disinfectant and Disinfection By-Products. Washington, D.C.: Drinking Water Standards Div., USEPA Office of Ground Water and Drinking Water.

———. 1992a. Notice of Intent to Form an Advisory Committee to Negotiate the Drinking Water Disinfection By-Products Rule and Announcement of Public Meeting. *Fed. Reg.*, 57(179):42533.

———. 1992b. Establishment and Open Meeting of the Negotiated Rulemaking Advisory Committee for Disinfection By-Products. *Fed. Reg.*, 57(220):53866.

———. 1993. National Primary Drinking Water Regulations; Analytical Techniques; Trihalomethanes; Final Rule. *Fed. Reg.*, 58(147):41344–41345.

———. 1994a. Monitoring Requirements for Public Drinking Water Supplies; *Cryptosporidium, Giardia*, Viruses, Disinfection By-Products, Water Treatment Plant Data, and Other Information Requirements. Proposed Rule. *Fed. Reg.*, 59(28):6332.

———. 1994b. National Primary Drinking Water Regulations: Disinfectants and Disinfection By-products; Proposed Rule. *Fed. Reg.*, 59(145):38668–38829.

———. 1994c. National Primary Drinking Water Regulations; Enhanced Surface Water Treatment Requirements; Proposed Rule. *Fed. Reg.*, 59(145):38832–38858.

———. 1995. USEPA/ASDWA Joint Guidance on Sanitary Surveys.

———. 1996. Information Collection Rule. Final. *Fed. Reg.*, 61(94):24354.

———. 1997a. US Environmental Protection Agency Microbial/Disinfection By-Products (M/DBP) Federal Advisory Committee Agreement in Principle. Dated June 9, 1997. Signed July 15, 1997.

———. 1997b. National Primary Drinking Water Regulations; Disinfectants and Disinfection By-products; Notice of Data Availability; Proposed Rule. *Fed. Reg.*, 62(212):59387–59484.

———. 1997c. National Primary Drinking Water Regulations; Interim Enhanced Surface Water Treatment Rule; Notice of Data Availability; Proposed Rule. *Fed. Reg.*, 62(212):59485–59557.

———. 1998a. Disinfectants and Disinfection By-products. Notice of Data Availability; Proposed Rule. *Fed. Reg.*, 63(61):15674–15692.

———. 1998b. Disinfectants and Disinfection By-products; Final Rule. *Fed. Reg.*, 63(241):69390–69476.

———. 1998c. Interim Enhanced Surface Water Treatment; Final Rule. *Fed. Reg.*, 63(241):69478–69521.

USPHS. 1925. Report of the Advisory Committee on Official Water Standards. *Public Health Rept.*, 40:693.

CHAPTER • EIGHT

Control of Disinfection By-Product Formation Using Chloramines

GERALD E. SPEITEL, JR.

Chloramines are formed by combining appropriate quantities of chlorine and ammonia. Chloramines are inexpensive relative to chlorine, and in general form halogenated disinfection by-products (DBPs) to a considerably lesser extent than free chlorine. The chemistry of chloramines, especially in the presence of bromide, is more complicated than that of chlorine. The complexities of chloramine chemistry with respect to DBP formation are not fully understood; however, considerable information is available.

HALOAMINE CHEMISTRY

Chloramine chemistry is fairly well understood and can be summarized in its simplest form by the three reversible reactions listed below.

$$NH_3 + HOCl \leftrightarrow NH_2Cl + H_2O \qquad (8\text{-}1)$$

$$NH_2Cl + HOCl \leftrightarrow NHCl_2 + H_2O \qquad (8\text{-}2)$$

$$NHCl_2 + HOCl \leftrightarrow NCl_3 + H_2O \qquad (8\text{-}3)$$

The dominant chloramine species is a function of the Cl_2:N mass ratio and pH. At typical operating conditions, monochloramine is the dominant chloramine species in drinking water disinfection. The production of dichloramine is favored as the Cl_2:N ratio increases from low (e.g., 3:1) to high (e.g., 7:1) values and as the pH decreases below pH 7. The hydrolysis reactions (the reverse reactions in Eq 8-1–8-3) also are of considerable interest because they generate HOCl, which may be important in the formation of chlorine-substituted organics, HOBr, OBr^-, and bromamines. The brominated species, in turn, may produce bromine-substituted organics. Also, the monochloramine hydrolysis reaction liberates ammonia, which may play a role in haloamine formation and subsequent nitrification in the distribution system. At equilibrium, several percent, at most, of the monochloramine is hydrolyzed. In addition to hydrolysis and Eq 8-3, dichloramine decomposes by several other reactions, two of which are base catalyzed (Jafvert and Valentine 1992). Dichloramine decomposition accelerates as the pH increases and as the concentration of bases (e.g., alkalinity) increases; therefore, dichloramine is much less stable than monochloramine under most conditions of practical interest. The greater instability of dichloramine, however, does not necessarily mean that it is of little significance in DBP formation.

Monochloramine decomposes through conversion to dichloramine, with subsequent decomposition of dichloramine being primarily responsible for oxidant loss. The two major pathways for monochloramine decomposition are hydrolysis of monochloramine followed by subsequent reaction of the free chlorine with monochloramine (Eq 8-1 and 8-2) and general acid catalysis (Valentine and Jafvert 1988):

$$NH_2Cl + NH_2Cl + H^+ \leftrightarrow NHCl_2 + NH_3 + H^+ \qquad (8\text{-}4)$$

General acid catalysis also liberates ammonia. The rate of general acid catalysis (Eq 8-4) is a function of the concentration of proton donors and increases as their concentration increases and the pH decreases. Valentine and Jafvert (1988) note that the carbonate system, via carbonic acid and bicarbonate, may significantly increase the rate of acid-catalyzed decomposition at concentrations and pH values typical of many drinking waters. Thus, the decomposition rate of both monochloramine and dichloramine may increase as alkalinity increases.

The presence of bromide complicates the chemistry of the system considerably. The reactions of bromide with chlorine and the chloramines to produce bromamines are as follows (Rook 1980; Wajon and Morris 1980; Haag 1980):

$$Br^- + HOCl \leftrightarrow Cl^- + HOBr \quad (8\text{-}5)$$

$$NH_3 + HOBr \leftrightarrow NH_2Br + H_2O \quad (8\text{-}6)$$

$$NH_2Br + HOBr \leftrightarrow NHBr_2 + H_2O \quad (8\text{-}7)$$

$$NHBr_2 + HOBr \leftrightarrow NBr_3 + H_2O \quad (8\text{-}8)$$

$$NH_2Cl + Br^- \leftrightarrow NH_2Br + Cl^- \quad (8\text{-}9)$$

The speciation of bromamines is a function of pH and the Br_2:N ratio (Johnson and Sun 1975). Depending on the Br_2:N ratio, any of the three bromamines can be present at neutral pH. As the Br_2:N ratio increases, a greater degree of bromination occurs. At high pH (9–10), monobromamine is the dominant species. At low pH (<6) monobromamine formation is not expected, and dibromamine or tribromamine dominates. When both chloramines and bromamines are present, the dominant brominated species is not easily estimated, especially around neutral pH, because of the complexities in estimating the Br_2:N ratio.

Bromamines form rapidly, relative to chloramines, and are less stable, dissipating rapidly in the environment. For example, the half-lives of dibromamine and monobromamine are 30 minutes and 19 hours, respectively, at pH 8 (Wajon and Morris 1980), whereas a half-life of 50 to 100 hours is typical for monochloramine (Jafvert and Valentine 1992). Several pathways for dibromamine decomposition have been postulated. Some show HOBr as a decomposition product. Nitrogen gas is a typical product, although one pathway shows ammonia as a decomposition product (Jolley and Carpenter 1983). Decomposition of tribromamine and hydrolysis of monobromamine also yield HOBr as a product.

In the presence of ammonia and bromide, the chlorination of waters also may produce bromochloramine (Valentine 1986). Various reactions have been proposed for the production of bromochloramine, including (Wajon and Morris 1980; Gazda et al. 1993):

$$NH_2Cl + HOBr \leftrightarrow NHBrCl + H_2O \qquad (8\text{-}10)$$

$$NH_2Cl + Br^- \leftrightarrow NHBrCl + \text{other products} \qquad (8\text{-}11)$$

The reaction of hypobromous acid with monochloramine is at least three orders of magnitude faster at pH 6.5 than the reaction of bromide with monochloramine (Gazda et al. 1993).

Bromochloramine reacts rapidly with *N*,*N*-diethyl-*p*-phenylenediamine (DPD) indicating that the bromine atom of bromochloramine is very labile and reactive (Valentine 1986). In reactions with organic chemicals, bromochloramine should produce products similar to those produced by free bromine. Valentine (1986) produced bromochloramine by adding bromide to solutions of monochloramine at pH 6.5. The resulting solutions showed significant concentrations of only monochloramine and bromochloramine, and no measurable amounts of the bromamine species. Although the results apply to only one condition, they are suggestive of the possible importance of bromochloramine in DBP formation in bromide-containing waters.

In summary, a variety of DBP-producing chemicals may be present during chloramination, especially if bromide is present. Chloramines, bromamines, bromochloramine, free chlorine, and free bromine may all be present and participate in reactions forming DBPs.

DISINFECTION BY-PRODUCTS

Characterization of chloramination DBPs is incomplete, even more so than with chlorine; however, some general trends can be pointed out. Also, a reasonable amount of information is available for DBPs of current and near future regulatory concern. One key point is that small concentrations of free chlorine, and possibly free bromine, are always expected with chloramination; therefore, formation of DBPs observed in chlorination can likewise be expected, but at much smaller concentrations. Chloramines are weaker oxidants than chlorine, and have a greater tendency to participate in chlorine substitution reactions, rather than oxidation reactions, in comparison to chlorine. Substitution reactions are especially prevalent with organic nitrogen compounds. The identified DBPs largely consist of small chlorinated and brominated compounds, including halogen-substituted amines, amino acids, and peptides; cyanogen chloride and bromide; halogenated nitriles, ethers, alcohols, and alkanoic acids; and dihalomethanes (Symons et al. 1997). Other chloramination by-products have been identified in nonpotable water, i.e., in wastewater and in solutions of fulvic acid, amino acids, amines, and other organic compounds (Crochet and Kovacic 1973; Ingols et al. 1953; Minisci and Galli 1965; Kotiaho et al. 1991; Kanniganti et al. 1992). For example, chlorinated aldehydes, chlorinated acids, and chlorinated ketone by-products have been identified when fulvic acid is chloraminated (Kanniganti et al. 1992).

Numerous nitrogen-containing compounds exist in surface waters, including amino acids, peptides, proteins, humic materials, chlorophylls, and synthetic organic chemicals, such as herbicides, pesticides, and nitrophenols (LeCloirec et al. 1983). The total dissolved amino acid concentration in surface waters has been reported to range from 50 to 1,000 μg/L (Hureiki, Crove, and Legube 1994) and in some cases has increased after ozonation or biological filtration (LeCloirec et al. 1983). Snyder and Margerum (1982) and Isaac and Morris (1983) showed that monochloramine could transfer chlorine to nitrogenous organic chemicals by general acid catalysis reactions in a similar reaction scheme to Eq 8-4, except that one of the monochloramine molecules is replaced by an organic nitrogen-containing chemical. Significant reaction rates for many of the test chemicals were observed through pH values as high as 8 to 8.5. Snyder and Margerum (1982) concluded that general acid catalysis with monochloramine produced a very reactive chlorinating agent. As with monochloramine decomposition, a dependence of reaction rates on the concentration of proton donors would be expected; therefore, reaction rates may increase as alkalinity increases. Another plausible mechanism for chloramination reactions is via formation of an imine intermediate to yield a nitrile (LeCloirec and Martin 1985).

Trihalomethanes

The usefulness of chloramines in controlling THM formation is well-established (Symons et al. 1982). A survey of 543 utilities by McGuire and Meadow (1988) showed that 100 of them had switched their disinfectant to chloramines to meet the trihalomethane (THM) standard after it was promulgated in 1979. Trihalomethane concentrations of less than 20 µg/L are typical following chloramination (Symons et al. 1982, 1997). Trihalomethanes are formed to a limited extent by chloramination for virtually all chloramination conditions of practical significance. Trihalomethane formation decreases as the Cl_2:N ratio decreases and as the pH increases (Stevens, Moore, and Miltner 1989; Symons et al. 1997). The pH effect is opposite of that found in chlorination, in which base-catalyzed formation mechanisms are favored. In the presence of bromide, THM formation is enhanced somewhat, and the speciation shifts toward the bromine-substituted THMs (Symons et al. 1997).

Haloacetic Acids

Haloacetic acids (HAAs) can reach significant and potentially troublesome concentrations with respect to possible regulatory levels. Diehl et al. (1997) studied the effect of Cl_2:N ratio and pH on HAA formation in batch studies using preformed chloramines; six of the nine chlorine and bromine-containing HAAs were measured. As shown in Figure 8-1, HAA formation decreases as the pH increases and the Cl_2:N ratio decreases. As with other DBPs, HAA formation also has a water-specific character to it. The dihalogenated HAAs (dichloroacetic acid, bromochloroacetic acid, and dibromoacetic acid) were by far the most common species formed (>90 percent), in contrast to chlorination in which significant concentrations of both tri- and dihalogenated acetic acids are common. The presence of bromide shifted the speciation toward the bromine-substituted HAAs and tended to increase the total HAA mass concentration.

Cowman and Singer (1996) studied all nine bromine and chlorine-containing HAAs as a function of bromide concentration at a pH of 8 using preformed chloramines (Cl_2:N ratio of 4:1). In Palm Beach, Fla., extracts, the molar yield of HAAs was constant with bromide concentration. The speciation changed,

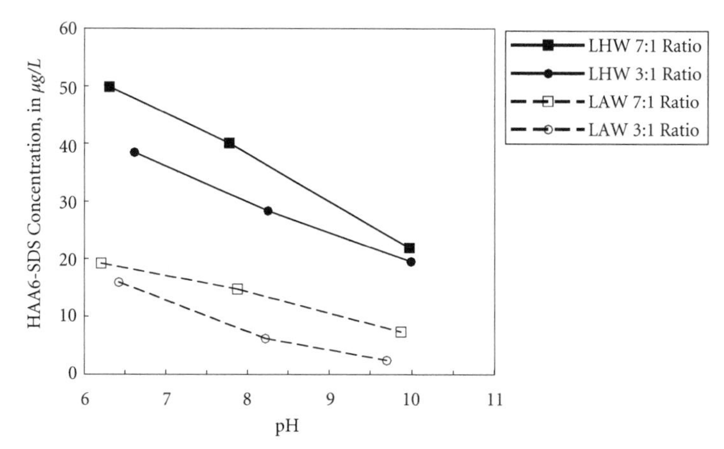

NOTE: LAW = Lake Austin Water; LHW = Lake Houston Water; Cl_2:N mass ratios = 7:1 and 3:1; 22°C; nominal 48-h residual = 2 mg/L as Cl_2; TOC concentration = 3.1 mg/L (LAW) and 9.3 mg/L (LHW)
Source: Diehl et al. 1997.

Figure 8-1 Haloacetic acid formation under simulated distribution system conditions

however, to the higher-molecular-weight, bromine-substituted species with increasing bromide concentration, thereby increasing the HAA mass concentration. On a mole fraction basis, approximately 60 percent of the HAAs were dihalogenated, 33 percent trihalogenated, and less than 10 percent monohalogenated. Thus, the dihalogenated species dominated, although not to the extent observed by Diehl et al. (1997).

As with other DBPs, HAA formation from chloramination is significantly less than that observed from chlorination. As shown in Table 8-1, HAA formation with concurrent addition of chlorine and ammonia is typically 5 to 20 percent of that observed from chlorination.

Cyanogen Halides

Krasner et al. (1989) found that cyanogen chloride was formed to a greater extent in chloraminated waters than chlorinated waters; therefore, although not now regulated, cyanogen halides (CNX) from chloramination may be of some concern. Diehl et al. (1997) studied the effect of Cl_2:N ratio and pH on CNX formation in batch

Table 8-1 Relative haloacetic acid formation with chloramination versus chlorination

Water Source	Relative HAA Formation From Chloramination*	Reference
Seymour Reservoir (B.C.)	23% of HAASDS†	Long, Logsdon, and Patterson (1992)
Capilano Reservoir (B.C.)	15% of HAASDS	Long, Logsdon, and Patterson (1992)
Quabbin Reservoir (Mass.)	6–20% of HAASDS	Zhu and Reckhow (1993)
Mississippi River (La.)	10% of HAASDS	Lykins, Koffsey, and Patterson (1994)
Palm Beach (Fla.) and Myrtle Beach (S.C.) extracts	5–10% of HAASDS	Cowman and Singer (1996)
Lake Austin (Texas)	3–20% of HAAFP‡	Symons et al. (1997)
Lake Houston (Texas)	7–17% of HAAFP	Symons et al. (1997)
Calif. State Project	5–30% of HAAFP	Symons et al. (1997)

*Chloramination HAA concentration/chlorination HAA concentration × 100.
†Haloacetic acid simulated distribution system concentration with chlorine.
‡Haloacetic acid formation potential with chlorine.

studies. Very low concentrations were observed at pH 10, because CNX is unstable at high pH, decaying by base-catalyzed hydrolysis. At lower pH (6 and 8), CNX concentrations as high as 18 µg/L were observed. Based on the work of Ohya and Kanno (1985), the CNX concentration is expected to increase with increasing Cl_2:N ratio up to a ratio of 8–9:1. As with the other DBPs, the presence of bromide favors increased formation of CNBr.

Total Organic Halides (TOX)

Chloramination produces lower, but still significant, concentrations of TOX in comparison to chlorine. As shown in Table 8-2, TOX production with chloramines ranges from 9 to 49 percent of that with chlorine for a variety of natural waters and humic and fulvic acids. Considering natural waters alone, TOX production seems to range from 10 to 20 percent of that observed with chlorine when chlorine and ammonia are added concurrently. At least some of the halogenated products are different than those found from chlorination. Jensen et al. (1985) found that the reaction of

Table 8-2 Relative TOX formation with chloramination versus chlorination

Water Source	Relative Chloramination TOX* (*percent*)	Reference
Midwest groundwater	9	Fleischacker and Randtke (1983)
Fulvic acid	13	Fleischacker and Randtke (1983)
Humic acid	20	Fleischacker and Randtke (1983)
Rhein River humic acid	28–49	Jensen et al. (1985)
Black Lake fulvic acid	8–14	Jensen et al. (1985)
Colorado River (Ariz.)	17	Nieminski and Shepard (1989)
Lake water	20	Jacangelo et al. (1989)
Ohio River (Cincinnati, Ohio)	20	Stevens et al. (1989)
Mississippi River (Jefferson Parish, La.)	11	Lykins et al. (1994)
Lake Austin (Texas)	16	Symons et al. (1997)
Lake Houston (Texas)	9	Symons et al. (1997)

*Chloramination TOX concentration/chlorination TOX concentration × 100

monochloramine with fulvic acid produced TOX that was more hydrophilic and had a larger molecular weight than chlorine-produced TOX. Limited data show that dichloramine gives a much greater production of TOX than monochloramine, not very much less than that associated with free chlorine (Fujioka, Tenno, and Loh 1983). The reactivity of bromamines in the formation of TOX is largely unknown.

Diehl et al. (1997) studied the effect of Cl_2:N ratio and pH on TOX formation in batch studies of three surface waters: Lake Austin (Texas), Lake Houston (Texas), and California State Project water. Most data followed the general trend of decreasing TOX formation with increasing pH, although some significant deviations from this pattern were observed in several instances. Bromide addition accentuated the production of TOX for nearly all conditions. Although the TOX analysis cannot differentiate between chlorinated and brominated organics, the increased production of TOX with bromide addition strongly suggests that brominated DBPs were formed. The effect of the Cl_2:N ratio on TOX formation was difficult to discern. On average, TOX formation decreased as the Cl_2:N ratio decreased; however, the data were quite scattered in this regard.

A relatively small fraction of the measured TOX can be accounted for by specific DBP measurements. Krasner et al. (1989) accounted for slightly less than 30 percent of the measured TOX by summing 21 measured DBPs in chloraminated water. Singer et al. (1995) accounted for approximately 36 percent of the measured TOX by summing 12 individual DBPs. Symons et al. (1997) accounted for 5 to 35 percent of the TOX by summing 12 individual DBPs. In general, a smaller fraction of the TOX was accounted for as the Cl_2:N ratio decreased from 7:1 to 3:1.

PILOT-PLANT AND FULL-SCALE EXPERIENCE

Jacangelo et al. (1989) conducted pilot-plant and full-scale studies on several surface waters. As shown in Table 8-3, switching from chlorination to ozonation followed by chloramination dramatically decreased THM, HAA, and TOX concentrations in the two water sources studied. Switching from chloramination to the combination of ozonation and chloramination also improved water quality in three water sources, but to widely varying extents. In one water, the large decrease in DBP formation can be attributed to the elimination of a 4-hour period of free chlorination in the chloramination treatment scheme. The large removal for the second water may have resulted from the relatively large ozone dose (1.45 mg O_3/mg total organic carbon [TOC]). A much smaller decrease in DBP concentrations occurred in the third water with an ozone dose of 0.51 mg O_3/mg TOC. As noted above, HAA formation is potentially problematic in chloramination. These data suggest that ozonation prior to chloramination may help control HAA formation. The CNCl concentration was 8 to 15 times greater with chloramines compared to chlorine in one water source. Also, preozonation prior to chloramination resulted in higher CNCl concentrations than with chloramines alone, which also was observed by Symons et al. (1997) with Lake Austin (Texas) water.

Table 8-3 Effect of changes in disinfectants on DBP formation

Disinfectant Change	TOC (*mg/L*)	O_3 (*mg/L*)	Cl_2:N Ratio	Br^- (*μg/L*)	pH	Percent Decrease in DBP Concentrations TTHM	HAA	TOX
Cl_2 to O_3 and NH_2Cl	2.46*	2.0	5.9:1	150	8.2	96	89	82
Cl_2 to O_3 and NH_2Cl	3.91†	2.0	4.2:1	320	7.8	95	89	82
NH_2Cl‡ to O_3 and NH_2Cl	2.46*	2.0	5.9:1	150	8.2	95	80	71
NH_2Cl to O_3 and NH_2Cl	5.51	8.0	5.0:1	230	9.3	84	64	59
NH_2Cl to O_3 and NH_2Cl	3.91†	2.0	4.2:1	320	7.8	23	16	8

*Same water source.
†Same water source.
‡Four-hour delay between chlorine and ammonia addition; no delay with ozonation. Chlorine and ammonia added concurrently for all other cases.
Source: Jacangelo et al. (1989).

Table 8-4 Jefferson Parish (La.) pilot-plant simulated distribution system DBPs

Disinfectant(s)	Incubation pH	TOC* (*mg/L*)	TTHM† (*μg/L*)	HAA6†‡ (*μg/L*)	TOX† (*μg/L*)
Pre- and postchlorine§	7–8	3.2	225	146	540
Pre- and postchloramines	7–8	3.2	9.4	14	59
Preozone, postchloramines	7–8	2.9	3.2	8.7	27

*Concentration in disinfection contact chamber effluent.
†Five-day simulated distribution system incubation time.
‡Ambient bromide concentration not reported, but HAA6 speciation suggests relatively low bromide concentration.
§Predisinfection occurred before sand filtration; postdisinfection after sand filtration.
Source: Lykins, Koffsey, and Patterson (1994).

Lykins, Koffsey, and Patterson (1994) studied different disinfection schemes for Mississippi River water that was coagulated, settled, predisinfected, sand filtered, and postdisinfected to simulate distribution system conditions. As shown in Table 8-4, dramatic reductions in DBP formation were observed in switching from chlorine to chloramines. The addition of ozonation, followed by chloramination after GAC filtration led to further reductions in DBP formation. Again note the effect of ozonation on subsequent HAA formation from chloramination.

Long, Logsdon, and Neden (1992) studied a limited range of DBPs, and showed a significant decrease in chloroform, dichloroacetic acid, and trichloroacetic acid concentrations by switching from chlorine to chloramines as the secondary disinfectant. Further improvement was achieved by switching from chlorine to ozone as the primary disinfectant, while retaining chloramines as the secondary disinfectant.

A pilot-plant study was conducted by Symons et al. (1997) on Lake Austin (Texas), which had a TOC concentration of 2.5 to 4.6 mg/L and a bromide concentration of 240 µg/L. Conventional lime softening, direct filtration with alum, and ozonation prior to conventional lime softening were tested. With conventional lime softening, DBPs from postchloramination with concurrent addition of chlorine and ammonia were well controlled under all operating conditions. TOX concentrations were 20 to 45 µg/L, and HAA6 concentrations were 10 to 20 µg/L. Small increases in THMs, TOX, and CNX occurred when the operation was switched to direct filtration. The addition of preozonation to conventional lime softening produced a significant reduction in HAA formation (HAA6 < 2 µg/L).

Conventional alum treatment, enhanced coagulation with alum, and conventional alum treatment with 0.5 mg/L bromide addition were tested at pilot scale on Lake Houston (Texas), which had a TOC concentration of 9 to 10 mg/L and a bromide concentration of 80 µg/L (Symons et al. 1997). Because Lake Houston has a relatively low ambient bromide concentration, the third operating condition was selected to simulate performance for a similar water with a higher bromide concentration. Using postchloramination with concurrent addition of chlorine and ammonia, THM formation was well controlled, but the HAA concentration approached levels that may be of future regulatory concern (20 to 26 µg/L). Enhanced coagulation had little effect on DBP precursor removal; in fact, the effect seemed to be an adverse one with respect to CNX, as the concentration of that species increased by 7 to 10 µg/L. As expected, bromide addition resulted in higher concentrations of all DBPs. The TOX concentration increased from approximately 100 to 200 µg/L, and the HAA6 concentration increased from approximately 20 to 40 µg/L with bromide addition. Therefore, water sources that experience time-varying bromide concentrations would also experience corresponding variations in DBP formation.

Table 8-5 Distribution system samples from chloramination plants

Location	TOC (*mg/L*)	Br^- (*µg/L*)	Cl_2:N Ratio	pH	TTHM (*µg/L*)	HAA6 (*µg/L*)	TOX (*µg/L*)	CNX (*µg/L*)
Mid-South water	3.0	1,500	3.8:1	7.1	40.3	16.2	72	4.7
Mississippi River	2.9	40	4:1	9	12.4	29.0	83	1.5
Northeastern creek	2.0	50	3:1	7.1	27.7	12.5	59	2.2
Pacific Northwest lake	1.4	7	4.7:1	6.7	10.6	37.8	85	3.4

Source: Symons et al. (1997).

A pilot-plant study was conducted by Symons et al. (1997) on California State Project water, which had a TOC concentration of 3.2 mg/L and a bromide concentration of 250 µg/L. Conventional alum treatment, ozonation prior to conventional alum treatment, and ozonation prior to conventional alum treatment in combination with biological filtration were tested. Chloramination with concurrent addition of chlorine and ammonia was practiced for all treatment conditions. As shown in several other waters, the introduction of ozonation significantly decreased HAA formation. The HAA6 concentration decreased from 15 µg/L to 3 µg/L with the addition of preozonation to conventional alum treatment. Biofiltration had little effect on DBP formation.

Symons et al. (1997) collected distribution system samples from four utilities across the United States (Table 8-5). All four used Cl_2:N ratios typical of normal chloramination practice (3:1 to 5:1). The THM, HAA, CNX, and TOX concentrations were similar to batch and pilot-plant chloramination data generated from other studies. As seen in these other studies, the HAA concentrations in two waters were at levels of possible future regulatory concern. The highest HAA concentration was associated with the Pacific Northwest Lake water, which had the highest Cl_2:N ratio and lowest pH of the four. Two of the four waters also had somewhat elevated THM concentrations, which suggests exposure to free chlorine, although the very high bromide concentration in the mid-South water might also be a factor. The TOX concentrations (59 to 85 µg/L) were within the typical range for chloramination.

SUMMARY

In general, DBP formation from chloramination diminishes as the pH increases and the Cl_2:N ratio decreases. The presence of bromide affects disinfectant speciation, DBP speciation, and DBP concentrations. As the bromide concentration increases, formation of bromine-substituted DBPs is favored, and an increase in the mass concentration of DBPs tends to occur. Trihalomethane concentrations less than 20 µg/L are possible under virtually all practical treatment conditions; therefore, THMs should not be of regulatory concern when practicing chloramination. Excessive concentrations of THMs following chloramination are indicative of a significant period of free chlorine exposure. Haloacetic acids, especially the dihalogenated HAAs, are not as well controlled in chloramination as other DBPs. Concentrations of HAA from about 10 to 40 µg/L are common, which may be of future regulatory concern in some waters. Concentrations of CNX are generally less than 10 µg/L. The highest concentrations are observed at or below neutral pH with larger Cl_2:N ratios (e.g., 5:1 or greater); lowest concentrations are observed at high pH because of CNX instability. Formation of TOX is generally less than 20 percent of that observed with free chlorine. Total organic halogen concentrations in the 10s to low 100s of µg Cl^-/L are typical in chloramination.

Specifically identified DBPs account for only a small fraction of the TOX (<35 percent); therefore, consideration of both individual DBP and TOX concentrations may be prudent in formulating DBP control strategies. Particular attention should be paid to system chemistry near neutral pH. Formation of DBPs may be very sensitive to the Cl_2:N ratio near pH 8, presumably because of the complex haloamine chemistry. With the possible exception of CNX, chloramination DBPs, especially HAAs, can be further controlled by using ozonation prior to chloramination. Overall, DBP formation from chloramination can be minimized by maintaining the distribution system pH as high as is practical and the Cl_2:N ratio as low as possible (consistent with good disinfection practice), by adding chlorine and ammonia simultaneously at the treatment plant in a well-mixed environment, and by using preozonation as needed.

REFERENCES

Cowman, G.A., and P.C. Singer. 1996. Effect of Bromide Ion on Haloacetic Acid Speciation Resulting from Chlorination and Chloramination of Aquatic Humic Substances. *Environ. Sci. & Tech.*, 30(1):16–24.

Crochet, R.A., and P. Kovacic. 1973. Conversion of 0-hydroxyaldehydes and Ketones into 0-hydroxyanilides by Monochloramine. *J. Am. Chem. Soc. Commun.*, 19:716–717.

Diehl, A.C., G.E. Speitel, Jr., J.M. Symons, S.W. Krasner, C.J. Hwang, and S.E. Barrett. 1998. Factors Affecting Disinfection By-Product Formation During Chloramination: Batch Studies. Denver, Colo.: *Jour. AWWA*, (submitted).

Fleischacker, S.J., and S.J. Randtke. 1983. Formation of Organic Chlorine in Public Water Supplies. *Jour. AWWA*, 75(3):132–138.

Fujioka, R.S., K.M. Tenno, and P.C. Loh. 1983. Mechanism of Chloramine Inactivation of Poliovirus: A Concern for Regulators? In *Water Chlorination: Chemistry, Environmental Impacts and Health Effects*, Vol. 4. Jolley, R.L., W.A. Brungs, J.A. Comming, J.S. Mattice, and V.A. Jacobs, eds. Stoneham, Mass.: Butterworth Publishers.

Gazda, M., L.E. Dejarmi, Z.K. Chowdhury, R.G. Cooks, and D.E. Margerum. 1993. Mass Spectrometric Evidence for the Formation of Bromochloramine and N-Bromo-N-Chloromethylamine in Aqueous Solution. *Environ. Sci. & Tech.*, 27(3):557–561.

Haag, W.R. 1980. Formation of N-Bromo-N-Chloramines in Chlorinated Saline Waters. In *Water Chlorination: Chemistry, Environmental Impacts and Health Effects*, Vol. 3. Jolley et al., eds. Ann Arbor, Mich.: Ann Arbor Science Publishers.

Hureike, L., J.P. Croué, and B. Legube. 1994. Chlorination Studies of Free and Combined Amino Acids. *Water Res.*, 28(12):2521–2531.

Ingols, R.S., H.A. Wyckoff, T.W. Kethley, H.W. Hodgden, E.L. Fincher, J.C. Hildebrand, and J.E. Mandel. 1953. Bactericidal Studies on Chlorine. *Ind. Eng. Chem.*, 45(5):996–1000.

Isaac, R.A., and J.C. Morris. 1983. Transfer of Active Chlorine and Chloramine to Nitrogenous Organic Compounds. 1. Kinetics. *Environ. Sci. & Tech.*, 17(12):738–742.

Jacangelo, J.G., N.L. Patania, K.M. Reagan, E.M. Aieta, S.W. Krasner, and M.J. McGuire. 1989. Ozonation: Assessing Its Role in the Formation and Control of Disinfection By-products. *Jour. AWWA*, 81(8):74–84.

Jafvert, C.T., and R.L. Valentine. 1992. Reaction Scheme for the Chlorination of Ammoniacal Water. *Environ. Sci. & Tech.*, 26:577–586.

Jensen, J.N., J.J. St. Aubin, R.F. Christman, and J.D. Johnson. 1985. Characterization of the Reaction Between Monochloramines and Isolated Fulvic Acid. In *Water Chlorination Chemistry, Environmental Impact and Health Effects*, Vol. 5. Jolley, R.L., R.J. Bull, W.P. Davis, S. Katz, M.H. Roberts, and V.A. Jacobs, eds. Chelsea, Mich.: Lewis Publishers.

Johnson, J.D., and W. Sun. 1975. Bromine Disinfection of Wastewater. In *Disinfection Water and Wastewater.* Johnson, J.D., ed. Ann Arbor, Mich.: Ann Arbor Science Publishers.

Jolley, R.L., and J.H. Carpenter. 1983. A Review of the Chemistry and Environmental Fate of Reactant Oxidant Species in Chlorinated Water. In *Water Chlorination: Chemistry, Environmental Impacts and Health Effects,* Vol. 4, Book 1. Jolley, R.L., W.A. Brungs, J.A. Comming, J.S. Mattice, and V.A. Jacobs, eds. Stoneham, Mass.: Butterworth Publishers.

Kanniganti, R., J.D. Johnson, L.M. Ball, and M.J. Charles. 1992. Identification of Compounds in Mutagenic Extracts of Aqueous Monochloraminated Fulvic Acid. *Environ. Sci. & Tech.,* 26(10):1998–2004.

Kotiaho, T., M.J. Hayward, and R.G. Cooks. 1991. Direct Determination of Chlorination Products of Organic Amines Using Membrane Introduction Mass Spectrometry. *Anal. Chem.,* 63:1794–1801.

Kotiaho, T., J.M. Wood, P.L. Wick, Jr., L.E. Dejarme, A. Ranasingh, R.G. Cook, and H.P. Ringhand. 1992. Time Persistence of Monochloramine in Human Saliva and Stomach Fluid. *Environ. Sci. & Tech.,* 26(2):302–306.

Krasner, S.W., M.J. McGuire, J.G. Jacangelo, N.L. Patania, K.M. Reagan, and E.M. Aieta. 1989. The Occurrence of Disinfection By-products in US Drinking Water. *Jour. AWWA,* 81(8):41.

LeCloirec, C., and G. Martin. 1985. Evolution of Amino Acids in Water Treatment Plants and the Effect of Chlorination on Amino Acids. In *Water Chlorination Chemistry, Environmental Impact and Health Effects,* Vol. 5. Jolley, R.L., R.J. Bull, W.P. Davis, S. Katz, M.H. Roberts, and V.A. Jacobs, eds. Chelsea, Mich.: Lewis Publishers.

LeCloirec, C., J. Morvan, and G. Martin. 1983. Forms of Organic Nitrogen in Surface Waters: Crude Waters or in Drinking Water Treatment Steps. *Rev. Fr. Sci. Eau.,* 2:25–39.

Long, B.W., G.S. Logsdon, and D.G. Neden. 1992. Options for Controlling Disinfection By-Products in Greater Vancouver's Water Supply. In *Proc. 1992 AWWA Water Quality Technology Conference, Part I.* Denver, Colo.: American Water Works Association.

Lykins, B.W., W.E. Koffsey, and K.S. Patterson. 1994. Alternative Disinfectants for Drinking Water Treatment. *Jour. of Env. Eng.,* 120(4):745–758.

McGuire, M.J., and R.G. Meadow. 1988. AWWARF Trihalomethane Survey. *Jour. AWWA,* 80(1):61–67.

Minisci, F., and R. Galli. 1965. A New, Highly Selective, Type of Aromatic Substitution, Homolytic Amination of Phenolic Ethers. *Tetrahedron Lett.,* 8:433–436.

Morris, J.C., and R.A. Issac. 1983. A Critical Review of Kinetic and Thermodynamic Constants for the Aqueous Chlorine-Ammonia System. In *Water Chlorination: Chemistry, Environmental Impacts and Health Effects,* Vol. 4. Jolley, R.L., W.A. Brungs, J.A. Comming, J.S. Mattice, and V.A. Jacobs, eds. Stoneham, Mass.: Butterworth Publishers.

Nieminski, E.C., and B.M. Shepherd. 1989. Pilot Scale Evaluation of Disinfection By-Product Formation with the Use of Ozone, Chlorine, and Monochloramine. In *Proc. 1989 AWWA Water Quality Technology Conference.* Denver, Colo.: American Water Works Association.

Ohya, T., and S. Kanno. 1985. Formation of Cyanide Ion or Cyanogen Chloride Through the Cleavage of Aromatic Rings by Nitrous Acid or Chlorine. VIII. On the Reaction of Humic Acid with Hypochlorous Acid in the Presence of Ammonium Ion. *Chemosphere,* 14(11/12):1717.

Rook, J.J. 1980. Possible Pathways for the Formation of Chlorinated Degradation Products During Chlorination of Humic Acids and Resorcinol. In *Water Chlorination: Chemistry, Environmental Impacts and Health Effects,* Vol. 3. Jolley, R.L., W.A. Brungs, R.B. Cumming, and V.A. Jacobs, eds. Ann Arbor, Mich.: Ann Arbor Science Publishers.

Singer, P.C., A. Obolensky, and A. Greiner. 1995. DBPs in Chlorinated North Carolina Drinking Waters. *Jour. AWWA,* 87(10):83–92.

Snyder, M.P., and D.W. Margerum. 1982. Kinetics of Chlorine Transfer from Chloramine to Amines, Amino Acids, and Peptides. *Inorganic Chemistry,* 21(7):2545–2550.

Stevens, A.A., L.A. Moore, and R.J. Miltner. 1989. Formation and Control of Non-Trihalomethane Disinfection By-products. *Jour. AWWA,* 81(8):54–60.

Symons, J.M., A.A. Stevens, R.M. Clark, E.F. Geldreich, O.T. Love, Jr., and J. DeMarco. 1982. *Treatment Techniques for Controlling Trihalomethanes in Drinking Water.* Denver, Colo.: American Water Works Association.

Symons, J.M., G.E. Speitel, Jr., C.J. Hwang, S.W. Krasner, S.E. Barrett, A.C. Diehl, and R. Xia. 1998. Factors Affecting Disinfection By-Product Formation During Chloramination. Denver, Colo.: American Water Works Association Research Foundation and American Water Works Association.

Valentine, R.L. 1986. Bromochloramine Oxidation of N,N-Diethyl-p-phenylenediamine in the Presence of Monochloramine. *Environ. Sci. & Tech.,* 20(2):166–170.

Valentine, R.L., and C.T. Jafvert. 1988. General Acid Catalysis of Monochloramine Disproportionation. *Environ. Sci. & Tech.,* 22(6):691–696.

Wajon, J.E., and J.C. Morris. 1980. Bromamination Chemistry: Rates of Formation of NH_2Br and Some N-Bromamino Acids. In *Water Chlorination: Chemistry, Environmental Impacts and Health Effects,* Vol. 3. Jolley, R.L., W.A. Brungs, R.B. Cumming, and V.A. Jacobs, eds. Ann Arbor, Mich.: Ann Arbor Science Publishers.

Zhu, Q., and D.A. Reckhow. 1993. Investigation of Pre-Ozonation and Chloramination on Haloacetic Acid Formation. In *Proc. of 1993 AWWA Annual Conference.* Denver, Colo.: American Water Works Association.

CHAPTER • NINE

Control of Disinfection By-Product Formation Using Ozone

DAVID A. RECKHOW

HISTORY OF OZONE USE

Ozone has been used in the treatment of drinking water since the end of the 19th century (Langlais, Reckhow, and Brink 1991). Most early installations were in Europe, especially France, Germany, and Switzerland. As engineers gained experience with this new technique for oxidation–disinfection, the number of ozone plants worldwide grew to several thousand. The United States has been slow to adopt ozone for drinking water treatment. However, since the 1970s, there has been a steady growth in the use of ozone in North America.

The earliest applications of ozone were for disinfection purposes. However, as field applications became more numerous, engineers began to notice that other benefits accompanied ozone's use. These include color removal, oxidation and subsequent precipitation of iron and manganese, and improvements in tastes and odors. In the mid-1960s it was noticed that ozonation could improve the

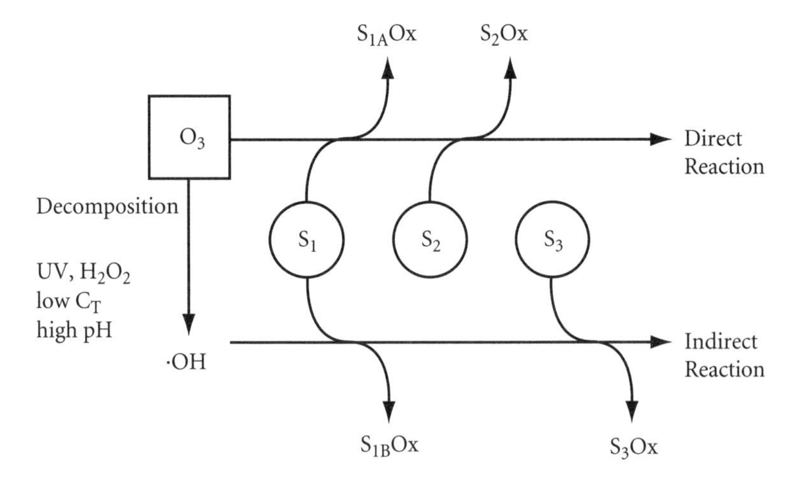

Figure 9-1 Dual nature of ozone: direct reaction as molecular ozone or indirect reaction as ·OH radical (S_1, S_2, and S_3 are solutes that form oxidized by-products S_1Ox, S_2Ox, and S_3Ox, respectively)

performance of subsequent coagulation, settling, and filtration. Later concerns over the presence of organic contaminants led to ozone's use as a means of controlling industrial solvents, pesticides, and other micropollutants. Since the mid-1970s, ozone has also been recognized as an important tool in controlling halogenated disinfection by-products. As will be discussed, the means by which this occurs can be site-specific and involve several mechanisms.

AQUEOUS OZONE CHEMISTRY

Ozone Decomposition

Ozone is a powerful oxidant capable of reacting with a wide range of organic and inorganic solutes in water. Ozone may also undergo a rapid, autocatalytic decomposition forming a variety of secondary oxidant species, most notably the hydroxyl radical, ·OH (see Figure 9-1). Because of the distinct chemistry of these two oxidants and their disparate concentrations in ozone contactors, some solutes in water will react only with ·OH radical (S_3), some will react only with molecular ozone (S_2), and some will react with both (S_1, possibly resulting in different oxidation products; $S_{1A}Ox$ and $S_{1B}Ox$).

Many chemical species affect the rate of ozone decomposition. In particular, high pH (i.e., hydroxide anion) will substantially accelerate the rate of decomposition of aqueous ozone. Ozone decomposition has been modeled as an autocatalytic chain reaction. It is autocatalytic because hydroxyl radicals participate in their own formation. It is a chain reaction because of the importance of initiation, promotion, and inhibition pathways. For example, iron (II), hydroxide, and certain organic moieties will initiate the decomposition by reacting with molecular ozone to form free radical species leading to the formation of the ·OH radical. Other solutes, such as HCO_3^-/CO_3^{-2}, some alcohols, and saturated alkyl moieties, can act as inhibitors by consuming ·OH radicals (Hoigné and Bader 1979; Staehelin and Hoigné 1983; Hoigné et al. 1985). These latter species tend to stabilize molecular ozone residuals. Addition of hydrogen peroxide or illumination with ultraviolet (UV) light also forces this decomposition reaction to occur more quickly. Applications that use both ozone and H_2O_2 or UV are referred to as advanced oxidation processes (AOPs). These systems generate large amounts of hydroxyl radicals, which are highly reactive and nonselective. In contrast, molecular ozone is far more selective in its reactions with organic and inorganic solutes. Because the reactivity and reaction mechanisms of these oxidants are different, it is important to know which types of reactions will predominate in any given water.

Reactions With Organic Substrates

Molecular ozone may react with substances in dilute aqueous solution by means of electrophilic attack or 1-3 dipolar cycloaddition (Bailey 1972; Decoret et al. 1984). Ozone reacts readily with olefinic bonds via the classical Criegee ozonolysis mechanism. The Criegee products are acids and aldehydes or ketones formed from cleavage of the molecule at the site of the double bond (Figure 9-2). Similarly, aromatic compounds that are activated by an electron-donating substituent react readily with ozone to yield ring-cleavage products. Other compounds that may quickly consume ozone include a variety of nitrogenous compounds and organic sulfides (Hoigné and Bader 1983). These organic bases are most reactive in their deprotonated form.

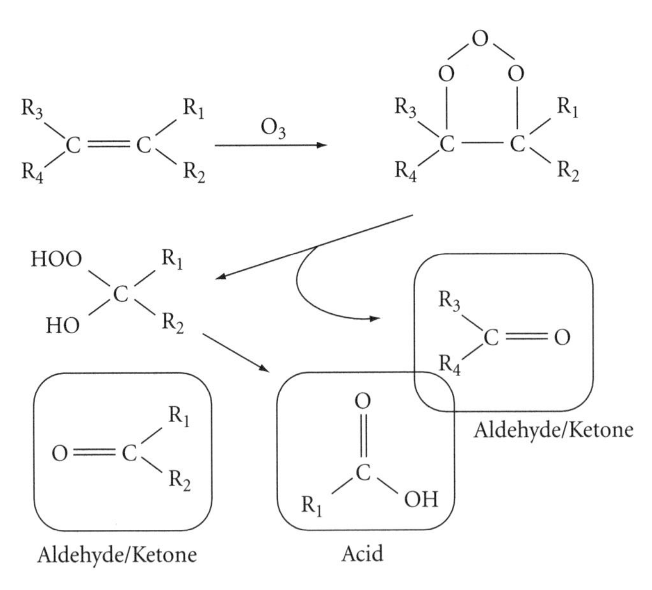

Figure 9-2 By-products of classical Criegee ozonolysis

Hydroxyl radicals will participate in at least the following three general types of reactions: hydrogen abstraction, electron transfer, and ·OH addition. Addition reactions occur when hydroxyl radicals add to an aromatic nucleus or an unsaturated aliphatic chain (Eq 9-1). An organic radical is formed that can proceed to react with other chemical species. Equation 9-1 is the preferred pathway for most aromatic compounds. This pathway is of particular interest because it can potentially lead to the production of THM precursors from non-THM precursors. Hydrogen abstractions may lead to the formation of another type of organic free radical and water (Eq 9-2). This is a major pathway for saturated organic compounds, alcohols, aldehydes, and ketones (Dorfman and Adams 1973).

$$\cdot OH + C_6H_6 \rightarrow \cdot C_6H_6OH \qquad (9\text{-}1)$$

$$\cdot OH + CH_3OH \rightarrow \cdot CH_2OH + H_2O \qquad (9\text{-}2)$$

Reactions With Other Disinfectants and Bromide

Both hypochlorite and chlorine dioxide will react with molecular ozone under conditions typical of drinking water treatment such that the oxidants cannot coexist for more than a few minutes. For hypochlorite, the reaction products are chloride and chlorate. Chlorate is the final product from the reaction of ozone with chlorine dioxide. Chloramines react with ozone somewhat slower than hypochlorite does, forming chloride.

In the absence of other more reactive species, molecular ozone will react slowly with bromide to form hypobromous acid. The hypobromous acid can then go on to react with more ozone, natural organic matter (NOM), or ammonia. If it reacts with NOM, brominated disinfection by-products (DBPs) will result. Subsequent reaction of hypobromous acid with ozone or hydroxyl radical can ultimately lead to bromate ion formation (Eq 9-3).

$$Br^- \xrightarrow{O_3} [HOBr \leftrightarrow OBr^-] \xrightarrow{O_3} BrO_3^- \qquad (9\text{-}3)$$

The second-order rate constant for the first step is 160 $M^{-1}s^{-1}$ (Haag and Hoigné 1983). This means that hypobromous acid will be slowly formed in an ozone contactor, and the extent of formation will depend on the ozone residual, time, and temperature. For example, bromide will be converted into hypobromous acid with a half-life of about 7 minutes in the presence of an aqueous ozone residual of 0.5 mg/L (at 20°C).

Reactions With Natural Organic Matter

Ozone reacts very rapidly with NOM, especially aquatic humic materials. It effects a quick decrease in color and UV absorbance (Figure 9-3). This is because ozone is selective for carbon–carbon double bonds and other sites with high electron density. Many of these are part of the same structures that are responsible for UV absorbance and color. In addition, ozone forms many new oxygenated functional groups. A large fraction of these will be carboxyl groups (i.e., increased acidity, Figure 9-3) that impart a negative charge to the NOM molecules. Accompanying these chemical changes will be bond cleavage, which ultimately leads to molecular fragmentation. The degree to which this occurs depends on such factors as the ozone dose and the initial molecular size of the NOM. A relatively small amount of the organic carbon in ozonated NOM is converted into inorganic carbon dioxide (i.e., small loss of total organic carbon [TOC]). Instead,

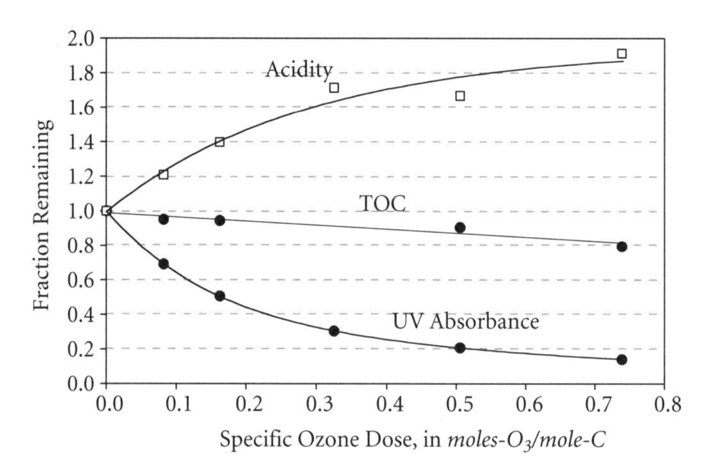

Source: Reckhow and Singer (1984).

Figure 9-3 Effect of ozone on various NOM characteristics

many of the reactions terminate in the formation of long-lived intermediates, such as oxalic acid. These are end products, because they are relatively nonreactive toward ozone.

Aquatic humic materials are believed to be composed of 20 to 40 percent aromatic carbon, much of which may be activated by phenolic groups. Because phenolic molecules are very rapidly attacked by ozone, it is generally assumed that ozone first attacks these moieties in aquatic humic materials. However, about half of the organic carbon in natural fresh waters is nonhumic organic matter. This material consists of the hydrophilic acids (volatile fatty acids, hydroxy-acids, and polymeric acids of very high charge density), polysaccharides, amino acids, proteins, tannins, hydrocarbons, and others. While many of these would not be expected to play an important role based on their low reactivity with ozone (e.g., hydrocarbons and polysaccharides; see Bose, Bezbarua, and Reckhow 1994), others may be very important (e.g., amino acids and tannins). In colored waters, humic substances will account for the vast majority of the ozone demand. However, in waters of low color, especially waters that have undergone coagulation, there will be a lower abundance of humic substances. In these cases, much of the ozone demand will occur due to reaction with the nonhumic material.

USE OF OZONE TO CONTROL HALOGENATED DBPs

Direct Destruction of DBP Precursors

The trihalomethanes. It has been known since the late 1970s that ozone can destroy some of the THM precursor sites in raw drinking waters (e.g., Umphres et al. 1979). It was also acknowledged early on that ozone could create new precursor sites, which may result in a net increase in subsequent THM formation under certain conditions. Many of these early studies also showed that chlorine demand could be reduced as a result of prior ozonation. Ozonation seems to produce the greatest net decrease in THM production when subsequent chlorination is at a low pH (Riley, Mancy, and Boettner 1978; Brunet et al. 1980; Reckhow, Legube, and Singer 1986). Systems that use high-pH chlorination (e.g., above pH 8.5) are more likely to see a net increase as a result of ozonation. However, there are a few reports of precursor enhancement when neutral chlorination was used (Legube et al. 1985).

In addition to low pH, high bicarbonate concentration can also help improve the effectiveness of ozonation for THM control (Legube et al. 1985; Reckhow, Legube, and Singer 1986). This may be due partly to the action of bicarbonate as a free radical scavenger. Scavengers consume hydroxyl radicals, slow the decomposition of ozone, and thereby maximize the exposure of solutes (e.g., THM precursors) to molecular ozone. On the other hand, bicarbonate may play a more direct role by forming bicarbonate radicals as a result of reaction with hydroxyl radicals. These moderately reactive free radical species are selective oxidants themselves, and data suggest that they may help to destroy DBP precursor sites (i.e., they are reactive with phenolic compounds). Figure 9-4 summarizes the data from an early study on the ozonation of natural aquatic fulvic acids. These results have been supported by contemporary studies conducted in different laboratories using NOM from other sources (Malley, Edzwald, and Ram 1986; Legube et al. 1985). All waters were ozonated and chlorinated between pH 7 and pH 7.5. The fact that all three studies gave similar results suggests a degree of uniformity in the reaction of NOM with ozone. One carefully documented exception to this behavior is that of Legube et al.'s Les Landes fulvic acid from

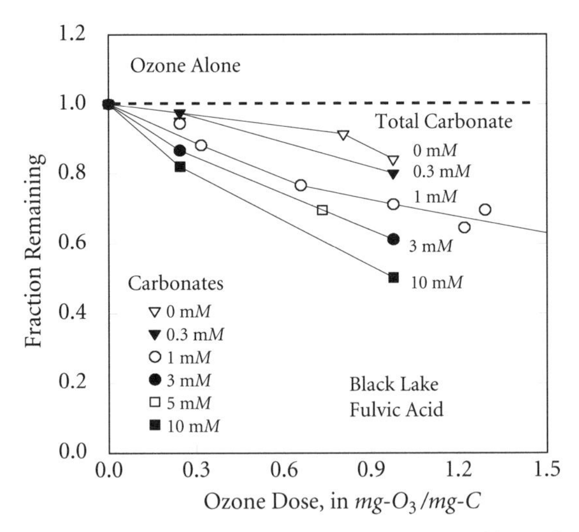

NOTE: Conditions: 4.1 mg/L TOC; Black Lake fulvic acid; 20 mg/L chlorine dose; 72-h reaction time; 20°C
Source: Reckhow et al. (1986).

Figure 9-4 Effect of ozone dose and bicarbonate concentration on the destruction of THM precursors

Bordeaux (France). In this case, a general increase in trihalomethane formation potential (THMFP) was observed with increasing ozone dose, especially in the absence of bicarbonate.

Analysis of precursor destruction in full-scale ozonation plants has generally supported the laboratory data. Figure 9-5 shows data from six full-scale plants superimposed over the general data pattern from the bench-scale work. Given neutral pHs and moderate levels of bicarbonate, one might expect THMFP reductions of 5 to 20 percent at the ozone doses most commonly applied in drinking water treatment. Data from Bay Metro (not shown) exhibit a broad range of precursor destruction (Grasso, Weber, DeKam 1989), which generally encompasses the range shown here. For reasons not well understood, these authors reported higher percent precursor destruction during the summer.

Waters with high bromide concentration will sometimes show elevated levels of brominated THMs as a result of preozonation (Miltner, Shukairy, and Summers 1992; Krasner 1996). This effect can only be observed after subsequent chlorination, and it

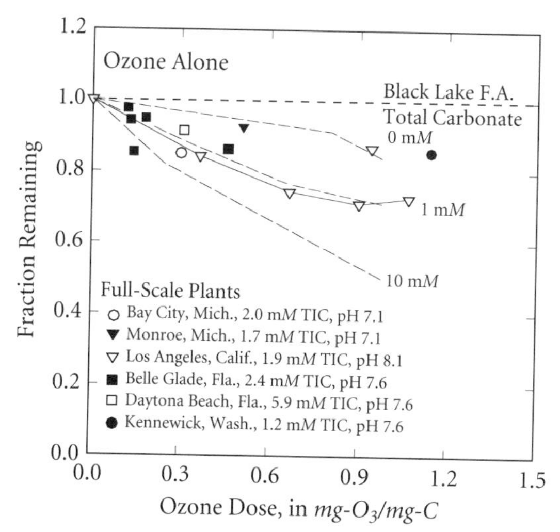

Source: Singer and Chang (1989); Georgeson and Karimi (1988).

Figure 9-5 Full-scale experience with destruction of THM precursors by ozone

should not be confused with the bromoform formation that may occur as a direct result of ozonation (discussed later). This ozone-induced shift in THM speciation is especially evident when low doses of ozone are applied, and chlorine dose is reduced to match the reduced chlorine demand typically observed for ozonated waters. Under these conditions, the chlorine-to-bromide ratio during subsequent chlorination is lower, and the bromine (produced from reaction with chlorine) is better able to compete for the now diminished precursors. At higher ozone doses, there is substantial bromate formation, and therefore less bromide is available for subsequent reaction.

In addition to altering precursor content, ozonation will change the relative rate at which chlorinated DBPs form (Riley, Mancy, and Boettner 1978; Hubbs and Holdren 1986; Reckhow et al. 1995). Bench-scale studies have shown that ozone destroys a disproportionate amount of the fast-reacting THM precursors (Weiner 1995). These precursors are thought to be largely composed of activated aromatic structures that are also highly susceptible to attack by molecular ozone. Trihalomethane formation is therefore slower following ozonation even when the ultimate

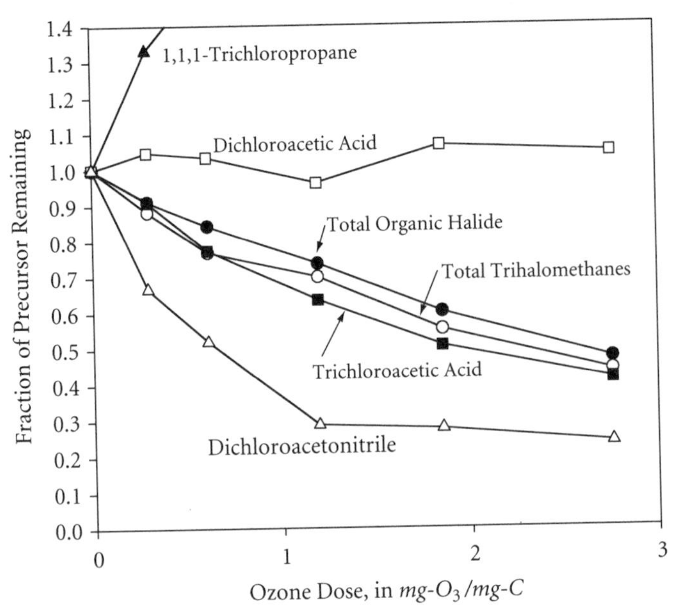

NOTE: Conditions: 4.1 mg/L TOC; Black Lake fulvic acid; 20 mg/L chlorine dose; 72-h chlorine contact time; 20°C
Source: Reckhow and Singer (1984).

Figure 9-6 Effect of ozone dose on chlorination by-product precursors

precursor content is unchanged. This results in a situation in which the percent reduction of THMs by ozonation appears greater as the subsequent chlorine contact time is decreased.

Other Chlorination By-Products

Ozonation will affect precursors to total organic halogens (TOX) and other halogenated by-products much as it affects THM precursors (e.g., Reckhow and Singer 1984; Legube et al. 1985; Lykins, Koffskey, and Miller 1986; Grasso, Weber, and DeKam 1989). Among the haloacetic acids (HAAs), trichloroacetic acid (TCAA) precursors have been shown to be quite susceptible to destruction by ozonation (Figure 9-6). In contrast, dichloroacetic acid (DCAA) precursors are frequently unaffected by ozonation. Some semi-mechanistic justification for these effects has been postulated (Reckhow and Singer 1985). However, definitive mechanisms for the ozonation of any given DBP precursor cannot be established until the exact chemical nature of the various DBP precursors is determined.

Some compounds, such as the halogenated ketones and aldehydes (e.g., 1,1,1-trichloropropanone and chloral hydrate) will form at greater concentrations as a result of prior ozonation (Reckhow and Singer 1984). It has been proposed that ozone produces large amounts of aldehydes and ketones, and thereby increases the likelihood that halogenated derivatives of these compounds will be formed during subsequent chlorination. Similar increases have been observed with chloropicrin (Becke, Maier, and Sontheimer 1984; Hoigné and Bader 1988). The reasons for this are not well understood.

Earlier it was noted that ozone changes the relative rate at which a water forms THMs. This also appears to be true for TOX formation (Hubbs and Holdren 1986) and HAA formation (Reckhow et al. 1995). This means that ozone should reduce TOX and total haloacetic acids (THAAs) as well as THMs at the consumer's tap to a greater extent than would be predicted based on reductions in the respective formation potentials.

Ozone as an Alternative to Chlorine

The most effective role for ozone in controlling chlorination DBPs is its use as a replacement for chlorine (i.e., as an alternative primary disinfectant–oxidant). Ozone is as effective or more effective than chlorine for nearly all of the treatment objectives for which one would use chlorine. It will not, however, provide a lasting disinfectant residual in the distribution system. For this reason, it is not generally used as an alternative secondary disinfectant.

There are several important advantages of applying ozone in place of chlorine at the head of the treatment plant. First, the formation of by-products from free chlorination is delayed or entirely avoided (especially when chloramines are used as a secondary disinfectant). Second, indirect beneficial effects of preozonation on downstream processes may be realized, such as increased biodegradation. There may also be beneficial effects that are not directly related to DBPs, such as improved control of tastes and odors, and better removal of turbidity or better filtration performance.

Improvements in DBP control for systems that apply ozone as a replacement for chlorine are very strongly linked to the exact placement of the ozonation step. A quantitative discussion of this topic appears later in this chapter.

Ozone as a Coagulant Aid

Ozonation can affect subsequent coagulation and filtration processes in some subtle, yet important ways (Reckhow, Singer, and Trussell 1986). Observations of the beneficial effect of preozonation include

- a shift in particle-size distribution towards larger sizes
- the formation of colloidal particles from "dissolved" organic matter
- the improved removal of TOC or turbidity during subsequent settling, flotation, or filtration
- a decrease in coagulant dose necessary for achieving desired effluent turbidity or TOC
- an increase in floc settling velocities
- extended filter run length due to slower head-loss buildup or delayed breakthrough

Although benefits from preozonation have been widely cited, there are a nearly equal number of cases in which detrimental effects have been reported, many exhibiting the exact opposite behavior from that indicated above. This has led to the general conclusion that preozonation may be either beneficial or detrimental, depending on some ill-defined set of circumstances. Despite the numerous conflicting reports, there are some reliable trends. First, preozonation often improves performance by subsequent filtration as measured by turbidity and particle control. This may be especially evident when cationic polymers are being used (e.g., Reckhow, Edzwald, and Tobiason 1993; Edwards, Benjamin, and Tobiason 1994). In this case, reduced coagulant demands are also observed. In contrast, there is a strong tendency for preozonation to cause a deterioration in subsequent coagulation of NOM (e.g., Tobiason et al. 1992; Tobiason, Reckhow, and Edzwald 1995). This is generally reflected in poorer removal of TOC and DBP precursors.

From many years of research on the subject, it has become evident that the underlying causes of these effects are numerous and complex. For example, there is evidence that with some waters the intrinsic affinity of NOM for aluminum hydroxide surfaces is not affected by prior ozonation. Instead, preozonation increases the negative zeta potential of the hydroxide–NOM particles, thereby inhibiting their agglomeration into settleable floc (Edwards and Benjamin 1992). On the other hand, some studies have shown some types of NOM to be less adsorbable to preformed alum floc after ozonation (e.g., Bose 1994). To date there is no quantitative way of reliably predicting the effects of preozonation on coagulation and filtration. Each new water must be tested in bench scale, or preferably pilot scale, in order to evaluate the full effect of preozonation.

The generally accepted reasons for ozone's detrimental effects on NOM removal during coagulation are closely linked to its tendency to fragment NOM molecules and increase their oxygen content (i.e., carboxyl, carbonyl, and hydroxyl groups). When this occurs, the organic matter is rendered more highly charged and more hydrophilic. Both properties have a negative effect on removal by hydrolyzing metal coagulants. The coagulating effects described here and the phenomena discussed should be viewed as separate from the direct effects of the oxidation of NOM by ozone (i.e., the bleaching of color, the increase in biodegradability, and the direct oxidation of THM precursors).

BIOLOGICAL TREATMENT

Biodegradability of NOM

It is well accepted that ozonation will increase the biodegradability of NOM. Evidence for this lies in the elevated levels of biodegradable dissolved organic carbon (BDOC) (e.g., Miltner, Shukairy, and Summers 1992), and assimilable organic carbon (AOC) (e.g., van der Kooij et al. 1989) in ozonated waters. These are supported by observations of greater levels of bioactivity in filters that receive ozonated water and greater bacterial regrowth in ozonated waters that are not subject to biological treatment (Cipparone, Diehl, and Speitel 1997). Recent results with waters of relatively low humic content have shown that rapidly biodegradable organic matter ($BDOC_{rapid}$) is preferentially formed by ozonation (Carlson and Amy 1997). Concentrations of the more slowly biodegradable organic material ($BDOC_{slow}$) are less affected by ozonation. Specific chemical analysis of known biodegradable compounds (e.g., acids and aldehydes) also shows sharp increases after ozonation. The specific identifiable ozonation by-products may account for as much as half of the $BDOC_{rapid}$, assuming that the by-products are completely mineralized on biodegradation. While other oxidants (e.g., chlorine, chloramines, and chlorine dioxide) will cause many of these same changes, none is quite as effective as ozone at the doses commonly employed.

Ozone Combined With Biological Filtration

Because ozonation increases the biodegradability of NOM, it also stimulates growth in downstream processes. Biodegradation may occur in downstream sedimentation tanks or treated water reservoirs (e.g., Miltner and Summers 1992); however, filtration processes are especially susceptible to effects from prior ozonation. This synergism tends to result in lower DBP concentrations (see chapter 16). The reasons for this include

- direct biodegradation of chlorination by-product precursors
- direct biodegradation of ozonation by-products
- reduction in chlorine-demanding substances, allowing the use of lower chlorine doses
- removal by biodegradation of potential sources of carbon and energy for microorganisms in the distribution system (e.g., BDOC and AOC), which can reduce biological growth, and thereby reduce formation of new DBP precursor compounds

Because it is important to control organic nutrients (e.g., BDOC and AOC) in finished water, ozonation should generally be followed by biologically active filtration (see the section on staging of ozonation). For more information on how biological filtration can be used to control DBPs, refer to chapter 16.

OZONATION BY-PRODUCTS

Organic By-Products

Although many researchers have applied gas chromatography/ mass spectrometry (GC/MS) techniques for the purpose of identifying organic ozonation products of raw waters and extracted humic materials, the identifiable yields are generally very small (i.e., <5 percent of the initial TOC). Among the products reported from the ozonation of drinking waters are simple aldehydes, low-molecular-weight aliphatic acids, some keto-acids, hydroxy-acids, organic peroxides, and benzene polycarboxylic acids. Brominated by-products (e.g., bromoform and tribromoacetic acid) are formed in waters with elevated levels of bromide ion. From this group of identified ozonation by-products, perhaps only the peroxides, bromo-organics, and aldehydes are suspected of posing human health hazards. In particular, formaldehyde seems to be a ubiquitous product of drinking water ozonation. Some of these by-products (e.g., peroxides and formaldehyde) may become depleted through slow reaction with other constituents in the water.

When waters with high levels of bromide ion are ozonated, there will be some formation of brominated organic by-products. This results from the oxidation of bromide to hypobromous acid, and subsequent reaction of the HOBr with NOM. Song and co-workers (1997) found that about 7 percent of the raw water bromide was converted to total organic bromine (TOBr) under typical conditions of pH and ozone dose. The formation of TOBr was found to be enhanced at low pH, high ozone dose, and high bromide levels. Utilities wishing to minimize bromate ion formation may want to ozonate at low pH (Krasner et al. 1993) (see bromate discussion later in this chapter), which could produce the undesired effect of increasing the formation of TOBr. Compounds that compete with organic precursors for active bromine, such as ammonia, will depress the formation of TOBr. However, compounds that encourage the complete reaction between ozone and bromide ion, such as free radical scavengers (e.g., bicarbonate), may enhance TOBr formation.

Earlier it was pointed out that ozone is not the only oxidant capable of producing elevated levels of biodegradable by-products

Table 9-1 Mass balance of ozonation by-products and various measures of biodegradability in California State Project water

Parameter	Concentration (*μg/L*)	Percentage of DOC	Percentage of BDOC	Percentage of $BDOC_{filter}$	Percentage of AOC
DOC	2,150	100			
BDOC	520	24	100		
$BDOC_{filter}$	330	15	63	100	
AOC	286	13	55	87	100
Aldehydes	12	0.6	2	4	4
Ketoacids	20	0.9	4	6	7
Carboxylic acids	73	3.0	14	22	26
Total by-products		4.5	20	32	37

Source: Krasner (1996).

(e.g., AOC). The same can be said for the formation of aldehydes, ketones, and acids. For example, chlorine is known to produce all of these by-products from reaction with NOM in water (e.g., Norwood et al. 1983; Lykins, Koffskey, and Miller 1986; Jacangelo et al. 1989). These compounds are more often associated with ozone because ozonation generally produces higher levels, and because there are no known chlorinated by-products of ozonation.

Comprehensive studies of ozonated waters have demonstrated that simple ozonation by-products may constitute a substantial fraction of the total biodegradable material (e.g., Krasner 1996). Table 9-1 shows that the three major groups of ozonation by-products in California State Project water can account for 20 percent of the measureable BDOC. This number rises to 32 percent when only considering the organic matter that can be removed across a biologically active filter ($BDOC_{filter}$). Finally, the major ozonation by-products accounted for 37 percent of the assimilable organic carbon (AOC), a parameter thought to represent the most highly biodegradable constituents.

Recent research has shown that most specific ozonation by-products will increase in concentration with increasing ozone dose. At some point, however, the reaction becomes less productive, and by-product formation plateaus or even declines. Figure 9-7 shows a summary of ozonation by-product yields at ozone doses before the

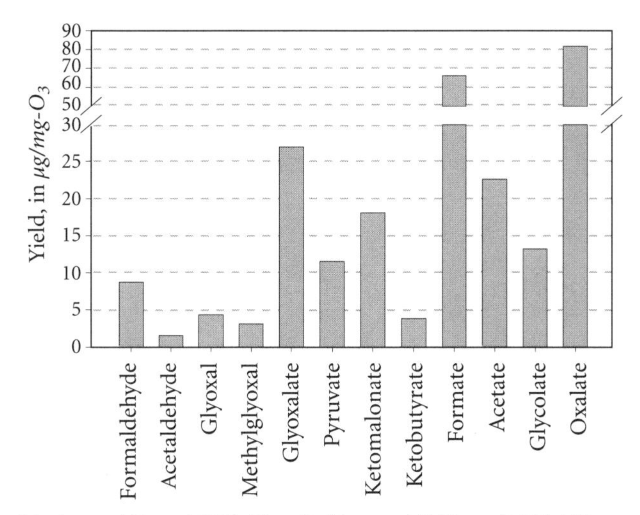

Source: Schechter and Singer (1995); Zhou, Reckhow, and Tobiason (1993); Miltner, Shukairy, and Summers (1992); Booth et al. (1996); Krasner (1996).

Figure 9-7 Summary of average ozonation by-product yields from published literature

point where yield diminishes. Simple acids dominate the group, followed by keto-acids, and then aldehydes.

Because ozone dissipates quickly in most waters, reaction time is much less important to the formation of these by-products. Studies focusing on the major aldehydes have shown that formation of these by-products is minimized at high pH (Zhou, Reckhow, and Tobiason 1993; Schechter and Singer 1995). There are indications that aldehyde formation is also enhanced in waters with substantial nonhumic character (Paode et al. 1997).

Bromate

Bromate ion formation occurs through a complex web of pathways, with several bromine-containing intermediates that undergo reactions with both molecular ozone and hydroxyl radicals (Figure 9-8). Many of the individual steps have been well characterized, but a comprehensive and quantitative model of the entire system has not yet been validated. There are many reports of empirical and semiempirical studies that cite the following factors as being important (Haag and Hoigné 1983; Siddiqui and

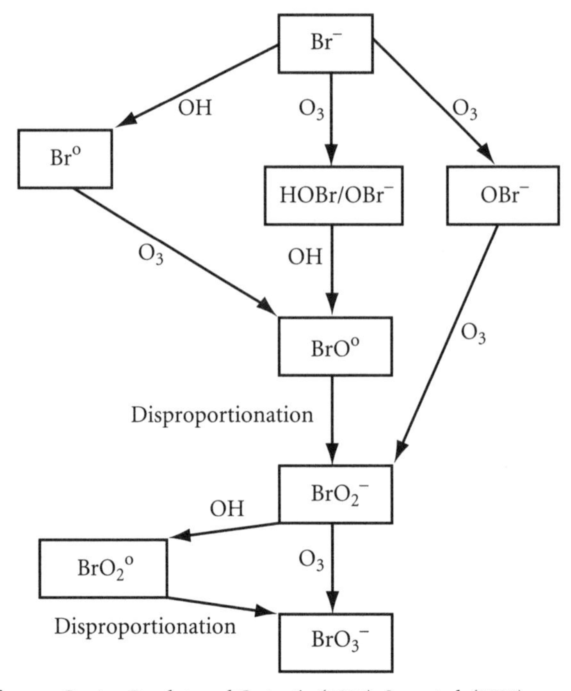

Source: After von Gunten, Bruchet, and Costentin (1996); Song et al. (1997).

Figure 9-8 Multiple pathways to the formation of bromate ion

Amy 1993; von Gunten and Hoigné 1994; Croué, Koudjonou, and Legube 1996):

- ozone dose
- pH
- alkalinity
- bromide ion
- nature and concentration of NOM
- ammonia
- addition of H_2O_2

Many of these have been incorporated in empirical power function models (e.g., Song et al. 1996; Ozekin and Amy 1997).

Above a certain threshold ozone dose (e.g., 0.5 mg-O_3/mg-C), the formation of bromate is approximately linear with ozone dose. However, it is inversely related to DOC at a fixed ozone dose. This is presumably due to competition between NOM and bromide for ozone. Increasing the pH of ozonation generally causes higher bromate formation. When ammonia is present, it is thought to preferentially react with active bromine, forming bromamines. While some beneficial effect of ammonia has been observed, it does not seem to be a viable bromate-control strategy by itself. Bicarbonate can stimulate bromate formation, presumably through stabilizing the ozone residual.

Addition of hydrogen peroxide may increase bromate formation at low dose ratios (below 0.3 moles H_2O_2/mole O_3), and decrease it at higher ratios. Studies from five full-scale water plants suggest that addition of hydrogen peroxide can substantially decrease bromate formation when ozone dose is not increased (von Gunten, Bruchet, and Costentin 1996). However, when a constant ozone residual is desired, peroxide addition results in an increase in bromate formation. This is due to a greater exposure to both molecular ozone and hydroxyl radicals. Pilot studies in the United Kingdom have shown that addition of peroxide at a mass ratio of 0.35 mg/mg-O_3 can reduce bromate formation by 67 percent (Croll 1996). This may be partly attributable to the rapid reaction between hydrogen peroxide and hypobromous acid forming bromide (von Gunten, Bruchet, and Costentin 1996). Treatment studies using only UV/H_2O_2 have failed to show measurable bromate formation (Symons and Zheng 1997).

Lowering the pH and minimizing aqueous ozone residuals are among the most effective control strategies. However, low pH may encourage the formation of TOX, and minimizing residuals will decrease $C \times T$ credit for disinfection. When the ozone dose is distributed between two stages in a multistage ozone contactor, the overall aqueous residual can be reduced by adjusting the dose so that more ozone is added in the first stage, and less in the second. Pilot-plant data showed that bromate could be reduced by 40 percent using this approach, as compared to splitting the overall dose evenly between the two stages. Mathematical modeling of ozone reactions and reactor kinetics suggests that contactor flow patterns may also affect bromate formation (Roustan et al. 1996).

TREATMENT OPTIONS: DOSES AND STAGING

Ozone dose is an important treatment variable that is determined by the quality of the raw water (i.e., ozone demand) and the specific treatment objective for ozonation. Those plants that apply ozone for oxidation of iron and manganese or improved coagulation and filtration may only need a small transient residual. However, effective inactivation of protozoan cysts and oxidation of some synthetic organic compounds requires higher and longer-lived residuals. The residual-time product (i.e., $C \times T$) or "ozone exposure" will be directly proportional to the ozone dose and inversely proportional to the ozone demand and decomposition rate. Doses used in drinking water treatment are typically in the range of 1–5 mg/L (see Tate 1991). Waters of low color and low inorganic demand (e.g., iron, manganese, and sulfide) may only require a dose of 1 mg/L for most purposes, whereas highly colored waters may need 5 mg/L or more.

The optimum placement of ozonation in a treatment train is determined by the various treatment objectives (e.g., control of taste and odor, iron and manganese, removal, disinfection, color removal, oxidation of micropollutants, coagulant aid, or DBP control) as well as the need to remove oxidation by-products generated by ozone. Many of these objectives have conflicting requirements with regard to ozone application, doses, and staging. The needs and optimum conditions for ozone when applied for DBP control are just one group of factors to be considered.

Considerations Regarding Preoxidation–Predisinfection

The most pronounced reductions in DBP formation come from systems where ozone has been employed in place of chlorination. Pilot- and full-scale studies have shown that the use of preozonation in place of prechlorination can result in finished water THM concentrations that are as much as 85 percent lower (Georgeson and Karimi 1988; Jacangelo et al. 1989; Singer, Robinson, and Elefritz 1990; Croll 1996). In addition, preozonation has been found to result in lower THM concentrations than preoxidation with either chloramines or chlorine dioxide (Croll 1996). In many of these studies, HAA and TOX levels were similarly reduced. These decreases may be less pronounced with systems that practice postchloramination (Jacangelo et al. 1989). Furthermore, as noted, use of preozonation under these conditions generally leads to increases in haloketones, aldehydes, chloral hydrate, and chloropicrin, and it causes shifts in DBP speciation toward brominated forms.

The addition of a preozonation step in a treatment process not previously employing preoxidation can also have substantial effects. When using postchloramination, very large reductions in finished water THM and HAA concentrations have been observed, while other chlorinated by-products are less strongly affected (Jacangelo et al. 1989). As before, aldehydes may be substantially increased. Pilot studies have shown that if the use of preozonation permits substitution of postchlorination with postchloramination, the greatest reductions in chlorinated DBPs can be realized (Jacangelo et al. 1989).

Considerations Regarding Coagulation and Biological Filtration

Because of the synergism that exists between ozonation and coagulation, as well as ozonation and biologically active granular media filtration, attempts to find the optimal placement for ozonation must include consideration of these two processes. Three commonly studied configurations include (1) ozonation upstream of coagulation and filtration (i.e., preozonation); (2) ozone after settling, but before filtration (intermediate ozonation, for conventional plants only); and (3) ozone after filtration (postozonation).

Preozonation provides an opportunity to completely eliminate prechlorination in plants that need a preoxidation step. However, it also means that ozone is applied to the water with the highest TOC, causing high ozone demands. Preozonation may also adversely affect the physicochemical removal of NOM by diminishing the formation of particulate organic carbon.

The second option (intermediate ozonation) avoids much of the high oxidant demand that occurs with preozonation. However, it doesn't allow for some of the various benefits of a powerful preoxidation step (e.g., removal of iron and manganese in the clarifier sludge).

The third alternative results in the lowest ozone demands, but does not generally provide a means for removal of the organic ozonation by-products and subsequent biological stability. A possible solution combining some of the best features of these three options is to apply ozone after filtration and follow it with a second filtration step (e.g., a GAC contactor). Although more costly, this approach has been used in several large European water treatment plants. Decisions regarding the placement of ozonation are best made after long-term pilot testing with a parallel train as a control. Studies of this sort have shown that the optimal placement depends on many factors that can be highly site-specific (e.g., Tobiason et al. 1993; Tobiason 1994; Yannoni, Zhu, and Clark 1997).

REFERENCES

Bailey, P.S. 1972. Organic Groupings Reactive Toward Ozone. Evans, F.L., ed. *Ozone in Water and Wastewater Treatment* (pp. 29–59). Ann Arbor, Mich.: Ann Arbor Science Publishers.

Becke, C., D. Maier, and H. Sontheimer. 1984. Herkunft von Trichlornitromethan in Trinkwasser. *Vom Wasser*, 62:125–135.

Booth, S.D.J., G.A. Gagnon, S. Peldszus, D. Mutti, F. Smith, and P.M. Huck. 1996. Formation and Removal of Carboxylic Acids in a Full-Scale Treatment Plant Using Ozone. In *Proc. Annual Conference of the American Water Works Association*. Denver, Colo.: American Water Works Association.

Bose, P. 1994. *Selected Physico-chemical Properties of Natural Organic Matter and Their Changes due to Ozone Treatment: Implications for Coagulation using Alum.* Unpublished PhD dissertation. Amherst, Mass.: University of Massachusetts at Amherst.

Bose, P., B.K. Bezbarua, and D.A. Reckhow. 1994. Effect of Ozonation on Some Physical and Chemical Properties of Aquatic Natural Organic Matter. *Ozone: Science and Engineering*, 16(2):89–112.

Brunet, R., M. Bourbigot, B. Legube, and M. Doré. 1980. Evolution Chimique des Matieres Organiques au Cours de l'Ozonation d'une Eau de Surface Filtree. *Aqua*, 4:76–81.

Carlson, K., and G. Amy. 1997. The Formation of Filter-Removable Biodegradable Organic Matter During Ozonation. *Ozone: Science and Engineering*, 19(2):179–199.

Cipparone, L.A., A.C. Diehl, and G.E. Speitel, Jr. 1997. Ozonation and BDOC Removal: Effect on Water Quality. *Jour. AWWA*, 89(2):84–97.

Croll, B.T. 1996. The Installation of GAC and Ozone Surface Water Treatment Plants in Anglian Water, UK. *Ozone: Science and Engineering*, 18(1):19–40.

Croué, J., B.K. Koudjonou, and B. Legube. 1996. Parameters Affecting the Formation of Bromate Ion During Ozonation. *Ozone: Science and Engineering*, 18(1):1–18.

Decoret, C., J. Royer, B. Legube, and M. Doré. 1984. Experimental and Theoretical Study of the Mechanism of the Initial Attack of Ozone on Some Aromatics in Aqueous Solution. *Environmental Technology Letters*, 18:207–218.

Edwards, M., and M.M. Benjamin. 1992. Effect of Preozonation on Coagulant–NOM Interactions. *Jour. AWWA*, 84(8):63–72.

Edwards, M., M.M. Benjamin, and J.E. Tobiason. 1994. Effects of Ozonation on Coagulation of NOM Using Polymer Alone and Polymer/Metal Salt Mixtures. *Jour. AWWA*, 86(1):105–116.

Georgeson, D.L., and A.A. Karimi. 1988. Water Quality Improvements With the Use of Ozone at the Los Angeles Water Treatment Plant. *Ozone: Science and Engineering*, 10(3):255–276.

Grasso, D., W.J. Weber, Jr., and J.A. DeKam. 1989. Effects of Preoxidation with Ozone on Water Quality: A Case Study. *Jour. AWWA*, 81(6):85–92.

Haag, W.R., and J. Hoigné. 1983. Ozonation of Bromide-Containing Water: Kinetics of Formation of Hypobromous Acid and Bromate. *Environ. Sci. & Tech.*, 17(5):261–267.

Hoigné, J., H. Bader, W.R. Haag, and J. Staehelin. 1985. Rate Constants of Reactions of Ozone With Organic and Inorganic Compounds in Water-III: Inorganic Compounds and Radicals. *Water Res.*, 19(8):993–1004.

Hoigné, J., and H. Bader. 1988. The Formation of Trichlonitromethane (Chloropicrin) and Chloroform in a Combined Ozonation-Chlorination Treatment of Drinking Water. *Water Res.*, 22(3):313–319.

———. 1979. Ozonation of Water: Selectivity and Rate of Oxidation of Solutes. *Ozone: Science and Engineering*, 1(1):73–85.

———. 1983. Rate Constants of Reactions of Ozone with Organic and Inorganic Compounds in Water. II. Dissociating Organic Compounds. *Water Res.*, 17(2):185–194.

Hubbs, S.A., and G.C. Holdren. 1986. *Chloro-Organic Water Quality Changes Resulting From Modification of Water Treatment Practices.* Denver, Colo.: American Water Works Association Research Foundation and American Water Works Association.

Jacangelo, J.G., N.L. Patania, K.M. Reagan, E.M. Aieta, S.W. Krasner, and M.J. McGuire. 1989. Ozonation: Assessing Its Role in the Formation and Control of Disinfection By-products. *Jour. AWWA*, 81(8):74–84.

Krasner, S.W. 1996. The Effects of Ozonation, Biofiltration, and Secondary Disinfection on DBP Formation. In *Proc. of the Water Quality Technology Conference.* Denver, Colo.: American Water Works Association.

Krasner, S.W., W.H. Glaze, H.S. Weinberg, P.A. Daniel, and I.N. Najm. 1993. Formation and Control of Bromate During Ozonation of Waters Containing Bromide. *Jour. AWWA*, 85(1):73–81.

Langlais, B., D.A. Reckhow, and D.R. Brink. 1991. *Ozone in Water Treatment: Application and Engineering.* Chelsea, Mich.: Lewis Publishers.

Legube, B., J.P. Croué, D.A. Reckhow, and M. Doré. 1985. Ozonation of Organic Halide Precursors: Effects of Bicarbonate and Bromide. In *Proc. of the International Conference, "The Role of Ozone in Water and Wastewater Treatment,"* (pp. 73–86). London: Selper, Ltd.

Lykins, B.W., Jr., W.E. Koffskey, and R.G. Miller. 1986. Chemical Products and Toxicologic Effects of Disinfection. *Jour. AWWA*, 78(11):66–75.

Malley, J.P., Jr., J.K. Edzwald, and N.M. Ram. 1986. Preoxidant Effects on Organic Halide Formation and Granular Activated Carbon Adsorption of Organic Halide Precursors. In *Proc. 1986 AWWA Annual Conference.* Denver, Colo.: American Water Works Association.

Miltner, R.J., H.M. Shukairy, and R.S. Summers. 1992. Disinfection By-Product Formation and Control by Ozonation and Biotreatment. *Jour. AWWA*, 84(11):53–62.

Miltner, R.J., and R.S. Summers. 1992. A Pilot-Scale Study of Biological Treatment. In *Proc. 1992 AWWA Annual Conference.* Denver, Colo.: American Water Works Association.

Norwood, D.L., J.D. Johnson, R.F. Christman, and D.S. Millington. 1983. Chlorination Products From Aquatic Humic Material at Neutral pH. Jolley, R.L., W.A. Brungs, J.A. Cotruvo, R.B. Cumming, J.S. Mattice, and V.A. Jacobs, eds. *Water Chlorination: Environmental Impact and Health Effects*, Vol. 4. Chelsea, Mich.: Lewis Publishers.

Ozekin, K., and G.L. Amy. 1997. Threshold Levels for Bromate Formation in Drinking Water. *Ozone: Science and Engineering*, 19(4):323–337.

Paode, R.D., G.L. Amy, S.W. Krasner, R.S. Summers, and E.W. Rice. 1997. Predicting the Formation of Aldehydes and BOM. *Jour. AWWA*, 89(6):79–93.

Reckhow, D.A., J.K. Edzwald, and J.E. Tobiason. 1993. *Ozone as an Aid to Coagulation and Filtration.* Denver, Colo.: American Water Works Association Research Foundation and American Water Works Association.

Reckhow, D.A., B. Legube, and P.C. Singer. 1986. The Ozonation of Organic Halide Precursors: Effect of Bicarbonate. *Water Res.*, 20(8):987–998.

Reckhow, D.A., and P.C. Singer. 1985. Mechanisms of Organic Halide Formation During Fulvic Acid Chlorination and Implications with Respect to Preozonation. Jolley, R.L., R.J. Bull, W.P. David, S. Katz, M.H. Roberts, Jr., and V.A. Jacobs, eds. *Water Chlorination: Environmental Impact and Health Effects*, Vol. 5. Chelsea, Mich.: Lewis Publishers.

———. 1984. The Removal of Organic Halide Precursors by Preozonation and Alum Coagulation. *Jour. AWWA*, 76(4):151–157.

Reckhow, D.A., P.C. Singer, and R.R. Trussell. 1986. Ozone as a Coagulant Aid. In *Proc. AWWA Seminar on Ozonation: Recent Advances and Research Needs.* Denver, Colo.: American Water Works Association.

Reckhow, D.A., J.M. Weiner, A. MacNeill, and J.E. Tobiason. 1995. Effects of Preoxidation on the Rates of Formation of Chlorination By-products. *ACS 210th National Meeting, Environmental Chemistry Division Preprints*. American Chemical Society.

Riley, T.L., K.H. Mancy, and E.A. Boettner. 1978. The Effect of Preozonation on Chloroform Production in the Chlorine Disinfection Process. *Water Chlorination: Environmental Impact and Health Effects*, Vol. 2. Chelsea, Mich.: Lewis Publishers.

Roustan, M., J.-P. Duguet, J.-M. Lainé, Z. Do-Quang, and J. Mallevialle. 1996. Bromate Ion Formation: Impact of Ozone Contactor Hydraulics and Operating Conditions. *Ozone: Science and Engineering*, 18(1):87–97.

Schechter, D.S., and P.C. Singer. 1995. Formation of Aldehydes During Ozonation. *Ozone: Science and Engineering*, 17(1):53–69.

Siddiqui, M.S., and G.L. Amy. 1993. Factors Affecting DBP Formation During Ozone-Bromide Reactions. *Jour. AWWA*, 85(1):63–72.

Singer, P.C., and S. Chang. 1989. Impact of Ozone on the Removal of Particles, TOC, and THM Precursors. Denver, Colo.: American Water Works Association Research Foundation and American Water Works Association.

Singer, P.C., K. Robinson, and R.A. Elefritz. 1990. Ozonation at Belle Glade, Florida: A Case History. *Ozone: Science and Engineering*, 12(2):199–215.

Song, R., C. Donohoe, R. Minear, P. Westerhoff, K. Ozekin, and G.L. Amy. 1996. Empirical Modeling of Bromate Formation During Ozonation of Bromide-Containing Waters. *Water Res.*, 30(5):1161–1168.

Song, R., P. Westerhoff, R. Minear, and G.L. Amy. 1997. Bromate Minimization During Ozonation. *Jour. AWWA*, 89(6):69–78.

Staehelin, J., and J. Hoigné. 1983. Mechanism and Kinetics of Decomposition of Ozone in Water in the Presence of Organic Solutes. *Vom Wasser*, 61:337.

Tate, C.H. 1991. Survey of Ozone Installations in North America. *Jour. AWWA*, 83(5):40–47.

Tobiason, J.E. 1994. Optimal Use of Ozonation and GAC in Direct Filtration for Drinking Water Treatment. Klute, R., and H.H. Hahn, eds. In *Chemical Water and Wastewater Treatment III*. Berlin: Springer-Verlag.

Tobiason, J.E., J.K. Edzwald, D.A. Reckhow, and M.S. Switzenbaum. 1993. Effect of Preozonation on Organics Removal by In-line Direct Filtration. *Water Sci. & Tech.*, 27(11):81–90.

Tobiason, J.E., J.K. Edzwald, O.D. Schneider, M.B. Fox, and H.J. Dunn. 1992. Pilot Study of the Effects of Ozone and PEROXONE on In-Line Direct Filtration. *Jour. AWWA*, 84(12):72–84.

Tobiason, J.E., D.A. Reckhow, and J.K. Edzwald. 1995. Effects of Ozonation on Optimal Coagulant Dosing in Drinking Water Treatment. *Aqua*, 44(3):142–150.

Umphres, M.D., R.R. Trussell, A.R. Trussell, L.Y.C. Leong, and C.H. Tate. 1979. The Effects of Preozonation on the Formation of Trihalomethanes. *Ozonews*, 6:3, part 2:1–4.

van der Kooij, D., W.A.M. Hijnen, and J.C. Kruithof. 1989. The Effects of Ozonation, Biological Filtration and Distribution on the Concentration of Easily Assimilable Organic Carbon (AOC) in Drinking Water. *Ozone: Science and Engineering*, 11(3):297–311.

von Gunten, U., A. Bruchet, and E. Costentin. 1996. Bromate Formation in Advanced Oxidation Processes. *Jour. AWWA*, 88(6):53–65.

von Gunten, U., and J. Hoigné. 1994. Bromate Formation During Ozonation of Bromide-Containing Waters: Interaction of Ozone and Hydroxyl Radical Reactions. *Environ. Sci. & Tech.*, 28(7):1234–1242.

Weiner, J.M. 1995. Effects of Ozone on Trihalomethane Formation: Pilot Plant Processes and Kinetics. MS thesis. Amherst, Mass.: University of Massachusetts at Amherst.

Yannoni, C.C., Q. Zhu, and S.D. Clark. 1997. Pilot-scale Evaluation of Ozone, DAF, GAC Filtration and Chloramination on Removal of Particles and Disinfection Byproducts. In *Proc. 1997 Water Quality Technology Conference*. Denver, Colo.: American Water Works Association.

Zhou, X., D.A. Reckhow, and J.E. Tobiason. 1993. Formation and Removal of Aldehyde in Drinking Water Treatment Processes. In *Proc. Water Quality Technology Conference*. Denver, Colo.: American Water Works Association.

CHAPTER • TEN

Control of Disinfection By-Product Formation Using Chlorine Dioxide

ROBERT C. HOEHN
DON J. GATES

INTRODUCTION

As many as 500 water utilities use chlorine dioxide (ClO_2) full time in the United States and even more use it either on a short-term basis or seasonally. Reasons for its use are varied, ranging from taste-and-odor control to iron and manganese oxidation, disinfection, and disinfection by-product (DBP) control. For many years, its primary use was for control of odors caused by phenols, chlorophenols, and algal products (Aston 1947; Mounsey and Hagar 1946; Cardey 1953), but Dietrich et al. (1992) found that most utilities using ClO_2 today use it as part of their overall DBP-control strategy. More recently, about 13 percent of 340 utilities participating in the Information Collection Rule (ICR) reported using ClO_2 for pretreatment (Bissonette and

Allgeier 1997). Reports that ClO_2 is effective against *Cryptosporidium*, particularly when used in conjunction with chloramine and ozone (Liyanage, Finch, and Belosevic 1997; Liyanage et al. 1997), have fostered renewed interest in its use as a disinfectant.

In the late 1970s, USEPA (1977) listed ClO_2 as an alternative to chlorine that would aid utilities in controlling THMs, but they expressed concern over the potential health effects of the organic and inorganic ClO_2 by-products (Bull 1982). As a precaution, they recommended that the total residual concentrations of ClO_2, chlorite ion (ClO_2^-), and chlorate ion (ClO_3^-) be limited to 0.5 mg/L (*Federal Register* 1979). In 1983, the recommended level was increased to 1.0 mg/L (USEPA 1983). The Disinfectant/Disinfection By-Product (D/DBP) Final Rule (*Federal Register* 1998) established a maximum contaminant level (MCL) for ClO_2^- at 1.0 mg/L (average of three distribution-system concentrations) and a maximum residual disinfectant level (MRDL) for ClO_2 of 0.8 mg/L at the entrance to the distribution system. The MCL for ClO_2^- may be revised once data from comprehensive, now-completed, health-effects studies are evaluated (Gates and Harrington 1995), but it is not likely to be exceeded if the ClO_2 dosage at the treatment plant is below 1.0–1.4 mg/L. Lacking health-effects data, USEPA delayed recommending an MCL for ClO_3^- in the 1994 proposed D/DBP Rule, but studies are under way and an MCL is expected in the future.

Organic by-products of ClO_2 reactions have not yet been addressed in the regulations, but available data suggest that their concentrations are as much as 1,000 times lower than the inorganic by-product concentrations (Richardson 1998) and are not likely to pose health problems. Several researchers (Colclough et al. 1983; Richardson et al. 1994; Stevens, Seeger, and Slocum 1978; Werdehoff and Singer 1987) detected halogenated organic compounds, expressed as total organic halogen (TOX), in water treated with ClO_2, but concentrations were generally low. Lykins, Goodrich, and Hoff (1990) stated that TOX concentrations in ClO_2-treated waters are usually from 20 to 50 times lower than those in chlorinated water. According to Gordon (1993) and Masschelein (1979), however, organic halogen compounds formed during ClO_2 treatment most likely were products of reactions involving trace quantities of free chlorine in generated ClO_2 solutions.

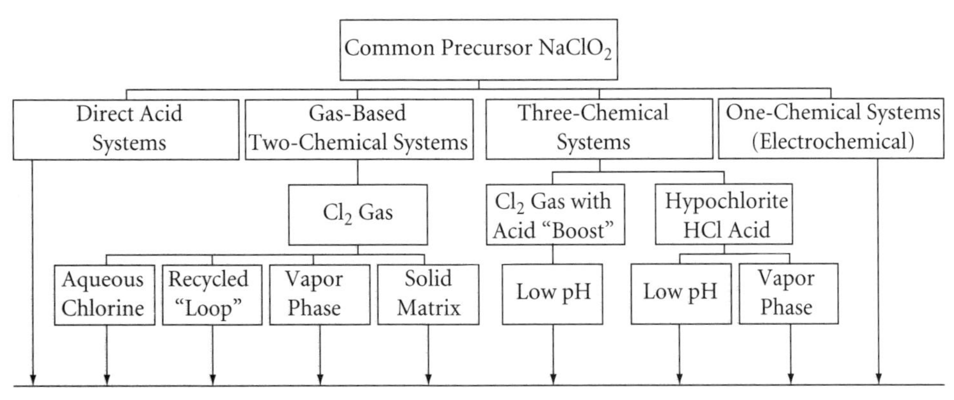

Source: After Gates (1998).

Figure 10-1 Chlorite-based chlorine dioxide generation systems

CHARACTERISTICS, GENERATION, AND CHEMISTRY OF CHLORINE DIOXIDE

In several respects, ClO_2 is unique as a drinking water oxidant and disinfectant in that (1) it is a gas at temperatures above 11°C and readily volatilizes in open basins, (2) its odor is offensively chlorinous at low concentrations, and (3) it decomposes rapidly in light, a characteristic that leads to rapid residual loss and ClO_3^- formation in uncovered basins (Zika et al. 1985).

In contrast to chlorine, which is readily compressible for shipping and storage, ClO_2 is unstable at extremely high concentrations and must be generated on site. Several generators that produce ClO_2 from sodium chlorite ($NaClO_2$) are available (Figure 10-1), but few can achieve high-percentage ClO_2 yields without excess free chlorine (Gordon and Rosenblatt 1995). Production of ClO_2 from $NaClO_2$ solutions occurs in most generators by reaction with either gaseous chlorine (Cl_2) or strong acid solutions of hypochlorous acid (HOCl) according to the following reactions (Gordon and Bubnis 1996):

$$2NaClO_2 + Cl_2(g) \rightarrow 2ClO_2(g) + 2NaCl \qquad (10\text{-}1)$$

$$2NaClO_2 + HOCl \rightarrow 2ClO_2(g) + NaCl + NaOH \qquad (10\text{-}2)$$

Optimal generator-operating conditions (e.g., concentration of generated ClO_2 and pH of the generator effluent mixture) vary considerably with the design of each particular generator. For instance, not all generating systems shown in Figure 10-1 operate like some of the earlier ones that required low pH and excess chlorine to overcome the basicity of $NaClO_2$ feedstock and caustic that is formed according to Eq 10-2. One particular manufacturer's generator reacts 25 percent aqueous $NaClO_2$ with stoichiometric quantities of $Cl_2(g)$ under partial vacuum to produce ClO_2 by reaction 10-1. Another manufacturer produces ClO_2 by reacting solid $NaClO_2$ with humidified $Cl_2(g)$ (Gordon and Rosenblatt 1995). Because $Cl_2(g)$ hydrolyzes in water and forms HOCl and HCl, conventional systems that prereact $Cl_2(g)$ with water acidify $NaClO_2$ directly, which accounts, in part, for their slightly reduced efficiency. The reaction that occurs in that type of system is

$$5NaClO_2 + 4HCl \rightarrow 4ClO_2(g) + 2H_2O + 5NaCl \quad (10\text{-}3)$$

During reactions 10-1 and 10-2, an intermediate ($[Cl_2O_2]$) is formed that becomes important in by-product formation. Gordon and Rosenblatt (1995) described the reaction as:

$$Cl_2(g) + ClO_2 \rightarrow [Cl_2O_2] + Cl^- \quad (10\text{-}4)$$

Early dilution of the $[Cl_2O_2]$ during the generation process will cause ClO_3^- to form, but if little dilution occurs and the concentrations of both reactants are high, the intermediate will give rise to ClO_2. Aqueous ClO_2 solutions are relatively stable when diluted to approximately 5 g/L or less with cold, demand-free water and kept in the dark (Kaczur and Cawlfield 1993).

Under conditions of either low reactant concentrations or excess chlorine, the intermediate $[Cl_2O_2]$ can form ClO_3^- according to the following equations (Gordon and Bubnis 1996):

$$[Cl_2O_2] + H_2O \rightarrow ClO_3^- + Cl^- + 2H^+ \quad (10\text{-}5)$$

$$[Cl_2O_2] + HOCl \rightarrow ClO_3^- + Cl_2^- + H^+ \quad (10\text{-}6)$$

According to Gordon and Bubnis (1996), the presence of excess ClO_2^- during the generation process favors ClO_2 production, but the presence of high chlorine concentrations favors ClO_3^- formation. Chlorine dioxide will disproportionate under highly basic conditions (e.g., pH 10–11 during precipitative softening) and form both ClO_2^- and ClO_3^-.

Because ClO_2 is highly soluble and hydrolyzes very slowly in water to form anionic species, it exists as a true undissociated, highly energetic molecule in aqueous solution. Its hydrolysis rate in water is approximately 10 million times slower than that of chlorine, and, as a result, its efficacy as a disinfectant, unlike chlorine, extends over a broad pH range (Gordon 1993). Several more detailed reviews of fundamental ClO_2 generation chemistry and ClO_2 reactions in water are available elsewhere (Gordon, Keiffer, and Rosenblatt 1972; Aieta and Berg 1986; Aieta and Roberts 1985; Gordon and Rosenblatt 1995).

INORGANIC BY-PRODUCTS OF CHLORINE DIOXIDE TREATMENT

When applied to surface water or groundwaters, ClO_2 reacts rapidly with naturally occurring, oxidizable organic material. As much as 70 percent of the dose is reduced to ClO_2^-, and the remainder is either ClO_3^- or Cl^- (Werdehoff and Singer 1987; Singer and O'Neil 1987; Aieta and Berg 1986). The major oxidant to which humans who drink ClO_2-treated water will be exposed, therefore, will be ClO_2^- rather than ClO_2 (USEPA 1994). While sensitive ClO_2^- and ClO_3^- analytical methods are currently available (e.g., ion chromatography), reliable and easy-to-use methods for analyzing low ClO_2 levels have yet to be developed, a fact that has caused concern among regulators (Fair 1992; Bellamy and McGuire 1994).

During the D/DBP negotiated rulemaking discussions in the early 1990s, technical workgroup members ascertained that the upper ClO_2^- exposure level in systems that use ClO_2 was in the 1.2–1.4 mg/L range, with a median of 0.6–0.7 mg/L (Krasner and Bubnis, personal communication). The mean and median ClO_2^- concentrations of data contributed by the Chlorine Dioxide Panel of the Chemical Manufacturers Association, representing 65 utilities serviced by a single generator manufacturer, were 0.58 and

Table 10-1 Finished water concentrations of chlorite and chlorate ions in the United States

Survey	ClO_2^- (*mg/L*)	ClO_3^- (*mg/L*)	Number of Utilities
2-year Running Quarterly Average, Single System (Bubnis 1997)	0.45–0.60	0.14–0.21	1
Texas Users' Average (Bubnis 1997)	0.21–0.86	0.08–0.48	18
Virginia Tech./AWWARF (Gallagher, Hoehn, and Dietrich 1994)	0.4–0.8	0.1–0.2	5
Chlorine Dioxide Panel, CMA (USEPA 1997)	0.01–5.36*	0.01–4.42†	65

*Mean 0.58 mg/L, median 0.52 mg/L.
†Mean 0.40 mg/L, median 0.23 mg/L.

0.52 mg/L, respectively; the 90th percentile was 1.02 mg/L, which is only slightly greater than the proposed MCL (USEPA 1997).

The purity of the ClO_2 added to water, in addition to the concentration of free chlorine added to ClO_2-treated finished water for disinfection, can influence ClO_2^- and ClO_3^- concentrations in tap water (Aieta et al. 1984). In addition, free chlorine and dilute ClO_2^- can react in the distribution system to produce low concentrations of both ClO_2 and ClO_3^- (Aieta and Berg 1986; Werdehoff and Singer 1987; Hoehn et al. 1990; Slootmaekers et al. 1990). Free ClO_2, even at low concentrations, can escape when customers open taps in their homes or business and cause strong chlorinous odors. If new carpet has recently been installed, the escaping ClO_2 can react with organic compounds released from the carpeting to produce offensive odors described as being similar to cat urine or kerosene (Hoehn et al. 1990). Sources of organic compounds other than carpeting have also been implicated.

Utilities in the United States generally apply ClO_2 at concentrations below 1.5 mg/L to ensure that they meet the USEPA-recommended 1.0 mg/L total residual oxidant concentration, and, generally, they are successful (Table 10-1). European countries tend to be more accepting of ClO_2 for water treatment, and some allow much higher ClO_2 (and, presumably, ClO_2^-) concentrations in finished water (Table 10-2) than are allowed in the United States.

Chlorate ion, once formed, cannot be removed from drinking water by conventional water treatment techniques; thus, its control

Table 10-2 Permissible pretreatment and post-treatment chlorine dioxide dosages in European countries

Country	Maximum Applied Dose (*mg/L*)	Acceptable Posttreatment Dose (*mg/L*)
Belgium	0.41	0.2–2
France	—	0.8–1.5
Germany	0.4	—
Italy	2	0.1–1.4
Netherlands	—	0.2–0.5
Portugal	—	0.1–1.4
Spain	2–18.5	0.2–3.0

Source: European Commission Directorate General Joint Research Committee (1996).

requires that steps be taken to minimize its formation. In contrast to ClO_3^-, ClO_2^- can be minimized either by optimizing generator operation or by reacting it with reducing agents such as ferrous iron (Iatrou and Knocke 1992; Hurst and Knocke 1997) and reduced-sulfur compounds (e.g., sulfite) (Slootmaekers et al. 1990).

Generator optimization is one of the simpler inorganic by-product control strategies. Unless generators are maintained properly and operated correctly, the generator effluents can contain excess chlorine, ClO_2^-, and ClO_3^-. If the pH is too low, conventional aqueous-chlorine/ClO_2 generators (Figure 10-1) can produce high ClO_3^- concentrations by reactions previously described. In some gas-phase generators (Figure 10-1), pH control is not essential.

The benefits of proper generator maintenance and operation were demonstrated in studies by Gallagher et al. (1995) at several well-run treatment facilities in the United States. Most (60–85 percent) of the ClO_3^- in finished water at these utilities did not appear until after final chlorination and, presumably, was produced by the reaction of residual ClO_2^- and HOCl according to Eq 10-6. The ClO_2^- was formed when ClO_2 was reduced during reactions with natural organic matter (NOM) in the raw water. Singer and O'Neil (1987) performed a mass balance on ClO_2 and its by-products and attributed the unaccounted-for ClO_2^- following ClO_2 addition and postchlorination to the formation of ClO_3^-.

Protection of ClO_2-treated water from light exposure is another by-product control strategy (Zika et al. 1985; Griese et al. 1991). When exposed to sunlight, or even fluorescent light, ClO_2 in water will be rapidly photolyzed, and ClO_3^- will be produced. Chlorine dioxide could be used more efficiently, therefore, if it were applied to water either in pipes or covered basins.

Iatrou and Knocke (1992) demonstrated that ClO_2^- could be rapidly reduced to Cl^- by ferrous iron at $Fe:ClO_2^-$ ratios of 3:1 (weight per weight [w/w]) in water at pH 5 to 7 and that the reaction was much faster than Fe^{2+} oxidation by dissolved oxygen. At pH 7 to 10, ratios ranging from 3.5:1 to 4:1 were required (Hurst and Knocke 1997). Because particulate ferric hydroxide is produced when Fe^{2+} is oxidized, ferrous salts should be added in the treatment train before solid–liquid separation processes, such as sedimentation and filtration.

Reduced sulfur compounds also can eliminate ClO_2^-, but reaction rates are slower and higher reductant-to-ClO_2^- ratios (10:1) are generally required (Dixon and Lee 1991; Slootmaekers et al. 1990; Griese et al. 1991). As is true for ClO_2^- reduction with Fe^{2+}, the rate of ClO_2^- reduction by reduced-sulfur compounds decreases markedly with increasing pH above pH 7.5. An advantage these compounds offer is that no solids are formed, so they can be added either immediately before or after filtration.

ORGANIC BY-PRODUCTS OF CHLORINE DIOXIDE REACTIONS

In the early 1980s, organic DBP production during ClO_2 treatment was discussed in *Treatment Techniques for Controlling Trihalomethanes* (AWWA 1982), and several preliminary observations were cited. Since then, organic by-product formation has been the subject of only a few investigations, and most research has focused on minimizing trihalomethane (THM) formation by substituting ClO_2 for chlorine early in the treatment process.

Raw water treatment with ClO_2 can reduce THM formation by oxidizing the organic precursors, but, in most instances, THM reduction has come about because utilities substituted ClO_2 for chlorine at the head of the treatment plant and did not chlorinate until some of the THM precursors were removed by coagulation, flocculation, and settling. During the 1980s, "delayed chlorination," either with or without ClO_2 pretreatment, became a popular THM-control practice, and many studies followed. Only a few, however, dealt specifically with THM-precursor oxidation.

Several of the earlier investigations of ClO_2 as an alternative to chlorine involved treatment of Ohio River water with ClO_2 (Stevens, Seeger, and Slocum 1978; Miltner 1976; Lykins and Griese 1986). Like most investigations that followed, the focus of the earlier studies was more on reduction in THM formation by delayed chlorination than on precursor oxidation by ClO_2. Miltner (1976), however, demonstrated that substituting ClO_2 for a portion of the final chlorine dose reduced the overall THM formation potential (THMFP) of the finished water.

Werdehoff and Singer (1987) found that ClO_2 treatment reduced the THMFP of fulvic acid solutions by 10 percent or more when the ClO_2-to-TOC ratio exceeded 0.4 (w/w). However, TOX concentrations increased slightly, a phenomenon they attributed to the formation of HOCl, when ClO_2 reacted with phenolic groups within the fulvic acid molecules. Wajon, Rosenblatt, and Burrows (1982) and Noack and Doerr (1978) also observed significant THM-precursor destruction following ClO_2 treatment, but they applied the oxidant in combination with chlorine at varying ratios and the percentage destruction increased with increasing ClO_2:Cl_2 ratios. Li et al. (1996) also demonstrated that the THMFP of solutions containing organic matter (commercial humic acid) could be reduced by treatment with mixtures of ClO_2 and Cl_2 in varying ratios.

Precursor concentrations may not always be reduced, however, following ClO_2 addition. Shelton (1996), for example, found no differences in the 48-hour THM and haloacetic acid (HAA) formation potentials in softened water that had been pretreated with ClO_2 prior to softening.

Effects of Bromide Ion and Light on THM Formation

Li and co-workers (1996) found that ClO_2 treatment of bromide-containing, humic-acid solutions kept in the dark produced bromoform ($CHBr_3$), which suggests either that ClO_2 can oxidize bromide ion to HOBr under certain conditions or that the ClO_2

solutions were contaminated with chlorine. They also showed that irradiation decreased the overall THM concentrations, most of which was $CHBr_3$.

Zika et al. (1985) demonstrated that light catalyzed the formation of $CHBr_3$ in three Floridian groundwaters containing Br^- treated with 10 mg/L ClO_2. They reported also that BrO_3^- was formed during reactions between ClO_2 and Br^- but only if the solutions were free of organic matter and exposed to light. The authors noted that bicarbonate and carbonate would quench halogen-radical formation by irradiation of ClO_2, thereby limiting $CHBr_3$ formation in natural water.

Several investigators have studied reactions between ClO_2 and numerous model organic compounds (e.g., Rice and Gomez-Taylor 1986; Masschelein 1979), but the reaction conditions seldom resembled those typically found at water treatment plants. For this review, therefore, the results, while interesting, are of questionable practical value. Only a few investigations of organic by-product formation following ClO_2 treatment of natural waters have been conducted. Other investigations have involved treatment of simulated natural water prepared by the addition of either fulvic acids recovered from natural waters or commercial humic acids to buffered distilled water.

Three studies (Stevens, Seeger, and Slocum 1978; Colclough et al. 1983; and Richardson et al. 1994) best illustrate the current state of knowledge regarding organic DBP formation resulting from ClO_2 reactions with natural organics (Table 10-3). Each investigation team reported recovering halogenated organic compounds, though the concentrations were always much less than those recovered from chlorinated water.

Stevens, Seeger, and Slocum (1978) were the first to detect aldehydes in Cl_2-treated water. Their studies involved treatment of Ohio River water in a USEPA pilot plant in Cincinnati, Ohio. Treated samples, along with appropriate control samples, were analyzed for organic DBPs by gas chromatography/mass spectroscopy (GC/MS).

Colclough et al. (1983) treated simulated natural water containing Black Lake (North Carolina) fulvic acid. Following ClO_2 treatment, samples were extracted with two organic solvents,

Table 10-3 Organic compound classes (excluding haloforms) recovered from ClO_2-treated natural waters or fulvic acid solutions

Compound Class	Stevens, Seeger, and Slocum (1978)	Colclough et al. (1983)	Richardson et al. (1994)
Aldehydes	X	—	X
Aromatic Compounds	—	—	X
Halogenated Aromatic Compounds	—	—	X
Carboxylic acids	—	X	X
Esters	—	—	X
Haloacids	—	X	—
Haloketones	—	—	X
Aromatic Compounds	—	—	X
Ketones	—	—	X

dried, and further concentrated. The concentrated samples were then methylated and analyzed by GC/MS. Recovered compounds included monomethyl and dimethyl esters of four carboxylic acid classes (benzenepolycarboxylic acids, aliphatic dibasic carboxylic acids, carboxyphenylgloxylic acids, and aliphatic carboxylic acids). In addition, trace amounts of two chlorinated compounds (dimethyl esters of monochloromalonic acid and monochlorosuccinic acid) and significant amounts of a third (dichloroacetic acid) were found. The authors did not detect chlorine by spectral analysis of the ClO_2 stock solution, so they concluded that the chlorine-containing products were ClO_2-reaction products and not the result of chlorine contamination.

Richardson et al. (1994) also studied organic DBP formation following ClO_2 treatment of Ohio River water in pilot plant studies, but their work was conducted at Evansville, Ind. Raw and treated water samples were concentrated on XAD resins (Rohm and Haas, Philadelphia, Pa.), extracted with two solvents, dried, and reconcentrated prior to analysis by GC/MS. Derivatization was purposefully avoided so that the researchers could determine directly the semivolatile compounds formed during treatment. The most abundant compounds isolated from the ClO_2-treated samples were carboxylic acids, including two maleic acids (as the anhydrides). Compounds found at lower levels included two unsaturated ketones, five aromatic compounds (styrene and naphthalene isomers), and one ester. One halogenated aromatic compound, tentatively identified as (1-chloroethyl)dimethyl benzene, and two haloketones also were isolated. According to Richardson (1998), these products were semiquantitatively determined to be in the 1 ng/L to 10 ng/L range.

SUMMARY

Chlorine dioxide will continue to play a significant role in the water treatment industry throughout the United States, Canada, and Europe, especially because it is primarily an oxidizing agent and produces few halogenated by-products. Past studies have clearly shown that applying ClO_2 and chlorine at different points in the treatment process can significantly reduce both the overall chlorine requirement and the halogenated DBP concentrations in finished water.

While inorganic rather than organic by-products appear to be the most significant in ClO_2-treated drinking water, they most likely will not pose any chronic or acute health risks to humans if the ClO_2 dosage is kept below 1.5 mg/L. By setting limits on ClO_2 and ClO_2^- (and, ultimately, on ClO_3^-), USEPA will satisfy a legal requirement that individual concentrations of contaminants, not their sum, be regulated.

If ClO_2 is applied to raw water, its concentration in finished water should be much below the proposed MRDL, though small amounts may be regenerated in the distribution system by reactions between chlorine and residual ClO_2^-. The ClO_2^- MCL will limit the practical ClO_2 dosage unless ClO_2^- removal is included as part of the treatment process. Chlorate cannot be removed once it is formed, but its concentration can be controlled by optimizing ClO_2 generation and by shielding ClO_2-containing water from light exposure.

Despite early indications that organic by-products appear only in extremely low concentrations in ClO_2-treated water supplies, the number of comprehensive studies is small and the types of waters examined have been few. Additional studies with natural waters of varying water-quality characteristics (e.g., different types of NOM at varying concentrations both with and without bromide ion) collected from a variety of geographical locations are needed for verification of existing data. If the recovered organic by-products are similar in nature and concentration (ng/L) to those already identified, one might safely conclude that the organic ClO_2 by-products are inconsequential insofar as their human health effects are concerned.

Sensitive, reliable, and inexpensive methods are needed for the analysis of ClO_2, ClO_2^-, and ClO_3^- in waters that have been treated with ClO_2 and then with either chlorine or chloramine. Existing methods available to water treatment plant operators are either unreliable for low-level analyses or require expensive equipment, considerable time, and considerable analytical skill. Newer, easy-to-use methods should be developed so that treatment plant operators can obtain real-time data needed for optimizing ClO_2 treatment and maintaining by-product concentrations at levels prescribed by the regulations.

Finally, ClO_2 is more effective against certain waterborne pathogens (e.g., *Cryptosporidium*) than either chlorine or monochloramine, though evidence is emerging that disinfectant combinations (e.g., ClO_2 followed by postchlorination or chloramination) are more effective than either of the disinfectants alone. Consequently, pathogen inactivation with ClO_2 may prove to be practical at utilities for which ozonation is not a realistic option. In such instances, however, ClO_2^- removal likely will be required because the ClO_2 dosages will have to be high enough to ensure pathogen inactivation. For that reason, plant-scale studies are needed for evaluating ClO_2^--removal techniques and for developing guidance for operators.

REFERENCES

Aieta, E.M., and J.D. Berg. 1986. A Review of Chlorine Dioxide in Drinking Water Treatment. *Jour. AWWA*, 78(6):62–72.

Aieta, E.M., and P.V. Roberts. 1985. The Chemistry of Oxo-Chlorine Compounds Relevant to Chlorine Dioxide Generation. In *Water Chlorination: Chemistry, Environmental Impact and Health Effects*, Vol. 5. Jolley, R.L., et al., eds. Chelsea, Mich.: Lewis Publishers.

Aieta, E.M., P.V. Roberts, and M. Hernandez. 1984. Determination of Chlorine Dioxide, Chlorite and Chlorate in Water. *Jour. AWWA*, 76(1):64–70.

Aston, R.N. 1947. Chlorine Dioxide Use in Plants on the Niagara Border. *Jour. AWWA*, 40:1207.

AWWA. 1982. *Treatment Techniques for Controlling Trihalomethanes in Drinking Water*. Denver, Colo.: American Water Works Association.

Bellamy, W.D., and M.J. McGuire. 1994. Report from Expert Workshop on Microbial and Disinfection By-product Research Needs. Sponsored by American Water Works Association Research Foundation and American Water Works Association Water Quality Council.

Bissonette, E., and S. Allgeier. 1997. Summary of Disinfection Application Practices, Evaluated Through the Analysis of ICR Plant Schematics and Small Systems CPRE Data. Presented at the Disinfectant/Disinfection By-Products Rule Meeting of FACA. Cincinnati, Ohio: USEPA Office of Ground Water and Drinking Water, Technical Support Center.

Bubnis, B. 1997. NOVATEK. Personal communication.

Bull, R.J. 1982. Health Effects of Drinking Water Disinfectants and Disinfectant Byproducts. *Environ. Sci. & Tech.*, 16(10):554–559.

Cardey, F. 1953. Treatment of Residual Phenolated Waters with Chloroperoxide. *Eau*, 40:75.

Colclough, C.A., J.D. Johnson, R.F. Christman, and D.S. Millington. 1983. Organic Reaction Products of Chlorine Dioxide and Natural Aquatic Fulvic Acids. In *Water Chlorination: Environmental Impact and Health Effects*, Vol. 4. Jolley, R.L., ed. Ann Arbor, Mich.: Ann Arbor Science Publishers.

Dietrich, A.M., M.P. Orr, D.L. Gallagher, and R.C. Hoehn. 1992. Tastes and Odors Associated With Chlorine Dioxide. *Jour. AWWA*, 92(6):82–88.

Dixon, K., and R. Lee. 1991. The Effect of Sulfur-Based Reducing Agents and GAC Filtration on Chlorine Dioxide By-Products. *Jour. AWWA*, 83(5):48–55.

European Commission Directorate General Joint Research Committee, Environment Institute. 1996. An Assessment of the Presence of Trihalomethanes (THMs) in Water Intended for Human Consumption and the Practical Means to Reduce Their Concentrations Without Compromising Disinfection Efficiency. Report to Directorate General for Environment, Nuclear Safety, and Civil Protection (DG XI) of the European Commission.

Fair, P. 1992. Status Report on Analytical Methods. Washington, D.C.: USEPA, Office of Ground Water and Drinking Water.

Federal Register. 1979. National Interim Primary Drinking Water Regulations; Control of Trihalomethanes in Drinking Water. *Fed. Reg.*, 44(231): 68624–68707.

———. 1997. National Primary Drinking Water Regulations: Disinfectants and Disinfection By-Products Notice of Data Availability, Proposed Rule. *Fed. Reg.*, 62:59388.

———. 1998. National Primary Drinking Water Regulations: Disinfectants and Disinfection By-products Final Rule. *Fed. Reg.*, 63(241):69390–69476. 40 CFR Parts 9, 141 and 142.

Gallagher, D.L., R.C. Hoehn, and A.M. Dietrich. 1994. Sources, Occurrence, and Control of Chlorine Dioxide By-Product Residuals in Drinking Water. Denver, Colo.: American Water Works Association Research Foundation and American Water Works Association.

Gates, D.J. 1998. The Chlorine Dioxide Handbook. Denver, Colo., American Water Works Association.

Gates, D.J., and R.M. Harrington. 1995. Neuro-Reproductive Toxicity Issues Concerning Chlorine Dioxide and the Chlorite Ion in Public Drinking Water Supplies. In *Proc. 1995 AWWA Water Quality Technology Conference.* Denver, Colo.: American Water Works Association.

Gordon, G., and B. Bubnis. 1996. Chlorine Dioxide Chemistry Issues. In *Chlorine Dioxide: Drinking Water, Process Water and Wastewater Issues.* Proc. of the Third International Symposium, Chemical Manufacturers Association, American Water Works Association Research Foundation and US Environmental Protection Agency. Denver, Colo.: American Water Works Association.

Gordon, G., R.G. Keiffer, and D.H. Rosenblatt. 1972. The Chemistry of Chlorine Dioxide. In *Progress in Inorganic Chemistry,* Vol. 15. Lippard, S.J., ed. New York: Wiley Intersciences.

Gordon, G. 1992. Water Chemistry of the Oxy-Chlorine Species. *Chlorine Dioxide: Drinking Water Issues.* Second Int'l Symposium. Denver, Colo.: American Water Works Association.

Gordon, G., and A.A. Rosenblatt. 1995. Gaseous Chlorine-Free Chlorine Dioxide for Drinking Water. In *Proc. 1995 AWWA Water Quality Technology Conference.* Denver, Colo.: American Water Works Association.

Griese, M.H., K. Hauser, M. Berkemeier, and G. Gordon. 1991. Using Reducing Agents to Eliminate Chlorine Dioxide and ClO_2^- Ion Residuals in Drinking Water. *Jour. AWWA,* 83(5):56–61.

Hoehn, R.C., A.M. Dietrich, W.S. Farmer, M.P. Orr, R.G. Lee, E.M. Aieta, D.W. Wood III, and G. Gordon. 1990. Household Odors Associated with the Use of Chlorine Dioxide. *Jour. AWWA,* 82(4):166–172.

Hurst, G.H., and W.R. Knocke. 1997. Evaluating Ferrous Iron for Chlorite Ion Removal. *Jour. AWWA,* 89(8):98–105.

Iatrou, A., and W.R. Knocke. 1992. Removing Chlorite by Addition of Ferrous Iron. *Jour. AWWA,* 84(11):63–68.

Kaczur, J.J., and D.W. Cawlfield. 1993. Chlorous Acid, Chlorites and Chlorine Dioxide (ClO_2, $HClO_2$). *Kirk-Othmer Encyclopedia of Chemical Technology,* 4th ed. 5:968–997.

Krasner, S.W., and B. Bubnis. Metropolitan Water District of Southern California and NOVATEK, respectively. Personal communication.

Li, J.W., Z. Yu, X. Cai, M. Gao, and F. Chao. 1996. Trihalomethane Formation in Water Treated with Chlorine Dioxide. *Water Res.,* 30(10):2371–2376.

Liyanage, L.R., G.R. Finch, and M. Belosevic. 1997. Sequential Disinfection of *Cryptosporidium* Parvum by Ozone and Chlorine Dioxide. *Ozone: Science and Engineering*, 19(5):409–423.

Liyanage, L.R.J., L.L. Gyürék, M. Belosevic, and G.R. Finch. 1997. Effect of Chlorine Dioxide Preconditioning on Inactivation of *Cryptosporidium* by Free Chlorine and Monochloramine: Process Design Requirements. In *Proc. 1995 AWWA Water Quality Technology Conference.* Denver, Colo.: American Water Works Association.

Lykins, B.W., and M.H. Griese. 1986. Using Chlorine Dioxide for Trihalomethane Control. *Jour. AWWA*, 78(6):88–93.

Lykins, B.W., J.A. Goodrich, and J.C. Hoff. 1990. Concerns of Using ClO_2 Disinfection in the USA. *Jour. Water SRT-Aqua*, 39:366–376.

Masschelein, W.J. 1979. *Chlorine Dioxide: Chemistry and Environmental Impact of Oxychlorine Compounds.* Rice, R.G., ed. Ann Arbor, Mich.: Ann Arbor Science Publishers.

Miltner, R.J. 1976. *The Effect of Chlorine Dioxide on Trihalomethanes in Drinking Water.* MS Thesis. Cincinnati, Ohio: Univ. of Cincinnati.

Mounsey, R.J., and M.C. Hagar. 1946. Taste and Odor Control with Chlorine Dioxide. *Jour. AWWA*, 38:1051.

Noack, M.G., and R.L. Doerr. 1978. Reactions of Chlorine, Chlorine Dioxide, and Mixtures Thereof with Humic Acid: An Interim Report. In *Water Chlorination: Environmental Impact and Health Effects*, Vol. 2. Jolley, R.L., H. Gorchev, and D.H. Hamilton, Jr., eds. Ann Arbor, Mich.: Ann Arbor Science Publishers.

Rice, R.G., and M. Gomez-Taylor. 1986. Occurrence of By-Products of Strong Oxidants Reacting with Drinking Water Contaminants - Scope of the Problem. *Envir. Health Perspectives*, 69:31–44.

Richardson, S.D. 1998. Identification of Drinking Water Disinfection By-Products. In *Encyclopedia of Environmental Analysis and Remediation*, Vol. 3, pp. 1398–1421. Meyers, R.A., ed. New York: John Wiley & Sons.

Richardson, S.D., A.D. Thruston, Jr., T.W. Collette, K. Schenck-Patterson, B.W. Lykins, Jr., G. Majetich, and Y. Zhang. 1994. Multispectral Identification of ClO_2 Disinfection Byproducts in Drinking Water. *Environ. Sci. & Tech.*, 28(4):592–599.

Shelton, S.L. 1996. Assessment of the Use of Chlorine Dioxide at the Topeka, Kansas, Water Treatment Plant. MS thesis. Lawrence, Kan.: University of Kansas.

Singer, P.C., and W.K. O'Neil. 1987. Technical Note: The Formation of Chlorate from the Reaction of Chlorine and Chlorite in Dilute Aqueous Solution. *Jour. AWWA*, 79(11):75–76.

Slootmaekers, B., S. Tachiyashiki, D. Wood III, and G. Gordon. 1990. Minimizing Chlorite Ion and Chlorate Ion in Drinking Water Treated with Chlorine Dioxide. *Jour. AWWA*, 82(4):160–165.

Stevens, A.A., D.R. Seeger, and D.J. Slocum. 1978. Products of Chlorine Dioxide Treatment of Organic Materials in Water. In *Proceedings of a Conference on Ozone/Chlorine Dioxide Oxidation Products of Organic Materials.* Rice, R.G., and J.A. Cotruvo, eds. Cincinnati, Ohio: Ozone Press International.

USEPA. 1977. *Ozone, Chlorine Dioxide, and Chloramines as Alternatives to Chlorine for Disinfection of Drinking Water.* EPA 600/D-77-003. Cincinnati, Ohio: US Environmental Protection Agency.

———. 1983. *Trihalomethanes in Drinking Water: Sampling, Analysis, Monitoring, and Compliance.* Washington, D.C.: Office of Drinking Water, US Environmental Protection Agency.

———. 1994. *External Review Draft for the Drinking Water Criteria Document on Chlorine Dioxide, Chlorite and Chlorate.* Prepared for the Health and Ecological Criteria Division, Office Science and Technol., Office of Ground Water and Drinking Water, Washington, D.C.

———. 1997. *Final Draft: Occurrence Assessment for Disinfectants and Disinfection Byproducts in Public Drinking Water Supplies.* Prepared by Science Applications International Corporation (McLean, Va.) for Office of Ground Water and Drinking Water, Washington, D.C.

Wajon, J.E., D.H. Rosenblatt, and E.P. Burrows. 1982. Oxidation of Phenol and Hydroquinone by Chlorine Dioxide. *Environ. Sci. & Tech.*, 16:396–402.

Werdehoff, K.S., and P.C. Singer. 1987. Chlorine Dioxide Effects on THMFP, TOXFP, and the Formation of Inorganic By-products. *Jour. AWWA*, 79(9):107–113.

Zika, R.G., C.A. Moore, L.T. Gidel, and W.J. Cooper. 1985. Sunlight-Induced Photodecomposition of Chlorine Dioxide. In *Water Chlorination: Chemistry, Environmental Impact, and Health Effects*, Vol. 5. Jolley, R.L., et al., eds. Chelsea, Mich.: Lewis Publishers.

CHAPTER • ELEVEN

Control of Disinfection By-Product Formation Using Ultraviolet Light

JAMES P. MALLEY, JR.

Downes and Blount (1877) were the first to recognize the effects that ultraviolet (UV) light in sunlight has on bacteria. Basic UV technology was established by 1910, when the mercury-vapor lamp enclosed in a quartz sheath was developed. The quartz sheath is used to dampen the effects of temperature changes on the light.

Increasing concerns over microbial risks from drinking water supplies have sparked new interest in alternative disinfectants such as UV irradiation. The pending Ground Water Rule (GWR) will allow the use of conventional, continuous-wave UV irradiation for compliance. This chapter focuses on the use of conventional, continuous-wave, mercury-vapor UV lamps for disinfection and its implications on disinfection by-products (DBPs) and their precursors. The chapter concludes with a brief look at emerging, innovative UV technologies.

PRINCIPLES OF UV IRRADIATION

Ultraviolet irradiation is produced in commercially available lamps by converting electrical power to light and heat. Most lamps emit UV irradiation by striking an electrical arc between filaments in a pressurized gas or vapor. The irradiance of UV emitted by any given lamp is a function of the power (photons) per unit area and is commonly reported as mW/cm^2 in the United States or $Joule/s{\cdot}m^2$ in Europe. The irradiance of a single UV lamp can be directly measured either electrically using a radiometer with appropriate wavelength filters or chemically using an actinometer (Hoyer et al. 1992). The higher the lamp power the larger the amount of waste heat generated.

For engineering applications of UV irradiation, it is also important to measure the UV dosage, which is defined as the product of irradiance and exposure time. Ultraviolet dosage is commonly reported as $mW\text{-}s/cm^2$ or mJ/cm^2. Ultraviolet dosage is independent of scale as long as plug flow conditions are maintained. Thus, it is possible to extrapolate required UV dosages from bench to pilot to full-scale applications.

The study of UV irradiation's effect on DBP formation deals with the interaction between light energy and molecules. After a molecule in its original (ground) state absorbs electromagnetic energy from a light source, the bonds in the molecule are transformed to an excited state, and chemical and physical processes become thermodynamically possible. In the excited state, the molecule becomes a more effective electron donor or acceptor. The electromagnetic radiation produced from the light source has a certain amount of energy associated with it, with the amount depending on many properties. Equation 11-1, developed by Planck in 1901, describes the energy produced based on the light frequency and wavelength (Calvert and Pitts 1966).

$$E = h\nu = \frac{hc}{\lambda} \qquad (11\text{-}1)$$

Where:

E = quantum of radiation, typically in joules
h = Planck's constant, 6.63×10^{-34} J-sec
ν = frequency, in cycles/s
c = speed of light, $\approx 3.0 \times 10^{17}$ nm/s
λ = wavelength, in nm

At least 167 kilojoules of energy is needed to break most chemical bonds. Commonly used, low-pressure, mercury-vapor UV lamps emit at 254 nm with a frequency of 1.2×10^{15} cycles/s resulting in an energy of 470 kJ/Einstein.

Another important relationship when evaluating UV irradiation effects is the Beer–Lambert law

$$A = \log_{10} \frac{I_o}{I} = \varepsilon c \qquad (11\text{-}2)$$

Where:

I = number of quanta of light transmitted through the homogeneous absorbing system per unit time, in watts
I_o = number of quanta of monochromatic (homogeneous) light produced per unit time, in watts
ε = molar extinction coefficient, or molar absorptivity, in volume/mole/length
l = distance from light source to the homogeneous absorbing system, in length
C = concentration, in mole/volume
A = absorbance

The value for the molar extinction coefficient, ε, is a characteristic of each individual chemical species. If more than one species is present, the coefficients can be summed ($\varepsilon_1 c_1 + \varepsilon_2 c_2 + \ldots \varepsilon_n c_n$) and substituted in Eq 11-2 for εc.

The drinking water UV industry uses the following modified form of the Beer–Lambert law because it is impossible to specify a single chemical concentration in a complex drinking water matrix:

$$A = \alpha l \qquad (11\text{-}3)$$

Where:

α = the absorbance coefficient and is equivalent to 2.303(A/l) and all other terms are as defined previously.

α is an aggregate estimate of UV absorbance at 254 nm for the given water quality matrix because absorbance per centimeter (A/l) is measured as specified in Standard Method 5910 (APHA, AWWA, and WPCF 1992).

Organic matter typically found in highly colored water supplies will exert a UV absorbance at 254 nm of between 0.100 and 0.800 cm^{-1} ($\alpha = 0.23 - 1.84$). Treated surface waters and typical groundwaters have significantly lower absorbances, generally ranging from 0.005 to 0.050 cm^{-1} ($\alpha = 0.01 - 0.12$). As a general rule, when the value of α for a given water sample exceeds 0.43, pretreatment will be required for successful UV disinfection.

ULTRAVIOLET SYSTEMS USED FOR DISINFECTION

An excellent review of UV disinfection of drinking water can be found in DeMers and Renner (1992). International experience with the use of UV irradiation in drinking water treatment has been well summarized in several recent articles (von Sonntag and Schuchman, 1992; Kruithof et al. 1992; Karanis et al. 1992; Hoyer et al. 1992). Key issues relating to UV disinfection follow.

Typical applications of UV disinfection in wastewater have applied dosages of 15 to 30 mW-s/cm^2 to achieve coliform reductions of up to 4 logs. Point of entry (POE) and point of use (POU) drinking water applications have used similar dosages to achieve groundwater disinfection of bacteria. More recently, concern over viruses and the forthcoming groundwater disinfection rule requirement of 4-log viral inactivation suggests that dosages between 40 and 80 mW-s/cm^2 will be required. Day-to-day verification of UV dosage is achieved by the electronic UV sensors on each UV unit. UV sensor calibration is required periodically using the MS-2 bacteriophage bioassay method (Snicer et al. 1998).

Typically, UV-light generation involves the passing of an electrical arc through a mercury vapor. This creates a 254-nm wavelength that has about 85 percent of the germicidal effectiveness of the ideal 260- to 265-nm range (Black & Veatch 1995). Low-pressure lamps are currently the most common in disinfection systems and mimic the operation of conventional fluorescent lamps (DeMers and Renner 1992). The lamps contain mercury vapor at pressures of about 10^{-5} atm and require an optimum surface operating temperature of 40 to 50°C. About 85 percent of the light energy produced from these systems is at a wavelength of 254 nm and conversion of electricity to light is typically between 35 and 40 percent (Black & Veatch 1995). A low-pressure bulb can be expected to have a life span on the order of 7,500 to 8,800 hours. Shorter life can result with increased stopping and starting, high current, or voltage fluctuations (DeMers and Renner 1992; Thampi 1990).

Medium-pressure lamps are gaining popularity because of their higher-intensity output. Medium-pressure lamps operate at a high pressure (1 to 10 atm) and at substantially higher temperatures (500 to 800°C). Given that the output is higher, one medium-pressure lamp has the potential to replace up to 25 low-pressure lamps. The spectrum emitted by medium-pressure lamps is more broadband than that of low-pressure lamps. Operating at maximum output, bulb life ranges from 2,000 to 5,000 hours (Black & Veatch 1995).

The power required to deliver a given UV dose using a medium-pressure system is higher than an equivalent low-pressure system. Because there are fewer lamps in a medium-pressure system, the costs of equipment, lamp replacement, and cleaning are lower than that of low-pressure systems. These factors contribute to make the net present value of medium-pressure systems lower.

The key design parameter for UV reaction chambers is that they must be plug flow. Dispersion in the direction of flow must be kept to a minimum. This can be accomplished by maintaining a small cross-sectional area with respect to reactor length. Conversely, turbulence perpendicular to the direction of flow is necessary to provide sufficient mixing to overcome the nonuniform UV intensity field in the reactor (Montgomery Watson 1994). The intensity at the ends of the UV lamp will be significantly less than that in the center. This effect is minor in systems with long lamps, but maintaining turbulence within the reaction chamber, particularly between banks of lamps in series, will ensure adequate disinfection (Blatchley, Wood, and Schuerch 1995).

Basic data necessary for designing a UV disinfection reactor include flows, suspended solids, initial microbial density, particulate density, UV absorbance coefficient (a measure of the UV demand of the water), inactivation rate constant (a measure of microbial sensitivity to UV radiation), and required percent inactivation. These parameters can be found by direct testing or estimated based on experience (Rice and Hoff 1981).

Lamps can be arranged in the following two different patterns: contact (or annular) design and coaxial design. In the contact or annular design, water flows between closely spaced, transparent UV lamps enclosed in larger quartz sleeves submerged in the water channel or pipeline (Scheible 1987). In the coaxial design, water flows through thin-walled polytetrafluoroethylene (PTFE) or quartz conduits, which are transparent to the UV light produced by coaxial lamps. The conduit material will affect the dose, since PTFE and quartz have different transparencies to UV light. The contact designs are the most widely used at present (Scheible 1987).

The primary advantage of UV irradiation is that the actual detention time of the water in the UV reactor is seconds. Thus, money is saved in not having to construct large contact chambers and piping. Furthermore, the hydraulic gradient is not disrupted in the treatment system because the UV lamps can be placed in-line and produce negligible head loss. When a large contact chamber is used to provide adequate disinfectant contact time, additional pumping is then required to maintain pressure throughout the distribution system.

Several disadvantages to UV technology have also been reported. The most difficult problem in using UV irradiation is determining the actual UV dose the water receives, because measuring a residual is not possible. Modern UV system monitors have been improved to continuously monitor the UV light intensity and automatically control the operation of the system. Fouling of the lamp surfaces with mineral deposits or biofilms has been observed. Many utilities have begun to use new automated cleaning systems that can eliminate the fouling problem. Biological regrowth can occur in UV disinfection systems that do not provide for a residual disinfectant (Wolfe 1990). For many water supplies, the use of UV irradiation followed by low dosages of residual disinfectants will be more reliable and cost-effective than using higher dosages of chemical disinfectants alone. Hydraulic short-circuiting through the UV chambers may be a problem; baffles are normally installed to balance the flow through the reactor. Tracer studies for chemical reactors or the use of computational fluid dynamics (CFD) in pipeline reactors allow the designer to determine how well the reaction chamber is mimicking plug-flow conditions.

DISINFECTION BY-PRODUCTS FROM UV IRRADIATION

Little information exists in the literature on by-products formed by UV irradiation in water treatment applications. Montgomery Watson Inc. (1994) concluded in one study that UV irradiation of reclaimed wastewater did not form trihalomethanes (THMs), significantly reduced the formation of aldehydes compared to chlorine, and did not form other regulated DBPs.

Amy, Siddiqui, and Ozekin (1994) found that under certain conditions UV irradiation of surface waters may initiate a series of reduction reactions whereby bromate can be reduced to bromide. Since bromate only absorbs UV irradiation in the 200- to 240-nm range, these reactions would not be initiated by typical low-pressure mercury lamps as shown by Malley, Shaw, and Ropp (1995). However, medium-pressure lamps may cause such reactions.

Malley, Shaw, and Ropp (1995) conducted a study for the American Water Works Association Research Foundation (AWWARF) examining the effects of continuous-wave, low-pressure mercury vapor, UV irradiation on DBPs and their precursors. The study examined 20 groundwater and 10 surface water supplies from throughout the United States. The groundwater studies looked at DBPs from two perspectives. The first involved whether or not UV irradiation alone would cause the formation of DBPs. Of particular interest was determining if UV would form similar by-products to those formed by ozonation or advanced oxidation processes (AOPs). Therefore, the 20 samples were analyzed for aldehydes and ketones before and after UV irradiation. With the exception of one source, which contained up to 24 mg/L of dissolved organic carbon (DOC) and was highly colored, no by-product formation was observed. In the high-DOC source, low levels of formaldehyde (12 μg/L) were measured in duplicate experiments. Gas chromatography–electron capture detector (GC–ECD) chromatograms before and after UV irradiation were also compared for the 20 groundwaters and no significant shifts or unknown peak formations were observed.

The influence of UV on postdisinfectant DBP formation was also examined for the 20 groundwater samples using simulated distribution system (SDS) DBP tests with chlorine before and after

UV irradiation. The data (Figure 11-1) indicate that UV irradiation did not significantly alter the SDSDBP formation by chlorine.

The research also examined the effects of UV irradiation on nitrate (nitrite), bromide, iron, and manganese. Statistical comparison of these data suggested that UV irradiation did not significantly affect the nitrate, nitrite, bromide, iron, or manganese concentrations in the groundwaters tested. In none of the groundwaters was the conversion of bromide to bromate or brominated organics by UV irradiation observed. Low-pressure, mercury-vapor lamps emitting at 254 nm also did not affect the levels of bromate spiked into a limited number of samples before UV dosing.

UV irradiation was found to produce low levels of formaldehyde in the majority of the surface waters studied (Figure 11-2). The highest formaldehyde concentrations were found in the untreated surface waters. The photochemical literature (Ellis and Wells 1941; Murov 1973) suggests that UV irradiation of water can result in the formation of ozone or radical oxidants. Further, work by Glaze et al. (1989) and several other researchers has shown that formaldehyde is a common by-product of ozonation. Chromatographic comparisons of the surface water samples before and after UV irradiation showed no other significant changes.

Simulated distribution system DBP results using chlorine as the simulated distribution disinfectant for the 10 surface waters are shown in Figure 11-3. Because of the chlorine demand of these waters, higher chlorine dosages were required, which resulted in larger DBP concentrations than in the groundwaters studied. However, the overall effect of UV irradiation on SDSDBPs was insignificant. As in the groundwater studies, UV irradiation did not significantly alter the total concentration, the rate of formation, or the speciation of the DBPs (THMs, HAA5, HANs, or HKs).

Malley, Shaw, and Ropp (1995) also compared the effects of UV followed by chlorine versus chloramination. Chloramines produced significantly fewer DBPs than did chlorine in all cases. Ultraviolet irradiation had no significant effects on DBP formation with either chlorine or chloramines. By contrast, chloramines produced significantly more cyanogen chloride than did chlorine in all cases. However, the cyanogen chloride production prior to UV irradiation was not significantly different from its production after UV irradiation.

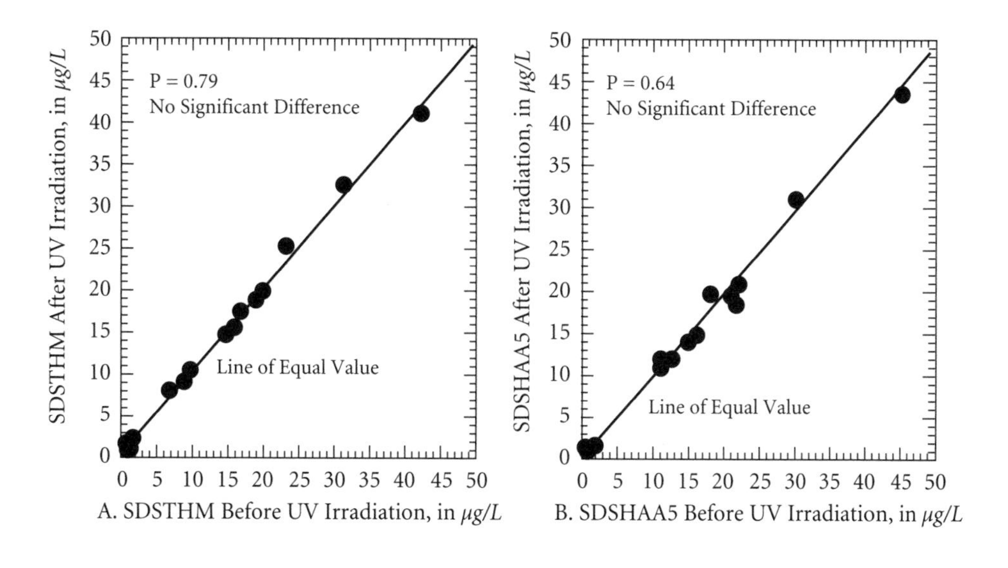

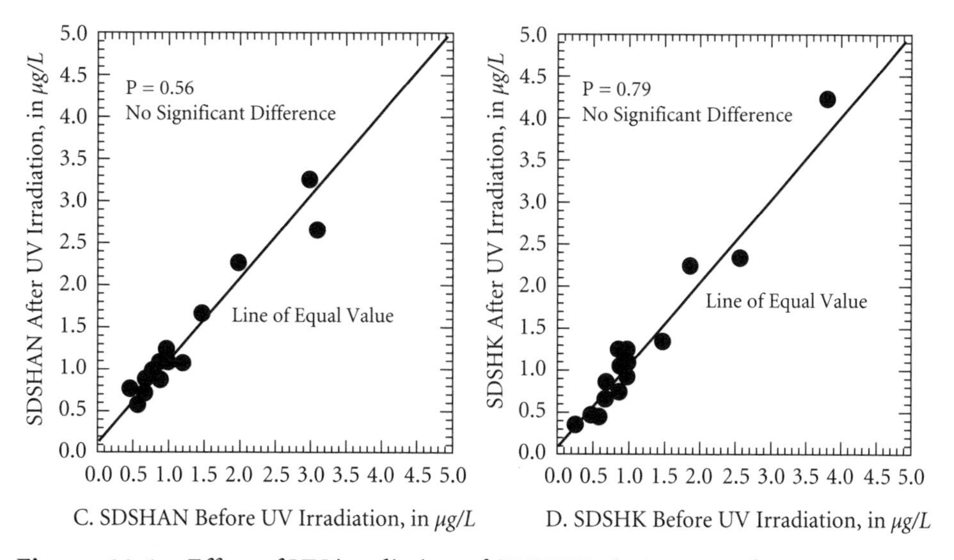

Figure 11-1 Effect of UV irradiation of SDSDBPs in 20 groundwaters. (A) SDSTHMs, (B) SDSHAA5, (C) SDSHANs, (D) SDSHKs. All studies were performed at a UV dosage of 130 mW-s/cm^2.

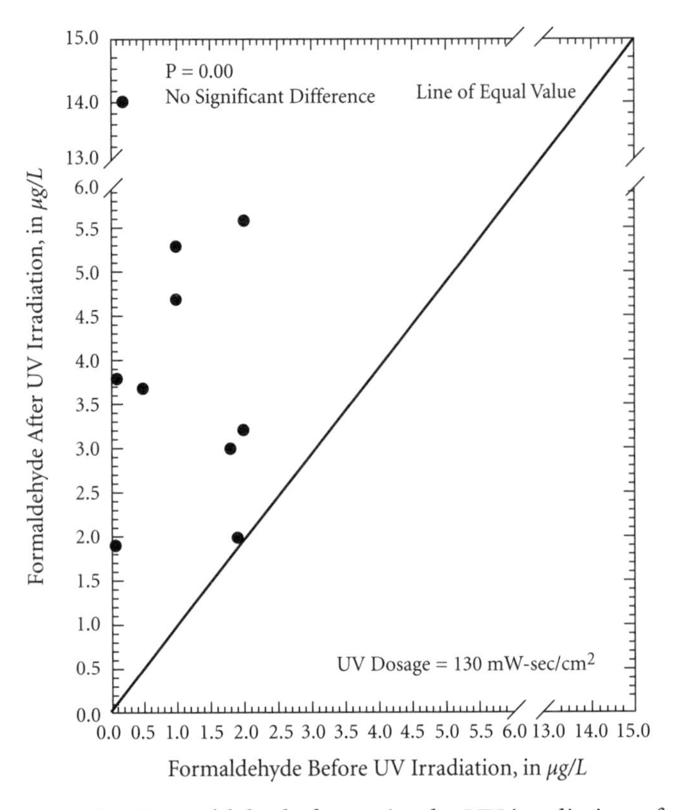

Figure 11-2 Formaldehyde formation by UV irradiation of surface waters

EMERGING UV TECHNOLOGIES AND RESEARCH NEEDS

The international search for effective means of inactivating *Cryptosporidium* combined with the commercialization of technologies spun off from the Strategic Defense Initiative (SDI) has resulted in numerous advanced electrotechnologies being tested in the drinking water field. Notably, medium-pressure UV lamps, prolonged surface UV irradiation (dosages of up to 8,000 mW-s/cm^2), and pulsed UV have all been introduced in the past two years. In each of these systems, the intensity and/or the wavelengths of UV light produced are significantly different than in conventional continuous-wave, low-pressure, mercury-vapor UV systems. As a result, the effects of these emerging UV systems on

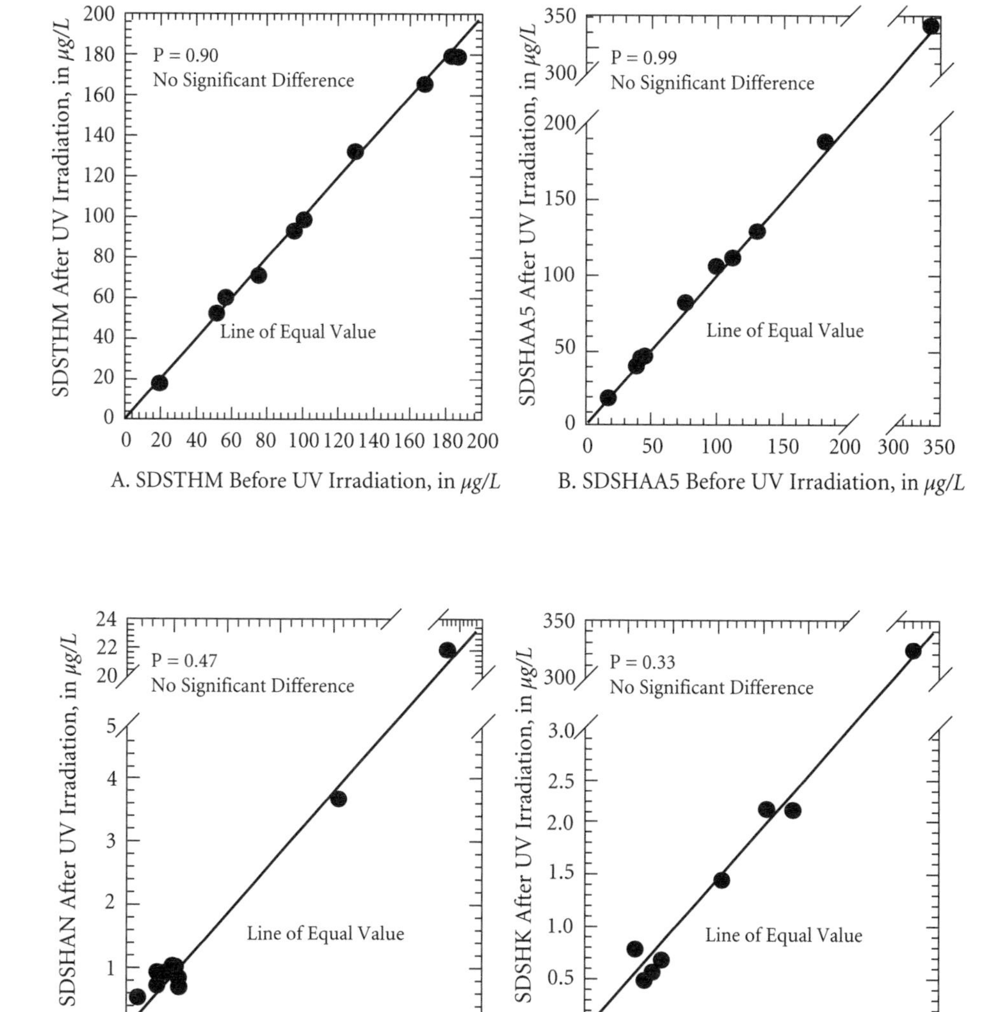

Figure 11-3 Effect of UV irradiation of SDSDBPs in 10 surface waters. (A) SDSTHMs, (B) SDSHAA5, (C) SDSHANs, (D) SDSHKs. All studies were performed at a UV dosage of 130 mW-s/cm^2.

DBPs and their precursors will require careful research prior to their widespread acceptance by the drinking water community.

SUMMARY

Conventional (continuous-wave, low-pressure mercury vapor) UV technology is a promising alternative primary disinfectant for utilities seeking to comply with the pending Ground Water Rule (GWR). Conventional UV technology did not produce any DBPs nor did it alter the rate or extent of DBP formation when followed by chlorination or chloramination. The effects of medium-pressure UV systems and emerging UV technologies, such as pulsed UV, on DBP formation have not been determined and warrant future research.

REFERENCES

Amy, G.L., M. Siddiqui, and K. Ozekin. 1994. *Removal of Bromate After Ozonation During Drinking Water Treatment.* Paper presented at the AWWA Annual Conference, June 19–23, 1994, New York.

American Public Health Association, American Water Works Association, and Water Pollution Control Federation. 1992. *Standard Methods for the Examination of Water and Wastewater*, 18th ed. Washington, D.C.: American Public Health Association.

Black & Veatch. 1995. Ultraviolet Radiation for the Disinfection of Water and Wastewater. Report for EPRI-CEC.

Blatchley, E.R., W.L. Wood, and P. Schuerch. 1995. UV Pilot Testing: Intensity Distributions and Hydrodynamics. *Jour. of Env. Eng.*, 121(3):258–262.

Calvert, J.G., and J.N. Pitts, Jr. 1966. *Photochemistry.* New York: John Wiley & Sons.

DeMers, L., and R.C. Renner. 1992. *Alternative Disinfection Technologies for Small Drinking Water Systems.* Denver, Colo.: American Water Works Association Research Foundation and American Water Works Association.

Downes, A., and T.P. Blount. 1877. The Germicidal Effects of Sunlight. In *Proc. of the Royal Society*, 26:488–503.

Ellis, C., and A.A. Wells. 1941. *The Chemical Action of Ultraviolet Rays.* New York: Reinhold Publishing Company.

Glaze, W.H., M. Koga, D. Cancilla, and K. Wang. 1989. Ozonation Byproducts. 2. Improvement of an Aqueous-Phase Derivatization Method for the Detection of Formaldehyde and Other Carbonyl Compounds Formed by the Ozonation of Drinking Water. *Environ. Sci. & Tech.*, 23:838.

Hoyer, O., R. Kryschi, I. Piecha, G. Mark, M.N. Schuchmann, H.P. Schuchmann, and C. von Sonntag. 1992. UV Fluence Rate Determination of the Low-Pressure Mercury Arc in the UV Disinfection of Drinking Water. *Aqua*, 41(2):75–81.

Karanis, P., W.A. Maier, H.M. Seitz, and D. Schoenen. 1992. UV Sensitivity of Protozoan Parasites. *Aqua*, 41(2):95–100.

Kruithof, J.C., R. Char, C.R. van der Leer, and W.A. Hijnen. 1992. Practical Experiences With UV Disinfection in the Netherlands. *Aqua*, 41(2):88–94.

Malley, J.P., Jr., J.P. Shaw, and J.D. Ropp. 1995. *Evaluation of By-Products Produced by Treatment of Groundwaters with Ultraviolet Irradiation.* Denver, Colo.: American Water Works Association Research Foundation and American Water Works Association.

Montgomery Watson, Inc. 1994. *A Comparative Study of UV and Chlorine for Wastewater Reclamation.* Pasadena, Calif.: Montgomery Watson, Inc.

Murov, S.L. 1973. *Handbook of Photochemistry.* New York: Marcel Dekker.

Rice, E.W., and J.C. Hoff. 1981. Inactivation of *Giardia lamblia* Cysts by Ultraviolet Irradiation. *Applied Environ. Micro.*, 42:546–547.

Scheible, O.K. 1987. Development of a Rationally Based Design Protocol for the Ultraviolet Light Disinfection Process. *Jour. WPCF*, 59:25–31.

Snicer, G.A., J.P. Malley, Jr., A.B. Margolin, and S.A. Hogan. 1998. Evaluation of Ultraviolet Technology in Drinking Water Treatment. Denver, Colo.: American Water Works Association Research Foundation and American Water Works Association.

Thampi, M.V. 1990. Basic Guidelines for Specifying the Design of Ultraviolet Disinfection Systems. *Pollution Engineering*, 22:65–69.

von Sonntag, C., and H.P. Schuchmann. 1992. UV Disinfection of Drinking Water and By-Product Formation: Some Basic Considerations. *Aqua*, 41(2):67–74.

Wolfe, R.L. 1990. UV Disinfection of Drinking Water. *Environ. Sci. & Tech.*, 24:768.

CHAPTER • TWELVE

Disinfection By-Product Precursor Removal by Coagulation and Precipitative Softening

STEPHEN J. RANDTKE

Coagulation and precipitative softening are not necessarily the most effective processes for removing precursors, but they tend to be relatively inexpensive compared to other precursor removal options, and the necessary facilities to meet this objective are already in place at most surface water treatment plants. Alternative technologies, such as granular activated carbon (GAC) adsorption and membrane processes, are usually considered only when coagulation or softening is unable to achieve the required degree of removal at a reasonable cost or when the source water does not require coagulation or softening (Jacangelo et al. 1995).

HISTORICAL PERSPECTIVE

Before THMs

Water utilities in North America began practicing coagulation primarily to prepare water for rapid sand filtration and disinfection. Those with less protected sources often found it helpful to prechlorinate the water to control growths of bacteria and algae inside the confines of the treatment facilities and to increase the contact time (CT) available for disinfection. Those with source waters high in color typically found it necessary to employ large coagulant dosages to achieve satisfactory removal of both turbidity and color. It gradually became recognized that the dissolved organic matter naturally present in water typically controls the required coagulant dosage, even for waters low in color. Hard source waters were (and are) commonly treated using precipitative (lime or lime-soda) softening.

Prior to the discovery of trihalomethanes (THMs) in drinking water, relatively few studies examined the removal of organic matter from water supplies by coagulation and softening. The majority of this research, including the classical studies of Black (Black and Willems 1961; Black et al. 1963) and Packham (Packham 1964; Hall and Packham 1965), focused on color removal. In an effort to better understand the origins, chemical characteristics, and behavior of color in water, early investigators (e.g., Black and Christman 1963a, b) drew on the knowledge of limnologists and soil chemists, who contributed to the water supply field the terminology of humic substances and various methods for characterizing them.

During the 1960s and 1970s, interest in physical–chemical treatment of municipal wastewater, for discharge or reuse, grew rapidly. Coagulation and softening were integral components of many proposed treatment schemes, and numerous investigations examined the ability of these processes to remove various types of organic contaminants, as well as pollutants such as metals, viruses, and phosphate. Efforts to characterize the nature of the organic matter removed from wastewater by coagulation and softening (e.g., Manka and Rebhun 1982), together with the efforts of those studying color and its removal from natural waters, were instrumental in developing the array of analytical tools available today for characterizing precursors (see chapter 4).

After THMs

By the mid-1970s, many practitioners were of the opinion that everything important to know about coagulation and softening of drinking water was already known. However, the discovery of THMs in drinking water (Rook 1974) and their subsequent regulation forced many water utilities to reexamine their treatment systems, and interest in precursor removal by coagulation and softening rapidly increased. High interest rates and an energy crisis provided strong incentives for utilities to avoid capital- and energy-intensive treatment processes, such as ozonation, GAC adsorption, and membrane filtration, and there were also a number of technical issues pertaining to these and other processes that had not yet been adequately addressed. Hence, utilities sought to reduce THM formation by modifying their existing facilities and treatment practices.

Encouraged by the results of a number of early studies demonstrating the ability of coagulation and softening to significantly reduce precursor concentrations (AWWA 1979; Symons et al. 1981; Randtke 1988), many more studies were undertaken to examine these processes in greater detail and to assess their application to other source waters. Ultimately, most utilities chose to control THMs by modifying their chlorination practices (McGuire and Meadow 1988). Relatively few found it necessary or cost-effective to modify the coagulation or softening process, or to employ processes such as GAC or membrane filtration.

The enhanced coagulation and softening requirements of the proposed Stage-1 Disinfectants/Disinfection By-Products (D/DBP) Rule (see Table 12-1), as well as the proposed maximum contaminant levels (MCLs) for THMs and haloacetic acids (HAAs), have renewed and intensified interest in precursor removal by coagulation and softening. Although it is too early to know exactly how many utilities will be affected, it appears that a significant number will need to modify the operation of their coagulation or softening facilities to achieve the required percentage of total organic carbon (TOC) removal or to reduce precursor concentrations to a level suitable for maintaining compliance with the new MCLs. Such efforts will be most effective if guided by knowledge of the applicable fundamental concepts.

Table 12-1 Percentage TOC removal requirements for enhanced coagulation and softening*

Raw TOC	Raw Alkalinity (*mg/L as $CaCO_3$*)		
(*mg/L*)	0–60	>60–120	>120†
>2–4	35	25	15
>4–8	45	35	25
>8	50	40	30

*The proposed rule will require most Subpart H systems (those treating surface water or groundwater under the influence of surface water) to control DBP precursors by removing the specified percentage of TOC by enhanced coagulation or softening. Some systems will be able to qualify for an exemption under one or more of criteria proposed in the Stage 1 D/DBP Rule. Most systems will be required to achieve precursor (TOC) removal prior to continuous addition of a disinfectant, but some exceptions will be permitted and others are presently under discussion. Systems not able to meet the TOC removal requirements shown, due to water quality characteristics or unique operating conditions, will be required to apply for alternative performance criteria (see White et al. 1997, for an analysis); the application must include, as a minimum, the results of bench- or pilot-scale tests. If certain criteria are met, the source water will be judged nonamenable to treatment by enhanced coagulation and softening and further reductions in TOC will not be required.

†Systems practicing precipitative softening must meet the TOC removal requirements shown in this column.

Source: Federal Register (1994).

FUNDAMENTAL CONCEPTS

The basic chemical reactions involved in coagulation and softening, as well as general process design considerations, are described in numerous reference books (e.g., AWWA 1990; AWWA and ASCE 1998) and will not be repeated here. The following paragraphs focus on the mechanisms whereby these processes remove disinfection by-product (DBP) precursors and on molecular characteristics influencing removal.

Precursor Removal Mechanisms

Coagulation, aided by flocculation and accompanied by a suitable process for removing particles (sedimentation, dissolved air flotation, or filtration) can remove DBP precursors by the following three mechanisms: (1) colloid destabilization, usually accomplished by charge neutralization or "enmeshment"

(AWWA 1990); (2) precipitation, i.e., the conversion of a dissolved substance to a solid; and (3) coprecipitation, primarily by occlusion or surface adsorption but conceivably by nonisomorphic inclusion as well (Randtke 1988).

Disinfection by-product precursors may be classified as particulate (e.g., plant debris and microorganisms) or dissolved based on their ability to pass through a filter (e.g., a 0.45-μm membrane filter). Except in very turbid rivers, where 50 percent or more of the precursors may be in particulate form, most of the precursors are usually dissolved. By definition, coagulation can remove particulate matter only by colloid destabilization (i.e., by particle removal mechanisms), while truly dissolved precursors can be removed only by precipitation or coprecipitation, i.e., by conversion to particles that can then be separated from the water. Since conventional surface water treatment plants are designed and operated to remove particles rather than dissolved substances, many efforts to increase precursor removal by coagulation and softening have focused on the removal of dissolved precursors measured in the form of dissolved organic carbon (DOC).

Humic substances (humic and fulvic acids) comprise a significant and variable fraction of the DOC present in a typical water supply. Because humic substances and other large molecules that may be present exhibit colloidal properties, they are considered by some scientists to be particulate in nature. However, natural waters typically contain much more fulvic acid than humic acid. Fulvic acids do not precipitate at pH 1 (by definition), nor do they precipitate at their isoelectric point. Because fulvic acids cannot be effectively destabilized by simple charge neutralization, yet are hydrophobic enough to be partially removed by metal-salt coagulants, many scientists consider them to exist in a dissolved state, i.e., in true solution. Due to uncertainty over the true physical state of the molecules comprising DOC, all three mechanisms listed above are invoked by those studying DOC removal by coagulation.

Dissolved organic carbon can be viewed as a complex mixture of polyprotic anions varying in molecular weight (MW), functionality, and hydrophobicity. These molecules interact with protons, metal hydrolysis products (both dissolved and precipitated), surfaces (including $CaCO_3$), and cations. These interactions are in turn influenced by the interactions of metal hydrolysis products with protons and certain anions able to alter metal hydrolysis product surfaces, especially their charge. These interactions are summarized in Table 12-2.

Molecular Characteristics Influencing Removal

Precursors removed by coagulation tend to be of higher MW, more hydrophobic, and more highly colored than nonremovable precursors. They typically exhibit greater specific ultraviolet (UV) absorbance (SUVA) values and a higher yield of DBPs, such that both SUVA and DBP yields decrease with increasing precursor removal. Edzwald et al. (1985) found SUVA to be a potentially useful indicator of precursor removability by coagulation, and other investigators (e.g., Moomaw et al. 1992; Edwards 1997) have developed models to predict precursor removal as a function of parameters such as DOC, SUVA, coagulant type, coagulant dosage, and pH. Presently, SUVA is being considered in defining the TOC removal requirements for enhanced coagulation (White et al. 1997).

Precursor removal generally follows one of two patterns as the coagulant dosage increases (Randtke and Jepsen 1981; Randtke 1988): (1) a sharp increase in removal at a particular dosage, stoichiometrically related to the precursor concentration; or (2) a relatively gradual increase in removal. The first pattern is indicative of a precipitation or charge-neutralization reaction and is likely to be observed when coagulating solutions of humic substances at relatively low pH values (e.g., 5 to 6). The second pattern is associated with precursor removal at higher pH values (by coagulation or softening) and with less removable, less humic, less highly colored, and more heterogenous precursor materials. As demonstrated by the results of White et al. (1997), TOC and DOC removal may follow different patterns at low coagulant dosages, at which DOC is converted to filterable particles that are not settleable. At higher coagulant dosages, removals of TOC, DOC, and UV absorbance plateau and SUVA values decline to a low level, e.g., about 1 to 2.5 L/mg-m.

Table 12-2 A conceptual model of DOC removal by coagulation and softening

Interaction(s)	Result(s)
DOC + H^+	Neutralization of acidic functional groups, making DOC molecules smaller, less soluble, and more hydrophobic; improved removal of DOC by coagulation, up to a point, beyond which protons begin to inhibit removal by outcompeting metal hydrolysis products for coordination sites; enhanced adsorption of DOC on metal hydroxide surfaces (and GAC, etc.); reduced removal of DOC by softening due to reduced dissociation of functional groups able to bind to calcium; precipitation of humic acids at $pH < 1$
DOC + metal hydrolysis products, polymers, and calcium carbonate	Precipitation of DOC molecules whose solubility is exceeded in the presence of dissolved metal hydrolysis products or polymers; coprecipitation (adsorption) of DOC with (on) metal hydroxide solids and $CaCO_3$, with interference by anions, e.g., OH^-, SO_4^{-2}, and CO_3^{-2}, able to compete with DOC for adsorption sites, to complex cations, or to render the surface charge more negative
DOC + cations	Precipitation of DOC molecules whose solubility is exceeded; reduced coagulant demand; enhanced coprecipitation (adsorption) of DOC with (on) metal hydrolysis products and $CaCO_3$
Metal hydrolysis products + H^+	A more positive surface charge, promoting increased removal by adsorption and coprecipitation during coagulation (but not softening)
DOC Characteristic	**Result(s)***
Greater hydrophobicity, color, or SUVA	Increased removal
More binding sites (functional groups) per molecule	Increased removal (unless overridden by increased solubility), greater coagulant demand, and a sharper increase in removal as coagulant dosage increases
Higher molecular weight	Increased removal (greater hydrophobicity and more binding sites per molecule)
Greater heterogeneity	A more gradual increase in removal as dosage increases
Treatment	**Result(s)**
Coagulation	Removes higher-MW and more hydrophobic DOC; remaining DOC molecules may adsorb more rapidly and go deeper into GAC pores, so a net increase in adsorptive capacity is possible; bed life may be greatly extended due to decreased DOC loading; can impede adsorption of DOC and other chemicals on powdered activated carbon (PAC) by depositing metal hydrolysis products in PAC pores, but can also increase adsorption on PAC by making molecules smaller in size and more hydrophobic and by removing large molecules; reduced concentration polarization (gel formation) and DOC precipitation on membranes; reduced oxidant demand
Softening	Removes much of the coagulable DOC, primarily by coprecipitation, with results similar to those listed for coagulation; DOC removal is usually improved by precipitation of Mg^{+2} as $Mg(OH)_2$, which has a positive surface charge and a much greater surface area than $CaCO_3$
Ozonation–Oxidation	Decreased hydrophobicity, decreased MW, but possibly an increased number of binding sites per molecule; can help or hinder DOC removal

*Assuming that all other characteristics remain unchanged.

APPLICATIONS

Most utilities coagulating surface water optimize the process for turbidity removal rather than precursor removal, with those treating source waters high in color or relatively low in alkalinity being common exceptions. Likewise, utilities practicing softening typically tailor the process to achieve the desired level of hardness removal, paying little attention to factors influencing precursor removal. However, even when no special effort is made to remove DBP precursors, coagulation and softening remove nearly all of the particulate precursor material and usually a significant fraction of the dissolved precursors. Hence, coagulation can be an effective means of controlling DBP formation, especially when the disinfection process is modified to minimize contact between the disinfectant(s) and the removable fraction of the precursors. Common modifications include the following: (1) eliminating or minimizing prechlorination; (2) using a different chemical for predisinfection and preoxidation, thereby reducing the formation of halogenated DBPs; and (3) adding ammonia to convert free chlorine to chloramines after a short period of time.

Utilities seeking to increase precursor removal, whether in response to new regulations, changing source water characteristics, or simply a desire to reduce the concentrations of DBPs in the finished water, may find that only a modest increase in removal is necessary to meet their objectives. In such cases, it may be possible to achieve the desired removal by modifying the coagulation or softening process, a practice generically referred to as *enhanced coagulation* or *enhanced softening* (proposed regulatory performance criteria are described in chapter 7 and Table 12-1). This is often an attractive option because capital expenditures and operator training requirements are typically minimal and the direction of future regulations remains uncertain.

Enhanced Coagulation

Precursor removal by coagulation can be improved in several ways, including increasing the coagulant dosage, optimizing the pH for precursor removal, changing coagulants, adding cations

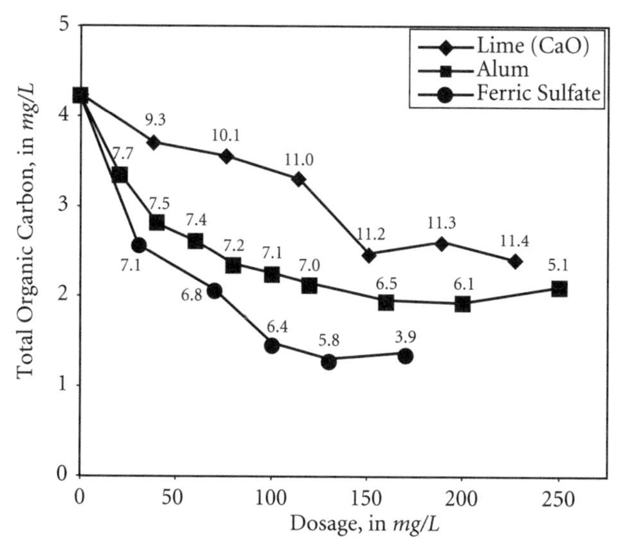

NOTE: Water drawn from Clinton Lake, Kan.; alkalinity = 112 mg/L as $CaCO_3$; calcium hardness = 100 mg/L as $CaCO_3$; magnesium hardness = 29 mg/L as $CaCO_3$, data points labeled with treated water pH values.

Source: Shorney and Randtke (1994).

Figure 12-1 Jar-test results showing the effect of increasing coagulant and lime dosage on TOC removal

(if the water is very soft), and, in some cases, by improving mixing or applying a moderate dosage of an oxidant.

Increasing the coagulant dosage. As the coagulant dosage is increased beyond the dosage required for satisfactory turbidity removal, precursor removal typically exhibits a gradual increase. If the pH is held constant, removal is expected to asymptotically approach a maximum value. However, if the coagulant is acidic or basic and the pH is not independently controlled, or if restabilization occurs, then removal will begin to decrease after reaching a maximum value, as illustrated in Figure 12-1. (Note that the coagulant dosages shown in Figure 12-1 are excessive. Much lower dosages would be needed to achieve similar results for a low-alkalinity water and, as discussed below, acid addition would significantly reduce the coagulant dosages needed to treat a high-alkalinity water such as this one.) Since precursor removal efficiency (milligrams of TOC removed per milligrams of coagulant) decreases as the dosage increases, there is a point where the increased costs (including chemical and sludge disposal costs) can no longer be justified.

Optimizing the pH. Maximum precursor removal by metal-salt coagulants and cationic polymers typically occurs at pH values of about 4.5 to 6, with the values generally being about 0.5 to 1 pH unit lower for iron salts than for alum. At these pH values, precursors receive little competition from hydroxyl ions for adsorption sites on the metal hydrolysis products, and the proton (H^+) concentration is sufficient to reduce coagulant demand without dissolving the metal hydrolysis products or neutralizing the acidic functional groups that enable precursors to coordinate with metal ions and cationic polymers.

If the source water is highly colored or low in alkalinity, it will generally be necessary to add a base (typically lime or sodium hydroxide) to keep the pH from falling below the optimum value. If the alkalinity of the source water is high, the optimum pH can be reached either by increasing the coagulant dosage (if the coagulant is acidic), by adding acid (usually sulfuric acid), or by purchasing a coagulant containing an excess of acid.

The potential benefits of acid addition are illustrated by the jar-test results presented in Figure 12-2 for a source water typically having an alkalinity of 150 to 200 mg/L as $CaCO_3$. One full-scale plant using this water as a source of supply coagulates it using an alum dosage of about 10 mg/L at a pH of about 8 to 8.5, resulting in excellent removal of turbidity and particulate organic carbon, but little or no removal of DOC. By feeding higher dosages of ferric chloride or ferric sulfate, the settled water TOC can be reduced from about 4 mg/L to less than 2 mg/L. Acid addition permits a given degree of TOC removal to be achieved using a lower dosage of coagulant and results in the production of a smaller quantity of residuals.

Changing coagulants. One coagulant may be significantly more effective than another for precursor removal, but reported differences must be carefully scrutinized to determine whether the comparison has been made on the basis of weight, chemical equivalents, or cost. It appears that the maximum removal achievable using iron salts generally exceeds that obtainable using alum (Randtke et al. 1994; Edwards 1997), but the dosages used

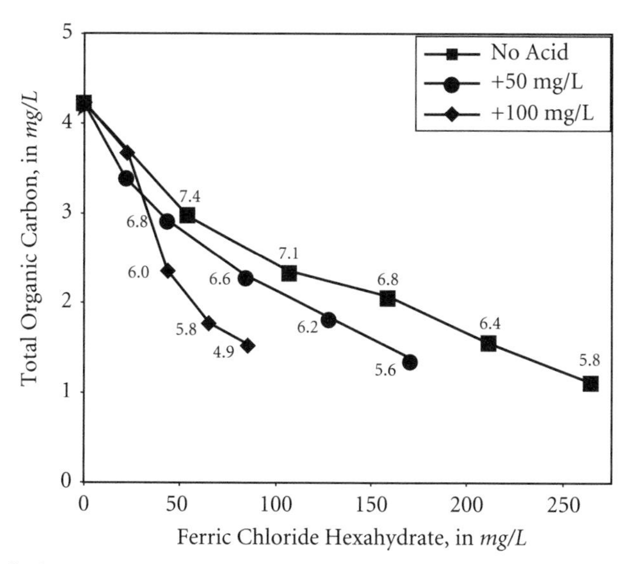

NOTE: The data points are labeled with the treated water pH values.
Source: Randtke et al. (1994).

Figure 12-2 Effect of sulfuric acid addition on ferric chloride coagulation of presettled Kansas River water

to obtain maximum removal far exceed those normally used in practice. When lower dosages are applied, alum may be more effective (Edwards 1997), but differences in the acidity of aluminum and iron salts tend to mask such differences.

Polymeric coagulants, both organic and inorganic, have also been found to effectively remove precursors. Low dosages of organic polymers may outperform low dosages of metal-salt coagulants, as illustrated in Figure 12-3, but higher dosages of metal salts have been found to achieve a greater degree of removal than organic polymers. This is likely due to the superior ability of precipitated metal hydrolysis products to adsorb dissolved precursors. Polymeric iron and aluminum products may outperform their metal-salt counterparts under some conditions (e.g., Dempsey et al. 1985), but their costs are greater on a metal-equivalent basis.

Changing coagulants may produce other benefits in addition to increased precursor removal. For example, the City of Tampa, Fla., found that not only did ferric sulfate remove significantly more natural organic matter (NOM) than alum but also that the iron-humate sludge could be applied to land, whereas land disposal

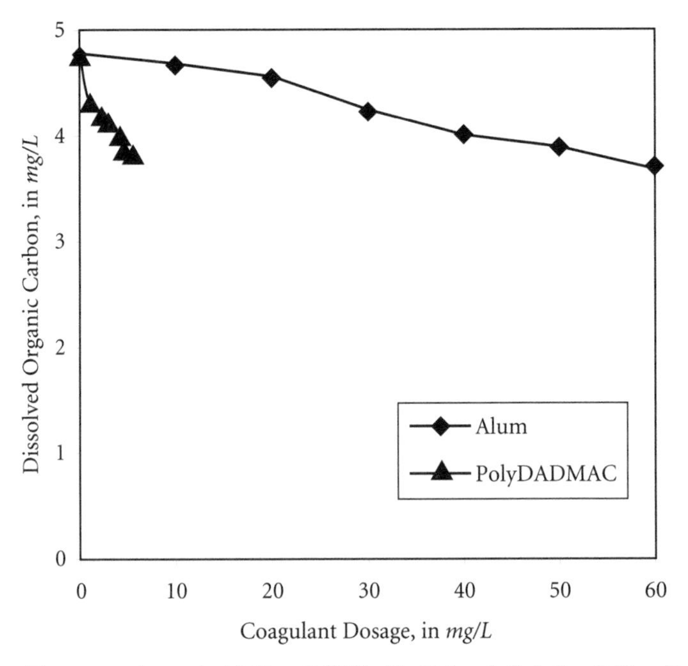

NOTE: The untreated water had 9.79 mg/L TOC, pH of 8.0, and alkalinity of 182 mg/L as $CaCO_3$. Following coagulant addition, the samples were flocculated for 15 min at 40 rpm, settled for 30 min, then filtered through a glass-fiber filter.

Figure 12-3 Dissolved organic carbon concentrations in Kansas River water treated with alum (MW = 600) or polyDADMAC undiluted

of the alum sludge was prohibited (Hook et al. 1992). Other considerations in selecting a coagulant include chemical costs (for both the coagulant and any chemicals required for pH adjustment or stabilization of the treated water); the presence of trace impurities (e.g., high dosages of some iron coagulants can create a dissolved manganese problem); the anionic composition of the treated water (e.g., chloride accelerates corrosion of steel); the quantity and characteristics of the residuals produced, as well as their cost of disposal; postprecipitation problems; and aesthetics (e.g., staining of basins).

Adding cations. The coagulant dosage required to achieve a given level of precursor removal from a very soft water (i.e., one having less than 50 mg/L of hardness as $CaCO_3$) can be significantly reduced by divalent cations, especially calcium (Randtke and Jepsen 1981; Randtke 1988; Dowbiggin and Singer 1989). Since such waters often have low alkalinity as well, lime addition may be helpful both for pH control and for lowering the required coagulant dosage.

Improving mixing. Mixing intensity may significantly influence the removal of particulate precursors, especially when the destabilization mechanism is charge neutralization (Amirtharajah and Mills 1982). Mixing intensity is not expected to exert a strong influence on the removal of dissolved contaminants that come into contact with the coagulant primarily by means of molecular diffusion, provided that the coagulant is uniformly dispersed into the water being treated in a relatively short period of time. In cases where mixing is inadequate, improved mixing may improve precursor removal or reduce the coagulant dosage required to achieve a given level of treatment.

Using a preoxidant. Preoxidation has sometimes been found to have beneficial effects on the coagulation of particulate matter (e.g., Edwards and Benjamin 1991). A mild dosage of a suitable preoxidant (e.g., ozone) may also improve precursor removal by increasing the number of acidic functional groups able to bind to metal hydrolysis products and by contributing to biodegradation of precursors within a biologically active filter. However, since oxidation of precursor molecules also renders them smaller and more soluble, higher oxidant dosages are generally found to adversely affect precursor removal by coagulation or to have little effect (e.g., Reckhow and Singer 1984; Tobiason et al. 1992; Edwards and Benjamin 1992a, b; Becker and O'Melia 1996).

Enhanced Softening

Coagulation and precipitative softening with lime are in some ways similar in regard to precursor removal, e.g., the same basic mechanisms are at work; but there are significant differences in process chemistry, the nature of the solids formed (Liao and Randtke 1985; Randtke 1988), and the options for disinfection and oxidation. Compliance with the MCL for THMs is especially difficult for utilities practicing precipitative softening because THM formation is very rapid in a high-pH environment. Trihalomethane formation can be reduced by substituting ozone or chlorine dioxide for chlorine, but both of these alternative disinfectants decompose rapidly and produce other undesirable by-products at the high pH values associated with precipitative softening, thus they usually cannot be used concurrently with precipitative softening. Many precipitative softening plants rely heavily on the use of chloramines to control DBP formation, and a number employ chlorine dioxide for preoxidation, allowing it to dissipate before adding lime.

The term *lime softening* encompasses the following three distinct processes: conventional lime softening, lime–soda softening, and excess lime softening. There are variations of these processes, e.g., split treatment (AWWA and ASCE 1998). Most lime softening plants practice conventional lime softening, i.e., they are operated to remove only calcium carbonate hardness. However, some use soda ash or caustic soda to remove noncarbonate hardness, and some add excess lime to raise the pH sufficiently to remove magnesium (and, in some cases, silica). Precursor removal can be increased by increasing the lime dosage (and pH), by precipitating magnesium, by adding a coagulant, and perhaps by operational changes; but some of these options can involve a fundamental change in the nature of the softening process.

Increasing the lime dosage. Precursor removal increases along with lime dosage. There are several reasons why this is so:

- If sufficient carbonate alkalinity is available, more $CaCO_3$ is precipitated, increasing the opportunity for precursors to be removed by coprecipitation.
- If carbonate alkalinity is insufficient, the excess calcium promotes precipitation, coprecipitation, and adsorption of precursors.
- The resulting increase in pH causes increased precursor removal, presumably by promoting stronger interactions between the precursors and calcium ions.
- The increase in pH may be sufficient to precipitate magnesium hydroxide, which strongly adsorbs precursors (Liao and Randtke 1985).

Typical jar-test results are shown in Figure 12-1. In contrast to the pattern observed for coagulation, precursor removal by lime softening often exhibits a second region of TOC decline corresponding to significant precipitation of magnesium hydroxide, as discussed below. Although lime softening does not remove precursors as effectively as very high dosages of metal-salt coagulants, as shown in Figure 12-1, its effectiveness is similar to that observed using relatively low dosages of coagulants. Lime dosages higher than those required to remove magnesium typically result in little if any additional precursor removal.

There are important practical considerations involved in significantly increasing the lime dosage at a softening plant. If the carbonate alkalinity is exhausted, addition of soda ash or caustic soda may be necessary for proper treatment, resulting in increased chemical costs and a higher sodium level in the finished water. Production of residuals will increase, and the handling and dewatering properties of the residuals may be significantly altered, especially if magnesium is removed. Increased chemical dosages will also be required to properly stabilize the treated water.

Precipitating magnesium. Unlike $CaCO_3$, which forms as dense cubic crystals that are negatively charged, freshly precipitated magnesium hydroxide exists in the form of light hydrous solids having a very high surface area and a positive surface charge, making it an excellent adsorbent for dissolved precursors and an effective coagulant for particles. Therefore, an increase in precursor removal commonly coincides with magnesium precipitation (Shorney and Randtke 1994; Randtke et al. 1994; Thompson et al. 1997). However, the increase may be negligible if another coagulant is applied or if the nature of the dissolved precursors is such that they are effectively removed by coprecipitation with $CaCO_3$.

Substantial removal of magnesium typically begins to occur at a pH of about 10.5 to 10.8, depending on the influent magnesium concentration, temperature, and other factors. Because measurable amounts of magnesium can be coprecipitated with $CaCO_3$, residual magnesium measurements do not provide a reliable indication of the onset of magnesium hydroxide precipitation. If no other coagulant is added, the onset of effective magnesium precipitation can be visually observed because its coagulative properties will be manifested by a marked increase in the clarity of the supernatant.

Adding a coagulant. A growing body of evidence shows that coagulant addition during lime softening can increase precursor removal, but it is not clear whether the increase is due only to improved capture of particulate matter or if improved removal of dissolved precursors also occurs. Some results (e.g., Shorney and Randtke 1994) indicate that the improvement is limited to low dosages of coagulant, i.e., those required to achieve a reasonable settled water turbidity, and that there is not a significant increase in precursor removal as the coagulant dosage is increased further. Since coagulant addition (or magnesium precipitation) is essential to the proper operation of a softening plant, it appears that softening plants may already be taking full advantage of the benefits of coagulant addition in regard to precursor removal. A limited amount of data indicates that the timing of coagulant addition significantly affects its performance.

Making operational changes. There is limited evidence (e.g., Liao and Randtke 1985) that precursor removal can be increased by reducing or eliminating solids recycling, thereby promoting coprecipitation of precursors rather than crystal growth, or by delaying the addition of soda ash to a subsequent stage of treatment, thereby avoiding competition between carbonate and precursors for adsorption sites. However, such measures should not be undertaken without careful consideration of their potentially significant practical implications and testing of their site-specific effectiveness.

Enhanced Presedimentation

Presedimentation is the first stage of treatment for many surface water supplies, especially those that are highly turbid or vulnerable to chemicals spills or sudden changes in quality. Even if no coagulant is added, a substantial fraction of the particulate precursors will be removed from a turbid water supply, and chlorine (or oxidant) demand will also be reduced. For example, Robinson (1974) reported that feeding a cationic polymer prior to presedimentation at Topeka, Kan., reduced chlorine demand and, by implication, DBP precursors, by 85 percent. Polymers may be just as effective as low dosages of metal-salt coagulants for this purpose, especially for high-alkalinity waters, and return of the settled solids to the water source is less likely to be prohibited when using polymers. Though polymers are widely believed to remove little or no DOC, this is not necessarily the case, as exemplified by the results shown in Figure 12-3 and those of many other investigators (e.g., Narkis and Rebhun 1975; Glaser and Edzwald 1979; Edzwald et al. 1986).

If the precursor removal requirements of the proposed Stage 1 D/DBP Rule can be met during presedimentation, subsequent stages of treatment can then be optimized for removal of turbidity or hardness without special attention to precursor removal. In some cases, precursor removal may be sufficient to permit the application of a predisinfectant or preoxidant, before subsequent stages of treatment, without violating the proposed MCLs for THMs and HAAs. This may be especially important for nutrient-rich waters typically found in agricultural areas, since some operators treating such waters report that they are unable to achieve adequate settled-water clarity or to control in-plant production of taste and odor without preapplication of a biocide.

Direct Filtration and Dissolved-Air Flotation

Like conventional (gravity) sedimentation, both direct filtration and dissolved-air flotation (DAF) are able to remove DBP precursors from properly coagulated water. Direct filtration plants typically employ low coagulant dosages (to minimize head-loss development and to produce filterable floc), so they generally do not remove DBP precursors to as great an extent as conventional facilities. They are not required to comply with the enhanced coagulation requirements of the proposed D/DBP Rule, but they do effectively remove particulate precursors and a portion of the dissolved precursors. Edzwald (1986) indicated that direct filtration is potentially feasible for treating waters having a mean annual TOC below 5 mg/L and a total trihalomethane formation potential (TTHMFP) below 400 μg/L, but conventional treatment should be considered for higher concentrations of precursors.

Dissolved-air flotation is an alternative to sedimentation that has historically enjoyed greater use on other continents but is now receiving more frequent consideration in North America, especially for algae-laden waters. Edzwald and his co-workers extensively studied DAF and delineated the important process design parameters (e.g., Edzwald 1995; Valade et al. 1996). Proper coagulation is essential in the DAF process, both for particle removal and for precursor removal. The ability of DAF to remove precursors has been demonstrated (e.g., Edzwald et al. 1992), and removal of organic matter has been found to correlate reasonably well with removal of *Cryptosporidium* oocysts (Plummer, Edzwald, and Kelley 1995). Precursor removal by coagulation is expected to be influenced by the same factors whether particle separation occurs by DAF or sedimentation.

Pretreatment for Subsequent Processes

Coagulation is often an excellent pretreatment process for subsequent processes needed to achieve a greater level of precursor removal or for other purposes. By reducing precursor loadings and removing interfering substances, the coagulation and subsequent solid–liquid separation process improves the performance and reduces the operating costs of downstream oxidation and disinfection (e.g., Robinson 1974), GAC adsorption (e.g., Semmens et al. 1986a, b; Randtke and Jepsen 1981; Hooper et al. 1996), and membrane processes (e.g., Weisner, Clark, and Mallevialle 1989; Laîné, Clark, and Mallevialle 1990).

LIMITATIONS OF COAGULATION AND SOFTENING FOR DBP PRECURSOR REMOVAL

Precursor removal by coagulation and softening has some important limitations and potential secondary effects that should be carefully considered.

Delaying addition of a disinfectant until after coagulation and solid–liquid separation will reduce the available disinfection contact time. It will also delay the response time available for correcting lapses in the disinfection process, and may therefore increase the risk of waterborne disease. Delaying the addition of an oxidant or biocide may cause an increase in settled water turbidity, promote in-plant growths of algae or bacteria (perhaps causing taste-and-odor problems), inhibit removal of iron and manganese, or preclude other benefits of preoxidation. If a higher disinfectant concentration, or a different disinfectant (e.g., free chlorine instead of monochloramine) is required to compensate for the delay, higher DBP levels may result.

Increasing the dosage of coagulant or lime, or removing more magnesium by lime softening, will increase chemical costs, including those associated with stabilization of the finished water; increase the quantity of residuals requiring disposal, as well as energy and environmental costs associated with residuals disposal; and, in all likelihood, adversely affect the dewaterability of the residuals (Knocke, Hamon, and Dulin 1987).

Water quality changes associated with increased chemical dosages (e.g., increased salt concentrations, reduced alkalinity, pH changes, or lower TOC concentrations) may, if not corrected, lead to increased lead and copper levels in the distribution systems and may accelerate corrosion of the treatment facilities, the distribution system, materials contacted by water during use, or the sewer system. Increased salinity may also interfere with water reclamation practices in water-short areas. Increased lime dosages may require the addition of soda ash or caustic soda, thereby increasing finished-water sodium levels. Precursor removal by coagulation or softening will increase the ratio of bromide to TOC and therefore may result in increased formation of brominated DBP species.

Increased coagulant dosages are likely to reduce floc size and density, leading to an increase in settled-water turbidity and shorter filter runs unless effective countermeasures, such as judicious use of polymers, are taken. High dosages of certain iron salts can cause manganese to increase to a problematic level. The decrease in pH associated with an increase in the dosage of a metal-salt coagulant may aggravate operational problems during cold weather, when a higher pH may be required for effective coagulation, or cause postprecipitation problems if the pH of the water is not properly adjusted prior to filtration.

SUMMARY

During the past 20 years, many utilities have modified their disinfection facilities and practices to control DBP formation without increasing the risk of microbial disease, but relatively few have significantly altered their coagulation or softening practices for this purpose. Proposed and future regulations are expected to cause many utilities to reexamine their coagulation and softening processes and perhaps to modify them to improve precursor removal or the performance of subsequent treatment processes. The challenge facing these utilities is to take advantage of the available benefits of enhanced coagulation and softening while avoiding potential problems, especially an increase in the risk of microbial disease. This will require careful analysis, judicious selection of treatment conditions, and maintenance of sound engineering and water treatment practices.

REFERENCES

Amirtharajah, A., and K.M. Mills. 1982. Rapid Mix Designs for Mechanisms of Alum Coagulation. *Jour. AWWA*, 74(4):210.

AWWA. 1990. *Water Quality and Treatment, 4th ed.* New York: McGraw-Hill.

AWWA and ASCE. 1998. *Water Treatment Plant Design, 3rd ed.* New York: McGraw-Hill.

AWWA Committee Report. 1979. Organics Removal by Coagulation: A Review and Research Needs. *Jour. AWWA*, 71(10):588.

Becker, W.C., and C.R. O'Melia. 1996. Ozone, Oxalic Acid, and Organic Matter Molecular Weight—Effects on Coagulation. *Ozone: Science and Engineering*, 18:311.

Black, A.P., and D.G. Willems. 1961. Electrophoretic Studies of Coagulation for Removal of Organic Color. *Jour. AWWA*, 53(5):589.

Black, A.P., and R.F. Christman. 1963a. Characteristics of Colored Surface Waters. *Jour. AWWA*, 55(6):753.

———. 1963b. Chemical Characteristics of Fulvic Acids. *Jour. AWWA*, 55(7):897.

Black, A.P., J.E. Singley, G.P. Whittle, and J.S. Maulding. 1963. Stoichiometry of the Coagulation of Color Causing Organic Compounds with Ferric Sulfate. *Jour. AWWA*, 55(10):1347.

Dempsey, B.A., et al. 1985. Polyaluminum Chloride and Alum Coagulation of Clay-Fulvic Acid Suspensions. *Jour. AWWA*, 77(3):74.

Dowbiggin, W.B., and P.C. Singer. 1989. Effects of Natural Organic Matter and Calcium on Ozone-Induced Particle Destabilization. *Jour. AWWA*, 81(6):77.

Edwards, M., and M.M. Benjamin. 1991. A Mechanistic Study of Ozone-Induced Particle Destabilization. *Jour. AWWA*, 83(6):96.

———. 1992a. Transformations of NOM by Ozone and Its Effect on Iron and Aluminum Solubility. *Jour. AWWA*, 84(6):56.

———. 1992b. Effect of Preozonation on Coagulant-NOM Interactions. *Jour. AWWA*, 84(8):63.

Edwards, M. 1997. Predicting DOC Removal During Enhanced Coagulation. *Jour. AWWA*, 89(5):78.

Edzwald, J.K., W.C. Becker, and K.L. Wattier. 1985. Surrogate Parameters for Monitoring Organic Matter and THM Precursors. *Jour. AWWA*, 77(4):122.

Edzwald, J.K., et al. 1986. Conventional Water Treatment and Direct Filtration Treatment and Removal of Total Organic Carbon and Trihalomethane Precursors. In *Organic Carcinogens in Drinking Water: Detection, Treatment, and Risk Assessment.* Ram, N.M., E.J. Calabrese, and R.F. Christman, eds. New York: John Wiley & Sons.

Edzwald, J.K., J.P. Walsh, G.S. Kaminski, and H.J. Dunn. 1992. Flocculation and Air Requirements for Dissolved Air Flotation. *Jour. AWWA*, 84(3):92.

Edzwald, J.K. 1995. Principles and Applications of Dissolved Air Flotation. *Water Sci. & Tech.*, 31:3–4:1.

Federal Register. 1994. National Primary Drinking Water Regulations; Disinfectants and Disinfection Byproducts; Proposed Rule. *Fed. Reg.*, 59(145):38668–38829.

Glaser, H.T., and J.K. Edzwald. 1979. Coagulation and Direct Filtration of Humic Substances With Polyethyleneimine. *Environ. Sci. & Tech.*, 13(3):299.

Hall, F.S., and R.F. Packham. 1965. Coagulation of Organic Color with Hydrolyzing Coagulants. *Jour. AWWA*, 57(9):1149.

Hook, M.A., J. Habraken, J. Gianatasio, and L.J. Hjerstad. 1992. Benefits of Enhanced Coagulation for Improved Water Quality and Beneficial Reuse of Residual Materials. In *Proc. Water Quality Technology Conference.* Denver, Colo.: American Water Works Association.

Hooper, S.M., R.S. Summers, G. Solarik, and D.M. Owens. 1996. Improving GAC Performance by Optimized Coagulation. *Jour. AWWA*, 88(8):107.

Jacangelo, J.G., J. DeMarco, D.M. Owen, and S.J. Randtke. 1995. Selected Processes for Removing NOM: An Overview. *Jour. AWWA*, 85(1):64.

Knocke, W.R., J.R. Hamon, and B.E. Dulin. 1987. Effects of Coagulation on Sludge Thickening and Dewatering. *Jour. AWWA*, 79(6):89.

Laîné, J.M., M.M. Clark, and J. Mallevialle. 1990. Ultrafiltration of Lake Water: Effects of Pretreatment on Organic Partitioning, THM Formation Potential, and Flux. *Jour. AWWA*, 82(12):82.

Liao, M.Y., and S.J. Randtke. 1985. Removing Fulvic Acid by Lime Softening. *Jour. AWWA*, 77(8):78.

Manka, J., and M. Rebhun. 1982. Organic Groups and Molecular Weight Distribution in Tertiary Effluents and Renovated Waters. *Water Res.*, 16(4):399–403.

Narkis, N., and M. Rebhun. 1975. The Mechanism of Flocculation Processes in the Presence of Humic Substances. *Jour. AWWA*, 67(2):101.

McGuire, M.J., and R.G. Meadow. 1988. AWWARF Trihalomethane Survey. *Jour. AWWA*, 80(1):61.

Moomaw, C., G. Amy, S. Krasner, and I. Najm. 1992. Predictive Models for Coagulation Efficiency in DBP Precursor Removal. In *Proc. 1992 AWWA Annual Conference*. Denver, Colo.: American Water Works Association.

Packham, R.F. 1964. Studies of Organic Color in Natural Water. *Proc. Soc. Water Treat. Exam.*, 13:316.

Plummer, J.D., J.K. Edzwald, and M.B. Kelley. 1995. Removing *Cryptosporidium* by Dissolved-Air Flotation. *Jour. AWWA*, 87(9):85.

Randtke, S.J., and C.P. Jepsen. 1981. Chemical Pretreatment for Activated Carbon Adsorption. *Jour. AWWA*, 73(8):411.

Randtke, S.J. 1988. Organic Contaminant Removal by Coagulation and Related Process Combinations. *Jour. AWWA*, 80(5):40.

Randtke, S.J., et al. 1994. A Comprehensive Assessment of DBP Precursor Removal by Enhanced Coagulation and Softening. In *Proc. 1994 AWWA Annual Conference*. Denver, Colo.: American Water Works Association.

Reckhow, D.A., and P.C. Singer. 1984. The Removal of Organic Halide Precursors by Preozonation and Alum Coagulation. *Jour. AWWA*, 76(4):151.

Robinson, C.N., Jr. 1974. Polyelectrolytes as Primary Coagulants for Potable Water Systems. *Jour. AWWA*, 66(4):252.

Rook, J.J. 1974. Formation of Haloforms During Chlorination of Natural Waters. *Water Treatment and Examination*, 23:234–243.

Semmens, M.J., et al. 1986a. Influence of pH on the Removal of Organics by Granular Activated Carbon. *Jour. AWWA*, 78(5):89.

———. 1986b. Influence of Coagulation on the Removal of Organics by Granular Activated Carbon. *Jour. AWWA*, 78(8):80.

Shorney, H.L., and S.J. Randtke. 1994. Enhanced Lime Softening for Removal of Disinfection By-Product Precursors. In *Proc. 1994 AWWA Annual Conference*. Denver, Colo.: American Water Works Association.

Symons, J.M., et al. 1981. Treatment Techniques for Controlling Trihalomethanes in Drinking Water. EPA-600/2-81-156. Cincinnati, Ohio: Municipal Environmental Research Laboratory, US Environmental Protection Agency.

Thompson, J.D., M.C. White, G.W. Harrington, and P.C. Singer. 1997. Enhanced Softening: Factors Influencing DBP Precursor Removal. *Jour. AWWA*, 89(6):94.

Tobiason, J.E., et al. 1992. Pilot Study of the Effects of Ozone and PEROXONE on In-Line Direct Filtration. *Jour. AWWA*, 84(12):72.

Valade, M.T., et al. 1996. Particle Removal by Flotation and Filtration: Pretreatment Effects. *Jour. AWWA*, 88(12):35.

White, M.C., et al. 1997. Evaluating Criteria for Enhanced Coagulation Compliance. *Jour. AWWA*, 89(5):64.

Wiesner, M.R., M.M. Clark, and J. Mallevialle. 1989. Membrane Filtration of Coagulated Suspensions. *Jour. of Env. Eng., ASCE*, 115(1):20.

CHAPTER • THIRTEEN

Adsorption of Disinfection By-Product Precursors

VERNON L. SNOEYINK
MARY J. KIRISITS
COSTAS PELEKANI

CHARACTERIZATION AND ADSORPTION PROPERTIES OF DISINFECTION BY-PRODUCT PRECURSORS

Effect of Natural Organic Matter Characteristics on Adsorption

Activated carbon is cited as one of the best available treatment (BAT) technologies in the US Environmental Protection Agency's (USEPA's) current and proposed regulations for trihalomethanes (THMs) and haloacetic acids (HAAs) (NPDWR 1994). Disinfection by-products (DBPs) themselves are usually less adsorbable than their organic precursors, so activated carbon is generally applied for the purpose of removing precursors before disinfection. Compared with other treatment processes (coagulation, oxidation, and ion exchange), activated carbon adsorption appears to be an effective treatment technology for water supplies with high THM precursor levels (Glaze and Wallace 1984), although it is usually best if the precursor concentration is reduced as much as possible by coagulation prior to adsorption.

The composition of natural organic matter (NOM) has a significant effect on its adsorption to granular activated carbon (GAC) or powdered activated carbon (PAC). Natural organic matter is a heterogeneous mixture of organic compounds consisting of nonadsorbing, weakly adsorbing, and strongly adsorbing compounds, which varies in type and concentration throughout the year. Seasonal data have shown that the amount of precursor or type of precursor that can react with disinfectants can quickly change. Chadik and Amy (1987) found that activated carbon was effective in removing NOM over a broad molecular weight range, consistent with the heterogeneity of NOM and the carbon surface.

Total organic carbon (TOC) or dissolved organic carbon (DOC) are commonly used to describe the adsorption of NOM. Figure 13-1 illustrates a typical adsorption isotherm for NOM, using a log–log scale. Isotherms of this nature are expected if some of the compounds are nonadsorbable and some are more strongly adsorbable (Randtke and Snoeyink 1983). These strongly and weakly adsorbing compounds consist primarily of intermediate-size molecules. Nonadsorbable compounds include molecules that are too large to access any pore volume (e.g., polysaccharides) and hydrophilic compounds, which prefer to be in solution rather than adsorbed to activated carbon (e.g., simple sugars and acids). The nonadsorbable fraction cannot be removed regardless of the adsorber design. Natural organic matter typically is larger than synthetic organic contaminants (SOCs), and the size affects the rate of approach to equilibrium. Intraparticle diffusion coefficients decrease as NOM molecular size approaches the size of the adsorbent pores. Thus, longer times are required to achieve equilibrium for NOM than for low-molecular-weight trace organic contaminants.

The performance analysis of GAC contactors commonly involves evaluation of a breakthrough curve. Figure 13-2 illustrates a representative normalized TOC breakthrough curve. The fraction that passes through the GAC column, typically 5 to 20 percent or 0.05 to 0.5 mg/L (McGuire et al. 1989), is comprised of

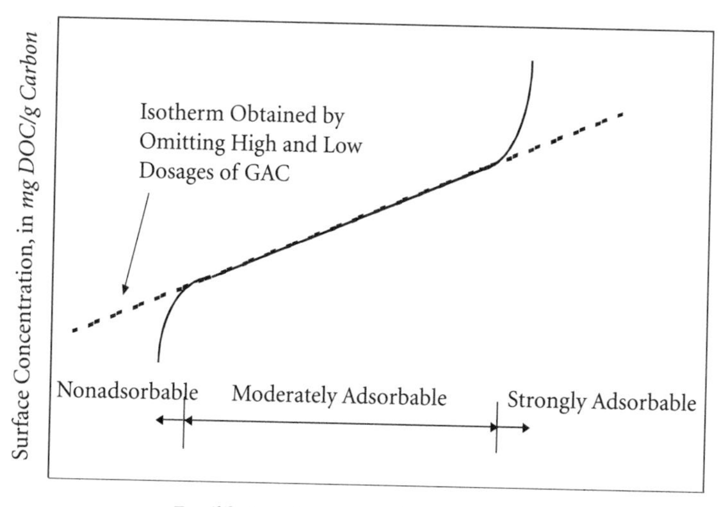

Figure 13-1 Typical DOC isotherm (log–log scale)

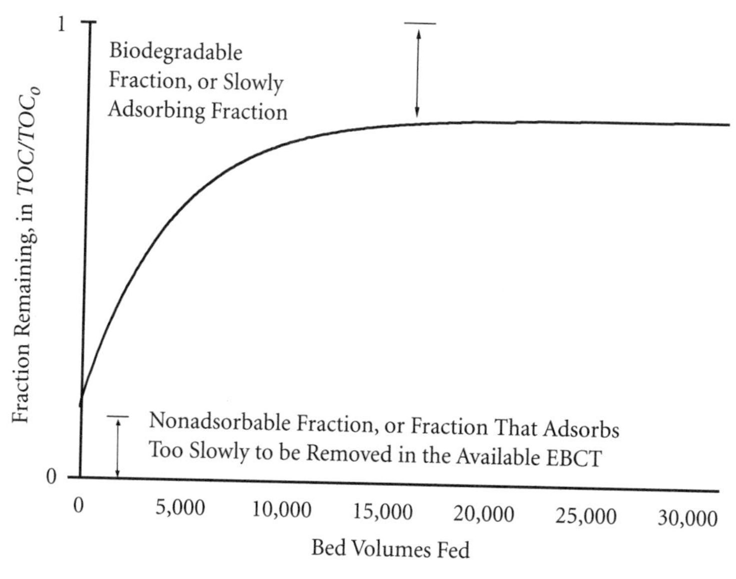

Figure 13-2 Typical breakthrough curve for TOC removal by granular activated carbon

nonadsorbable material or material that adsorbs too slowly to be removed in the available contact time. This results in the breakthrough curve starting at a normalized TOC level greater than zero.

A key parameter in the analysis of GAC contactors is empty-bed contact time (EBCT), which is defined as the volume of GAC in the contactor divided by the system flow rate. Shorter EBCTs result in elution of molecules with slow adsorption kinetics along with nonadsorbable NOM, thus increasing the initial value of TOC/TOC_0. As the EBCT increases, the initial TOC/TOC_0 approaches the nonadsorbable fraction. Over time, TOC/TOC_0 plateaus at a typical value of 0.6 to 0.8. The amount of TOC removed at this point can be attributed to slowly adsorbing TOC and to a change in the removal mechanism from adsorption to biodegradation. Granular activated carbon is a very good attachment media for microorganisms, but time is required to establish biological activity. By the time the adsorptive capacity of the GAC is approaching exhaustion, microorganisms have already acclimated to the system and can remove biodegradable NOM. Biological activity can significantly increase the life of the GAC adsorber, especially if only a small percentage of TOC removal is required, but this removal is highly dependent on temperature (Ventresque et al. 1991).

The shape of the breakthrough curve also depends on several physical properties of the GAC. Particle size is an important parameter because of its effect on the rate of adsorption. Particle size determines the time required for transport to the carbon surface (film mass transfer) and diffusion within pores to available adsorption sites. The smaller the particle, more rapid the approach to equilibrium. Pore size distribution and surface area are two important intrinsic properties of activated carbon. The maximum amount of adsorption is proportional to the amount of surface area within pores that are accessible to the adsorbate. Granular activated carbon with a low mesopore volume (pore volume in the 20 to 500 Å size range) also has a relatively low capacity for large NOM molecules. Lee, Snoeyink, and Crittenden (1981) showed that the quantity of humic substances of a given size that was adsorbed correlated well with certain mesopore volume fractions. Figure 13-3 illustrates the dramatic effect that these carbon properties can have on the TOC breakthrough curve (Carlson et al. 1994). The

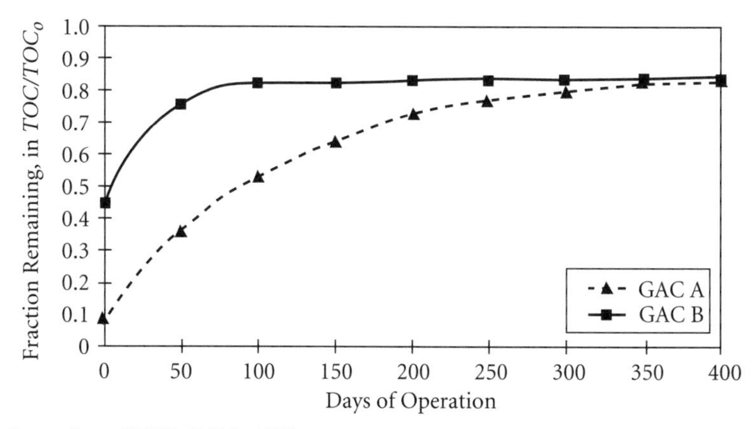

Source: Jour. AWWA *(1994), 86(3).*

Figure 13-3 TOC breakthrough curves for two different types of GAC

breakthrough curves for GAC A (effective size = 0.8–1.0 mm, pore volume = 0.85 cm^3/g) and GAC B (effective size = 0.8–1.0 mm, pore volume = 1.2 cm^3/g) were obtained using ozonated settled water. The EBCT was 14 minutes and the influent TOC varied in the range of 3 to 5 mg/L. The carbons show very different behavior, even though the effluent appears to plateau around the same TOC/TOC_0 ratio of 0.8. Even though GAC B has a higher pore volume, it reaches the plateau value more quickly; this is due, in part, to its pore size distribution.

Adsorption of Disinfection By-Product Precursors

Humic substances, present as part of NOM, have been demonstrated to be precursors for DBPs, such as THMs. Dissolved organic carbon (or TOC) is commonly used as a surrogate parameter for DBP formation potential (DBPFP). The use of DBPFP surrogate parameters for the assessment of GAC performance is important for water utilities primarily because of cost and ease of measurement. Total organic carbon has been demonstrated to be a good indicator of trihalomethane formation potential (THMFP) because a large fraction of the organic compounds present in natural waters form THMs. The measurement of TOC is rapid and accurate, compared to the analysis of

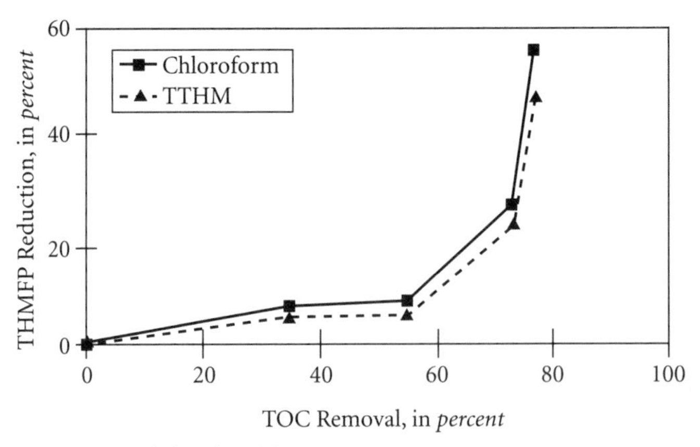

Source: Water Research *(1987), 21(5).*

Figure 13-4 Effect of GAC treatment on THMFP of humic acid

THMFP, which is more difficult and time-consuming. In full-scale applications, the routine measurement of THMFP would be impractical.

Due to the presence of nonadsorbable THMFP, the finished water THMFP after GAC adsorption can be in excess of the maximum level for some waters with elevated TOC concentrations, due to the immediate breakthrough of nonadsorbable DBP precursors (Jodellah and Weber 1985). El-Rehaili and Weber (1987) investigated carbon adsorption of THMFP using commercial purified Aldrich humic acid as a NOM source. Figure 13-4 shows that it was necessary to remove more than 70 percent of the TOC in order to achieve a reduction of approximately 20 percent in THMFP. Further analysis showed this to correspond to a reduction of less than 20 percent of organic matter in the low-molecular-weight range. Chadik and Amy (1987) have shown that the low-molecular-weight NOM fractions contribute significantly to THMFP. In studies with Colorado River water, the ratio THMFP (μg/L):TOC (mg/L) varied from 20:1 for the influent to 59:1 for the effluent (McGuire et al. 1989). These results demonstrate that the relationship between TOC and THMFP before and after GAC adsorption is not conserved as a result of selective removal of THMFP by GAC.

The presence of bromide (Br^-) in drinking water supplies has a significant effect on the THMFP speciation after GAC treatment.

The formation of proportionally higher levels of brominated DBPs in a sample of chlorinated GAC effluent, compared to a sample of chlorinated GAC influent, has been observed for several years (Symons et al. 1993). Conventional treatment and GAC adsorption do not remove Br^- to any significant extent, resulting in a higher Br^-:TOC ratio in the GAC effluent compared to the influent. As the Br^-:TOC ratio increases, the THMFP per unit TOC also increases, as well as the proportion of brominated DBPs (Sketchell, Peterson, and Christofi 1995; Jacangelo et al. 1995; Najm and Krasner 1995). Also, there is a higher $Br^-:Cl_2$ ratio in GAC treated water that is chlorinated because chlorine dosages decrease as a result of TOC removal. An increase in $Br^-:Cl_2$ ratio favors the formation of brominated DBPs. However, TOC adsorption decreases the total concentration of precursors and THMFP. Therefore, the total DBP concentration resulting from chlorination of GAC adsorber effluent is typically much less than if non-GAC treated water is chlorinated.

Effect of Pretreatment

The performance of activated carbon adsorption systems can be greatly affected by pretreatment processes prior to its application to activated carbon. The concentration of NOM may be reduced and the organic matrix composition may be changed. These changes can affect adsorbability and biodegradability. Furthermore, the inorganic composition may be altered in a way that affects adsorbability and the tendency of the activated carbon to become fouled. The removal of NOM by pretreatment reduces the organic loading that must be removed by adsorption, and this may result in lower GAC operating costs (Snoeyink 1990).

Chadik and Amy (1987) found that alum coagulation was effective in removing high-molecular-weight NOM but was less effective for the smaller molecular weight fractions. Hooper et al. (1996) investigated the effect of optimized coagulation on GAC performance when treating Ohio River water (Figure 13-5). A dramatic increase in the operating time was obtained, and similar results were obtained for other waters. It was hypothesized that enhanced coagulation not only yielded a reduction in the influent TOC concentration to the GAC column, but also improved the

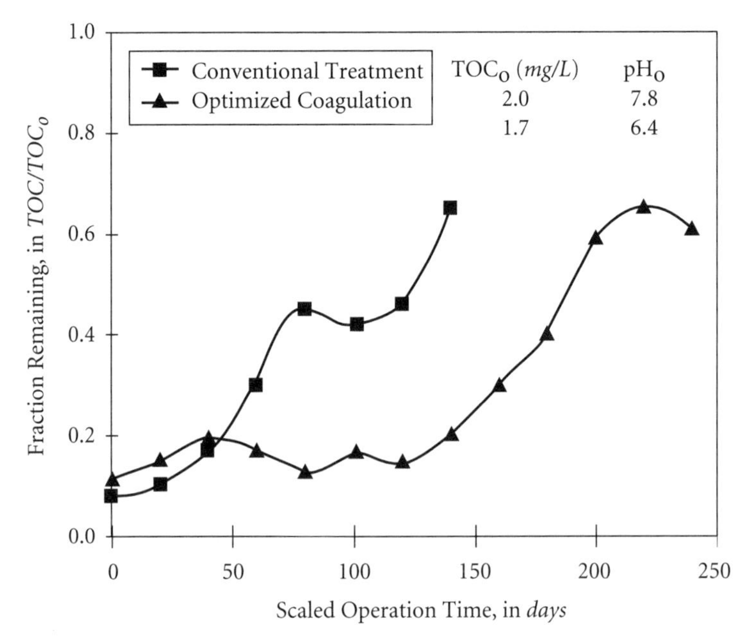

Source: Jour. AWWA. *(1996), 8(8).*

Figure 13-5 Effect of optimized coagulation on GAC performance for Ohio River water

adsorbability of the NOM remaining after coagulation. Enhanced coagulation decreased the pH of all the waters studied, and this lower pH was maintained through the GAC adsorber. At lower pH levels, NOM is less soluble and correspondingly more adsorbable.

Reactions of oxidative pretreatment chemicals, such as chlorine, ozone, chlorine dioxide, and permanganate, with the GAC surface or with NOM, can affect adsorption performance. These chemicals produce polar functional groups on the GAC that decrease its ability to adsorb certain compounds from water (Snoeyink 1990). Ozone can react with humic substances to produce more polar intermediates that are less adsorbable on GAC (Chen, Snoeyink, and Fiessinger 1980) but are usually more biodegradable (Sontheimer and Hubele 1987). If NOM is more biodegradable, increased removals by microbiological activity in the GAC contactor are expected.

If ammonium (NH_4^+) levels are high, nitrification may cause dissolved oxygen depletion in biologically active beds.

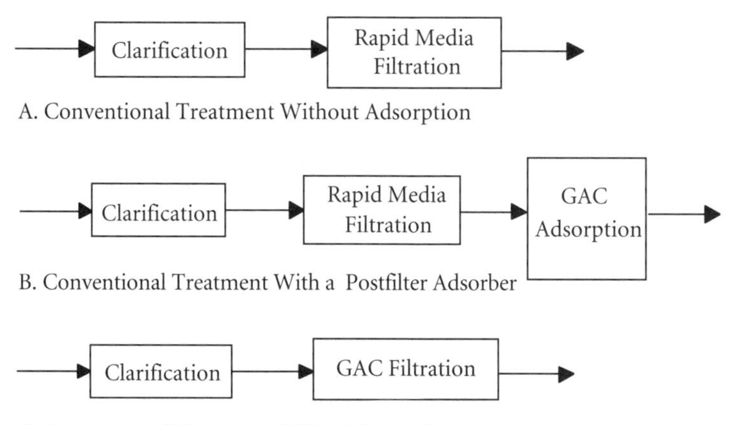

Figure 13-6 Alternative configurations for the application of GAC

Such depletion may cause a degradation of water quality through production of odorous compounds.

Pretreatment to prevent inorganic fouling of GAC is also important. A GAC influent supersaturated with respect to calcium carbonate, manganese dioxide, or iron hydroxide solids, for example, can lead to coating of the GAC particles and blockage of pores, thus reducing GAC adsorption capacity.

EFFECT OF PROCESS ARRANGEMENT ON GAC PERFORMANCE FOR DBPFP REMOVAL

Filter Adsorbers and Postfilter Adsorbers

Granular activated carbon contactors are classified according to their position in the treatment train. Figure 13-6 illustrates alternative configurations for the application of GAC contactors in water treatment. By convention, deep-bed contactors are placed at the end of the treatment train. In Europe, GAC is used in this manner almost exclusively. In the United States, replacing sand and anthracite with GAC in rapid media filters is more common. At issue, however, are the effect of solids and backwashing on adsorption efficiency, the EBCT that can be used, and the effect of media size on both turbidity removal and adsorption efficiency.

Filter adsorbers. The filter adsorber uses GAC to remove turbidity, adsorb dissolved organics, and biologically stabilize the treated water. A filter adsorber often is constructed simply by removing all or a fraction of the granular media from a rapid media filter and replacing it with GAC. When this is not appropriate, a new filter box and underdrain system for the GAC may be designed and constructed.

Hartman et al. (1991) found that filter adsorbers did not function well for removing less-strongly adsorbing compounds, such as THMs and NOM. The poor performance was attributed to a combination of slow adsorption kinetics and insufficient contact time. A short EBCT results in rapid breakthrough and the need for frequent regeneration or carbon replacement. However, good removal efficiency is expected for rapidly diffusing, strongly adsorbing compounds (e.g., pesticides and taste-and-odor-causing compounds).

Postfilter adsorbers. Granular activated carbon can also be used following conventional filtration as postfilter adsorbers, also called second-stage adsorbers. The sole objective of a postfilter adsorber is removal of organic matter.

Backwashing. Backwashing assisted by air scour is required to control the growth of zooplankton and higher organisms in the filter media and to prevent excessive head loss due to solids accumulation (Sontheimer 1983; Fiessinger 1983). Backwashing is necessary for filter adsorbers and often for postfilter adsorbers as well. Backwashing should be minimized, however, because of its possible effect on adsorption efficiency. Filter-bed mixing can take place during backwashing. If bed mixing occurs, GAC with adsorbed molecules near the top of the bed will move deeper into the bed where desorption is possible. Weakly adsorbing NOM may be partially desorbed in this new position, leading to a spreading of the mass transfer zone, and early breakthrough (Wiesner, Rook, and Fiessinger 1987).

The large uniformity coefficient (heterogeneous grain-size distribution) of most commercial activated carbons (1.6 to 1.9) promotes restratification after backwashing. However, if the underdrain system does not properly distribute the wash water, or if the backwash procedure is not performed in a way that promotes restratification, substantial mixing of the GAC filter can occur (Graese, Snoeyink, and Lee 1987). The uniformity coefficient is often kept smaller than 2 to minimize carbon losses during backwashing and to provide sufficient expansion and cleaning of the filter bed. By comparison, filters for turbidity removal achieve longer runs by using small uniformity coefficient media. Backwashing of postfilter adsorbers can result in adsorption characteristics approaching those of filter adsorbers.

Filter Adsorber Versus Postfilter Adsorber Performance

In most situations, sufficient NOM removal cannot be accomplished using filter adsorbers, but can be achieved using a postfilter adsorber. Generally, an EBCT greater than 15 minutes is required. The frequency of regeneration can vary widely, but 3- to 6-month intervals are most common (Singer 1994). Despite the superior efficiency of postfilter adsorbers for significant NOM removals, Wiesner, Rook, and Fiessinger (1987) concluded that the use of filter adsorbers for simultaneous turbidity and organic matter removal was more cost-effective than the use of postfilter adsorbers for TOC removal objectives less than 55 percent.

The ability of sand-replacement filter adsorbers to remove DBPs and DBP precursors is limited (Graese, Snoeyink, and Lee 1987). The EBCT and activated carbon depth in a sand replacement filter adsorber is usually limited by the existing filter design. The typical EBCT of a filter adsorber is about 9 minutes at operating flow rates and substantially less at design flow rates. Removal of weakly adsorbed organic matter with such short EBCTs would require regeneration or replacement of GAC after only a few weeks or months of operation if significant TOC removals are required. A high frequency of replacement or regeneration would be operationally cumbersome and costly.

The choice of an EBCT for a postfilter adsorber is more flexible than that for a filter adsorber, but capital cost considerations suggest a maximum of 20 minutes (DiGiano 1996). Size and operation are important determining factors. Postfilter adsorbers frequently use GAC with a small effective size and large uniformity coefficient to promote rapid adsorption of organic compounds and restratification to maintain the adsorption front.

Applications of PAC for THMFP Removal

Powdered activated carbon traditionally has been used in drinking water treatment for the removal of tastes and odors and periodic pesticide episodes. The doses used for these purposes are quite small, and DBPFP removals at these doses are usually minimal. Powdered activated carbon is not widely used for DBP precursor removal because of two factors. First, there usually is insufficient contact time for adsorption. Second, it is applied in conventional plants in a manner such that it tends to equilibrate at the effluent concentration of the process rather than at the influent concentration, as occurs with GAC. Examples of application points for PAC include at the head of the plant, to maximize contact time in the plant prior to sedimentation or filtration; after coagulation, to minimize treatment chemical interferences; or before filtration. All of these addition points have associated advantages and disadvantages (Snoeyink 1990; Snoeyink, Lai, and Young 1974).

Trihalomethane precursors are generally slowly adsorbing compounds, and long PAC contact times are necessary for good use of the carbon capacity. Such contact times are usually not available in a conventional treatment plant, or if they are available, such as by dosing to the filter influent, the dosage must be limited to ensure good removal of PAC in the filter. Various processes have been developed to improve the performance of systems employing PAC for precursor removal.

Powdered activated carbon and ultrafiltration systems. The application of PAC followed by ultrafiltration (PAC-UF) is considered to be an effective water treatment process (Adham et al. 1991) and can remove DBP precursors (Najm, Snoeyink, and Lykins 1991). A hollow-fiber or tubular-membrane configuration is commonly used in the PAC-UF system to allow cross-flow filtration with recirculation. The PAC is retained in the recycle loop long enough to allow the adsorption process to approach equilibrium. This maximizes the organic loading on the PAC and minimizes the carbon usage rate. The PAC also reduces the rate of fouling of the membrane surface (Adham et al. 1991).

Surface water usually contains significantly greater quantities of the types of organic matter that foul membranes than does groundwater, and thus coagulation and flocculation are commonly employed as pretreatments for PAC-UF systems with surface waters.

The system configuration studied by Schimmoller (1995) and Campos (1996) incorporated a floc-blanket reactor clarifier to the head of the PAC-UF system for coagulation and enhanced removal of NOM and organic micropollutants. This is achieved through improved adsorption efficiency by recycling the PAC from the UF system to the reactor clarifier where it can equilibrate at the concentration in the reactor clarifier effluent rather than the UF effluent. Using an Illinois surface water with an initial DOC of 6.5 to 6.8 mg/L and a PAC dose of 60 mg/L, a small pilot-scale system achieved 66 percent DOC removal, compared to 37 percent for the PAC-UF system alone (Campos 1996).

IMPORTANT DESIGN CONSIDERATIONS

Empty-Bed Contact Time and Carbon Usage Rate

The empty-bed contact time is defined as

$$\text{EBCT} = \frac{V}{Q} = \frac{D}{(Q/A)} = \frac{D}{u} \qquad (13\text{-}1)$$

Where:

V = volume of carbon in the bed, in L^3
Q = flow rate, in L^3/T
D = carbon bed depth, in L
A = bed area, in L^2
u = approach velocity, in L/T

The EBCT is proportional to the time that the adsorbate is in contact with the GAC. The molecular weight of the adsorbate and the size of the GAC determine how quickly adsorption will take place (Randtke and Snoeyink 1983). An increased EBCT generally facilitates increased adsorbate removal, as shown by the data of Glaze and Wallace (1984) in Figure 13-7. The GAC filter with the 24-minute EBCT demonstrated a slower THMFP breakthrough than the 12-minute filter. Jacangelo et al. (1995) note that EBCTs greater than 10 to 15 minutes are typically necessary to maintain acceptable precursor removal without excessive regeneration. For example, the GAC facility at the Cincinnati (Ohio) Water Works uses an EBCT of approximately 20 minutes.

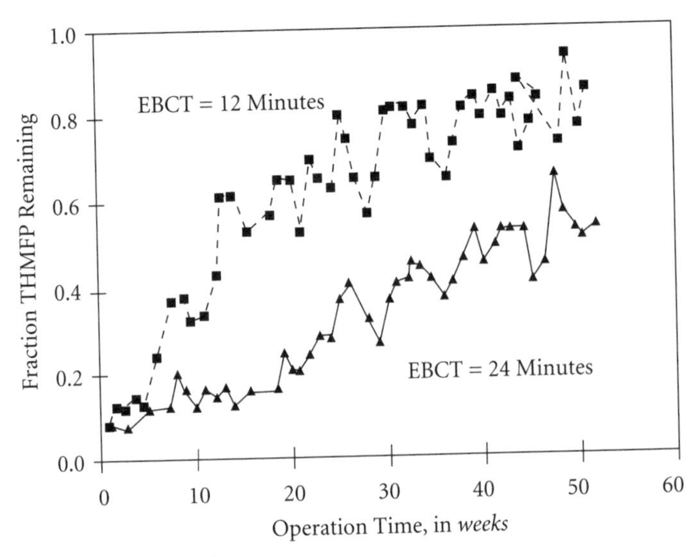

Source: Jour. AWWA *(1984).*

Figure 13-7 The effect of EBCT on THMFP breakthrough

Black, Harrington, and Singer (1996) examined TOC removal by GAC adsorption using the USEPA's Water Treatment Plant Simulation program (USEPA 1992). The predicted removal of TOC was translated to a reduction in the risk of cancer due to THM ingestion. The conversion between THM concentration and cancer risk was accomplished by using cancer potency factors for the various THM species. Figure 13-8 shows cancer risk reduction results obtained with an influent TOC concentration of 4 mg/L and EBCTs of 10 and 20 minutes.

Two important points can be derived from Figure 13-8. First, the use of the GAC adsorber with the 20-minute EBCT results in greater cancer risk reduction than the 10-minute adsorber. Second, as the bromide concentration increases, the removal of TOC by adsorption increases the number of cancer cases prevented. As mentioned earlier, brominated DBPs cause greater risk of cancer. By removing TOC through adsorption, brominated DBP formation potential is accordingly decreased; this results in a greater number of cancer cases prevented.

Carbon usage rate (CUR) is an important parameter related to the choice of an EBCT. In general, an increased EBCT allows

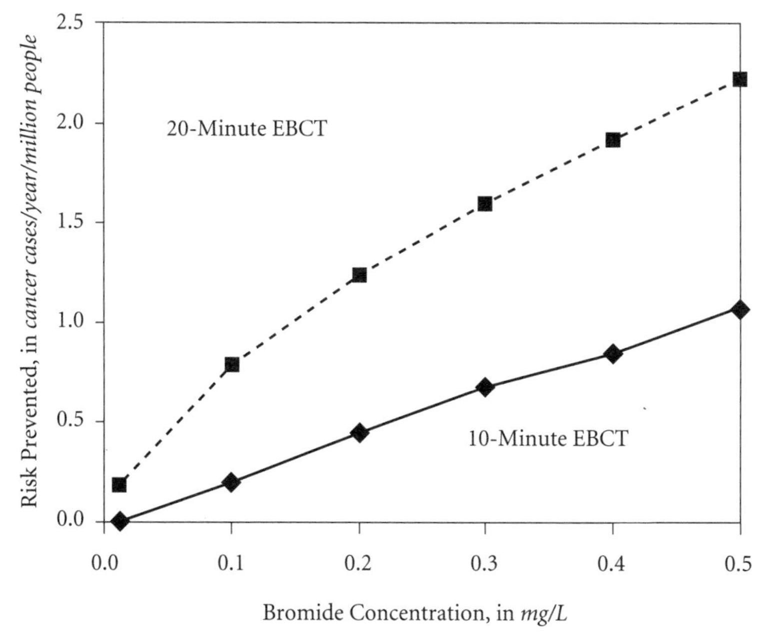

Source: Jour. AWWA *(1996).*

Figure 13-8 The use of GAC adsorption to reduce the number of cancer cases caused by THM ingestion

more of the adsorption capacity of the carbon to be utilized. This leads to a decreased carbon usage rate. The carbon usage rate is defined as

$$\text{CUR} = \frac{Mass_{GAC}}{WV} = \frac{\rho_{GAC}}{BV} \tag{13-2}$$

Where:

WV = volume of water treated to time of GAC replacement, in L^3

BV = volume of water treated to time of GAC replacement divided by carbon bed volume

ρ_{GAC} = carbon bed density, in M/L^3

As the EBCT increases, the CUR will plateau to a minimum value. Therefore, there exists a maximum EBCT beyond which no benefit to the CUR will be observed. This EBCT will vary depending on the source water and the effluent criteria. Using rapid small-scale column test (RSSCT) data (discussed later in this

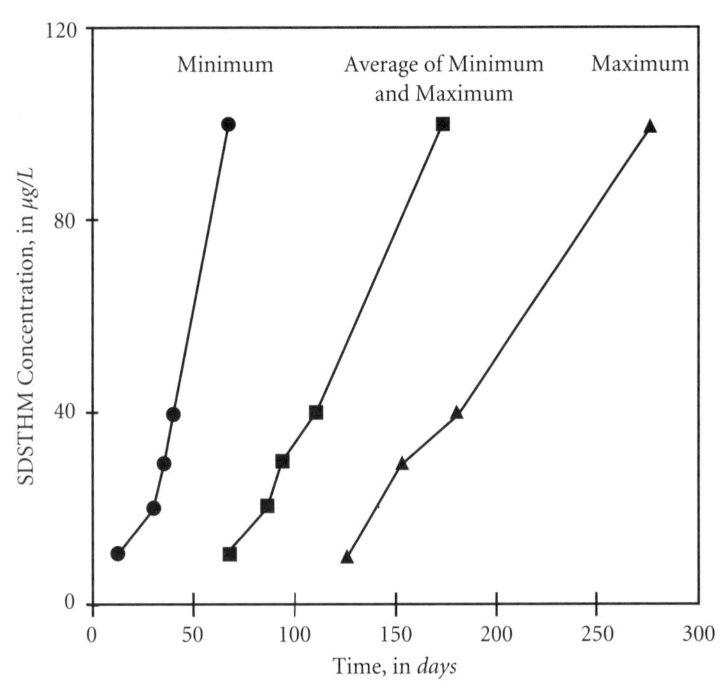

Source: Jour. AWWA *(1995).*

Figure 13-9 Variations in precursor removal by GAC at a single location at Cincinnati (Ohio) Water Works

chapter), Summers et al. (1995) examined carbon usage rates for Passaic River (Totowa, N.J.) water. In this study, the breakthrough concentration was defined as 80 percent of the stage 2 MCL for total trihalo-methanes (TTHMs) (32 μg/L). Carbon usage rates of 520 and 580 lb/mil gal were calculated for EBCTs of 10 and 20 minutes, respectively. Since doubling the EBCT did not have a dramatic effect on the carbon usage rate, it is reasonable to assume that the minimum CUR had been reached for this application.

Carbon usage rates, and therefore the life of the GAC bed, are highly variable. They depend on the influent and desired effluent contaminant concentrations and the particular contaminant of interest. For these reasons, it is difficult to pinpoint a typical range of carbon usage rates. For example, at the Cincinnati Water Works, the GAC bed life varied between 41

Table 13-1 Removal of DOC by GAC at various locations

Location	EBCT, in *min*	Bed Volumes	
		50 Percent Removal	70 Percent Removal
Cincinnati, Ohio	15	10,600	5,800
Jefferson Parish, La.	18.8	7,400	4,800
Miami, Fla.	18.6	2,900	1,300

Source: Jour. AWWA *(1995).*

and 182 days (2,800 to 12,500 bed volumes) for an effluent THM concentration of 40 μg/L (Jacangelo et al. 1995). The DOC removal varied from 8 to 48 percent and the maximum THMFP removal varied from 29 to 56 percent. This variability in conventional treatment causes fluctuations in the influent DOC concentration to the GAC filters. Figure 13-9 shows clearly how the GAC bed life can vary substantially as the influent DOC changes. In this figure, SDSTHM concentration is defined as the simulated distribution system trihalomethane concentration and is directly related to the DOC concentration.

There will be notable differences in GAC performance at different locations, since the concentrations of various NOM constituents will vary with the water source. Table 13-1 summarizes DOC removals at three water treatment plants. The number of bed volumes processed to 50 or 70 percent breakthrough can vary five-fold, depending on the water source.

The choice of EBCT is a trade-off between fixed capital costs and annual operating and maintenance (O&M) costs (see Figure 13-10). Fixed costs are related to the size of the contactor; as the contactor size increases, the capital costs likewise increase. The O&M costs are related to the CUR; as the EBCT increases, the CUR and O&M costs typically decrease. Therefore, an optimum EBCT can be determined that minimizes the total cost of the system.

It is again useful to refer to the data of Summers et al. (1995), where EBCTs of 10 and 20 minutes yielded CURs of 520 and 580 lb/mil gal, respectively. It is interesting to note that since the two filters had similar carbon usage rates, the 10-minute filter would have resulted in the lowest overall cost due to savings in capital expenditure. For example, for an average flow of 27 mgd, the unit cost of the 10-minute filter was determined to be $0.40/thousand gal, while the

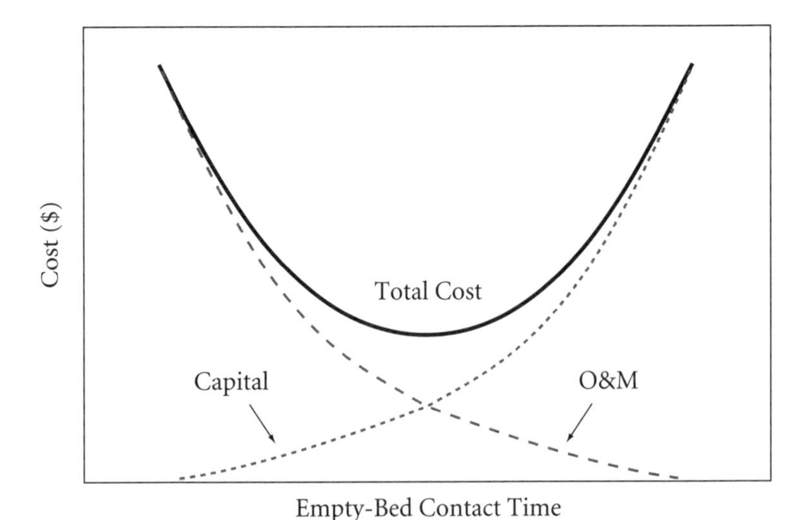

Figure 13-10 The effect of EBCT on system cost

unit cost of the 20-minute filter was calculated as $0.58/thousand gal. Summers et al. (1995) justly observed that it is economically beneficial to meet the D/DBP regulations with the lowest feasible EBCT.

Rapid Small-Scale Column Test

The RSSCT, developed by Crittenden, Berrigan, and Hand (1986), can be used to predict GAC column adsorption performance and carbon usage rate. Summers et al. (1995) observed that RSSCT data could be collected in about 1 to 15 percent of the time necessary to operate a corresponding pilot plant. Thus, the use of the RSSCT offers substantial savings of time and money.

As part of the Information Collection Rule (ICR), surface water utilities serving more than 100,000 people and having a raw water TOC concentration greater than 4.0 mg/L, and groundwater utilities serving more than 50,000 people, and having a finished water TOC concentration greater than 2.0 mg/L are required to operate bench- or pilot-scale experiments with GAC (or membranes) for precursor removal. The RSSCTs are a convenient tool for bench-scale assessments. The GAC studies must examine 10- and 20-minute EBCTs. The data obtained from these studies will provide a basis for cost estimates of precursor removal in a wide range of waters.

The RSSCT employs scaling equations to maintain similitude between full-scale contactors and the small bench-scale columns used in the analysis. Depending on whether intraparticle diffusivity is constant or dependent on carbon particle size, one of two different scaling equations can be used

$$\text{Constant Diffusivity } \frac{EBCT_{SC}}{EBCT_{LC}} = \left(\frac{d_{SC}}{d_{LC}}\right)^2 = \frac{t_{SC}}{t_{LC}} \quad (13\text{-}3)$$

$$\text{Proportional Diffusivity } \frac{EBCT_{SC}}{EBCT_{LC}} = \left(\frac{d_{SC}}{d_{LC}}\right) = \frac{t_{SC}}{t_{LC}} \quad (13\text{-}4)$$

Where:

LC = large column
SC = small column
t = operation time (T)
d = carbon particle diameter (L)

Equation 13-3 assumes that the diffusivity is constant and not dependent on the carbon particle size. Equation 13-4 assumes that the diffusivity is proportional to the carbon particle size. Equation 13-4 is most often used with TOC. In designing the RSSCT with the above equations, the small column typically has a length of about 10 cm and diameter less than 1 cm (DiGiano 1996).

Summers et al. (1995) examined a series of 30 experiments with four GAC types and nine different waters and concluded that the RSSCT-proportional diffusivity model (Eq 13-4) was most successful at predicting DOC breakthrough. Najm et al. (1996) compared the use of Eq 13-3 and 13-4 in conjunction with RSSCT data for predicting pilot-scale data with an EBCT of 15 minutes. The use of the proportional diffusivity model (Eq 13-4) resulted in a good prediction of the pilot-plant data, as shown in Figure 13-11. Note that while the proportional diffusivity model predicts the pilot-plant data very well for the first 25 days of the run, the RSSCT and pilot-plant data begin to diverge after this point. Empty-bed contact times of 30 and 60 minutes were also investigated in this study, and there were similar points of divergence where the RSSCT predicted greater TOC removals than actually observed in the pilot

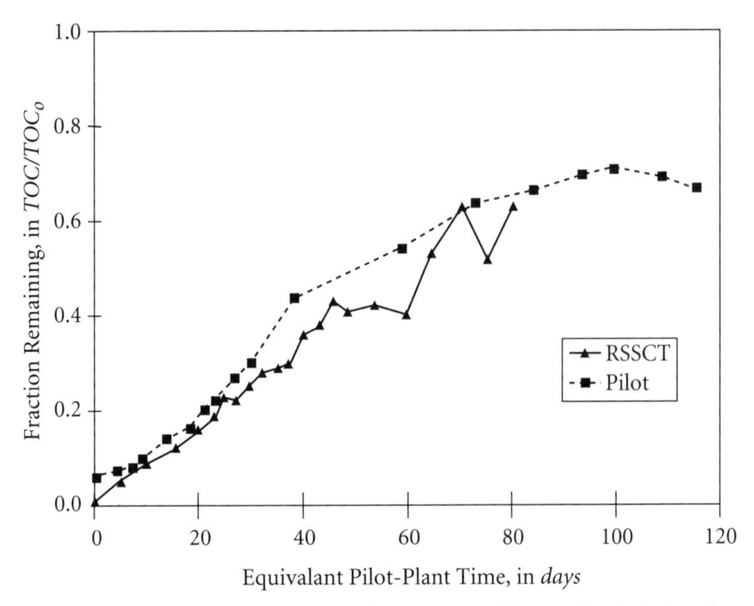

Source: NAJM et al., GAC, Membranes and the ICR: A Workshop on Bench-Scale and Pilot-Scale Evaluations *(1996).*

Figure 13-11 Fitting pilot-plant data with RSSCT data using the proportional diffusivity model for TOC breakthrough

plant. It was concluded that the RSSCTs simulated the pilot-plant data reasonably well. Clearly, the pilot-plant data were predicted well enough for an initial cost estimate.

There are limitations to the RSSCT. While RSSCTs are run with a single batch of water, the water quality influent to a full-scale plant will change over the course of a year. Fluctuations in the influent TOC concentration and changes in the adsorbability of the NOM present in the influent can effect the breakthrough curve (Summers et al. 1995). In order to minimize the affect of concentration fluctuations, Summers et al. (1995) recommend that the TOC concentration of the water used for the RSSCT be as close as possible to the average influent concentration of the pilot- or field-scale adsorber over the time scale of the experiment. The RSSCTs should be performed during different seasons in order to incorporate representative TOC concentrations and NOM adsorption characteristics (Dvorak 1996). Also, because RSSCTs are operated for relatively short periods of time, they will not be able to accurately predict biodegradation of NOM. In spite of these inherent shortcomings, Figure 13-11 shows that the RSSCT is a useful predictive tool.

It is important to consider how close the RSSCT prediction needs to be to the full-scale data to be useful. Harbold (1984) performed a sensitivity analysis that showed that the hydraulic loading rate and interest rate have more effect on GAC contactor cost than carbon loading. Even if a small error in the carbon usage rate is incurred from the RSSCT prediction, the effect of this error is negligible when compared to other factors.

Pilot-Plant Testing

The purpose of a pilot-plant study is to determine the optimum EBCT and the corresponding CUR for a particular system and to show if GAC performance could be improved with various pretreatment steps. Krasner, Westrick, and Regli (1995) observed that information concerning cost and scale-up are sometimes more easily interpreted from pilot-scale studies than from bench-scale studies. Consequently, some participants in the ICR volunteered to perform pilot-scale studies rather than bench-scale studies.

A pilot plant consists of suitable pretreatment operations followed by GAC column(s). Typical dimensions for a pilot column are a diameter of 2 to 4 in. and a length of at least 48 in. (Miller 1996). A general rule is that the column diameter should be at least 50 times the average particle diameter (Kornegay 1987). There are two ways to simulate multiple EBCTs. If only one GAC column is used, it can be equipped with several sampling ports along the column length. Alternatively, multiple GAC columns in series can be used.

The carbon mesh size in the pilot plant should be similar to that of the full-scale plant. It is a good idea to use a "typical" carbon as opposed to the "best" carbon. Avoid the situation in which only one company can provide the desired carbon. Such a restriction can lead to increased costs.

The materials used in the pilot plant may be plastic, glass, and/or stainless steel. Kreft et al. (1981) determined that no leaching or adsorption of volatile organic chemicals (VOCs) or TOC occurred from acrylic columns and polyethylene tubing. If organic contamination must be avoided, it is recommended to use stainless steel, polytetrafluoroethylene (PTFE), or glass.

Data Analysis

The effluent concentrations for each EBCT are plotted as breakthrough curves. Breakthrough curves can be plotted as $C_{effluent}$ or $C_{effluent}/C_{influent}$ versus time, number of bed volumes, or volume of water treated. The use of bed volumes normalizes the data so different EBCTs can be directly compared. When interpreting pilot-plant data, remember that reactivated carbon may not give the same results as virgin carbon.

Parallel column and series column data analysis. Pilot-plant data from one GAC column can be used to predict the breakthrough curve for n columns in parallel. Roberts and Summers (1982) found the following relationship between the fraction of organic matter in the effluent from contactor i, f_i, and the fraction of organic matter in the blended effluent, f_c:

$$f_c = \frac{1}{n}\sum_{i=1}^{n} f_i \tag{13-5}$$

Roberts and Summers (1982) found that the number of bed volumes to 50 percent TOC breakthrough could be increased two to three times by using multiple contactors in parallel. By using multiple parallel contactors, the regeneration frequency of the carbon could be decreased, thereby reducing the cost of the system. Example calculations for the parallel column analysis can be found in Snoeyink (1990).

When performing a pilot-plant study for DOC removal using multiple GAC contactors in series, consider that the breakthrough curves will be different for each of the columns. Some components of NOM will appear in the effluent earlier than others because of selective adsorption. Thus, the influent DOC composition to each column in a series will be different. Data from Owen, Amy, and Chowdhury (1993) showed that individual components of NOM have different breakthrough characteristics as compared to bulk NOM. In that study, using Salt River Project water, nonhumic and low-apparent-molecular-weight compounds achieved breakthrough first. Subsequently, humic materials and high-apparent-molecular-weight compounds achieved breakthrough. For this reason, it is desirable to operate the pilot columns long enough so that the lead column becomes saturated with respect to DOC, and the next column in series will be loaded with the same influent DOC concentration.

Contactors can be added to the series arrangement in order to obtain the lowest carbon usage rate. For example, if two contactors were placed in series, the leading contactor could be operated to exhaustion before regeneration. This, of course, would be subject to an economic analysis to determine whether or not the cost of building the additional contactor would be adequately compensated for by lower O&M costs.

SUMMARY

Depending on the nature and quantity of the NOM present in a natural water, GAC adsorption can provide an economical way to meet DBP regulations. Performance of GAC can be optimized by testing various process alternatives, including filter adsorbers, postfilter adsorbers, biologically active carbon, and series and parallel adsorbers. Rapid small-scale column tests and pilot plants are widely used experimental design tools for determining the effectiveness of GAC for a particular application.

REFERENCES

Adham, S.S., V.L. Snoeyink, M.M. Clark, and C. Anselme. 1991. Predicting and Verifying Organics Removal by PAC in an Ultrafiltration System. *Jour. AWWA*, 83(12):68.

Black, B.D., G.W. Harrington, and P.C. Singer. 1996. Reducing Cancer Risks by Improving Organic Carbon Removal. *Jour. AWWA*, 88(6):40.

Campos, C. 1996. Application of Powdered Activated Carbon to a Floc Blanket Reactor-Ultrafiltration Process for Removal of Natural Organic Matter and Atrazine. Masters thesis, University of Illinois.

Carlson, M.A., K.M. Heffernan, C.C. Ziesemer, and E.G. Snyder. 1994. Comparing Two GACs for Adsorption and Biostabilization. *Jour. AWWA*, 86(3):91.

Chadik, P.A., and G.L. Amy. 1987. Molecular Weight Effects on THM Control by Coagulation and Adsorption. *Jour. of Env. Eng.*, 113(6):1234.

Chen, A.S.C., V.L. Snoeyink, and F. Fiessinger. 1980. Activated Alumina Adsorption of Several Humic Substances. *Water Res.*, 14:151.

Clark, R.M., J.Q. Adams, and B.W. Lykins. 1994. DBP Control in Drinking Water: Cost and Performance. *Jour. of Env. Eng.*, 120(4):759.

Crittenden, J.C., J.K. Berrigan, and D.W. Hand. 1986. Design of Rapid Small-Scale Adsorption Tests for a Constant Diffusivity. *Jour. of Water Pollution Control Fed.*, 58:312.

DiGiano, F.A. 1996. Basic Concepts for the Use of GAC Technology in Drinking Water Treatment. *GAC, Membranes and the ICR: A Workshop on Bench-Scale and Pilot-Scale Evaluations.*

Dvorak, B.I. 1996. Practical Considerations in Designing RSSCTs: Seasonal Water Quality Variations and Pressure Drop. *GAC, Membranes and the ICR: A Workshop on Bench-Scale and Pilot-Scale Evaluations.*

El-Rehaili, A.M., and W.J. Weber, Jr. 1987. Correlation of Humic Substance Trihalomethane Formation Potential and Adsorption Behavior to Molecular Weight Distribution in Raw and Chemically Treated Waters. *Water Res.*, 21(5):573.

Fiessinger, F. 1983. Personal communication.

Glaze, W.H., and J.L. Wallace. 1984. Control of Trihalomethane Precursors in Drinking Water: Granular Activated Carbon With and Without Preozonation. *Jour. AWWA*, 76(2):68.

Graese, S., V.L. Snoeyink, and R.G. Lee. 1987. Granular Activated Carbon Filter Adsorber Systems. *Jour. AWWA*, 79(12):64.

Harbold, H. 1984. Granular Activated Carbon Cost Sensitivity Analysis. *Jour. of Env. Eng.*, 110(4):849.

Hartman, D.J., J. DeMarco, D.H. Metz, K.W. Himmelstein, and R. Miller. 1991. DBP Precursor Removal by GAC and Alum Coagulation. In *Proc. 1991 AWWA Annual Conference*. Denver, Colo.: American Water Works Association.

Hooper, S.M., R.S. Summers, S. Hong, G. Solarik, and D.M. Owen. 1996. Improving GAC Adsorption by Optimized Coagulation. *Jour. AWWA*, 88(8):107.

Jacangelo, J.G., J. DeMarco, D.M. Owen, and S.J. Randtke. 1995. Selected Processes for Removing NOM: an Overview. *Jour. AWWA*, 87(1):64.

Jodellah, M.A., and W.J. Weber, Jr. 1985. Controlling Trihalomethane Formation Potential by Chemical Treatment and Adsorption. *Jour. AWWA*, 77(10):95.

Kornegay, B.H. 1987. Determining Granular Activated Carbon Process Design Parameters. In *Proc. 1987 AWWA Annual Conference*. Denver, Colo.: American Water Works Association.

Krasner, S.W., J.J. Westrick, and S. Regli. 1995. Bench and Pilot Testing Under the ICR. *Jour. AWWA*, 87(8):60.

Kreft, P., A. Trussell, J. Lang, M. Kavanaugh, and R. Trussell. 1981. Leaching of Organics from a PVC-Polyethylene-Plexiglas Pilot Plant. *Jour. AWWA*, 73(10):558.

Lee, M.C., V.L. Snoeyink, and J.C. Crittenden. 1981. Activated Carbon Adsorption of Humic Substances. *Jour. AWWA*, 73(8):440.

Love, O.T., and R.G. Eilers. 1982. Treatment for the Control of Trichloroethylene and Related Industrial Solvents in Drinking Water. *Jour. AWWA*, 74:413.

McGuire, M.J., M.K. Davis, S. Liang, C.H. Tate, E.M. Aieta, I.E. Wallace, D.R. Wilkes, J. Crittenden, and K. Vaith. 1989. Optimization and Economic Evaluation of Granular Activated Carbon for Organic Removal. Denver, Colo.: American Water Works Association Research Foundation and American Water Works Association.

Miller, J.M. 1996. GAC: RSSCTs or Pilot Columns. *GAC, Membranes and the ICR: A Workshop on Bench-Scale and Pilot-Scale Evaluations.*

Najm, I.N., V.L. Snoeyink, and B.W. Lykins, Jr. 1991. Using Powdered Activated Carbon: A Critical Review. *Jour. AWWA*, 83(1):65.

Najm, I.N., and S.W. Krasner, 1995. Effects of Bromide and NOM on By-product Formation. *Jour. AWWA*, 87(1):106.

Najm, I., S. Liang, M. Davis, and E.M. Aieta. 1996. Comparison Between RSSCT and Pilot-Scale GAC Performance for TOC Removal. *GAC, Membranes, and the ICR: A Workshop on Bench-Scale and Pilot-Scale Evaluations.*

National Primary Drinking Water Regulations: Disinfectant and Disinfection By-Products: Proposed Rule. 1994. *Fed. Reg.*, 59:38668.

Owen, D.M., G.L. Amy, and Z.K. Chowdhury. 1993. Characterization of NOM and its Relationship to Treatability. In *Proc. 1993 AWWA Annual Conference*. Denver, Colo.: American Water Works Association.

Randtke, S.J., and V.L. Snoeyink. 1983. Evaluating GAC Adsorptive Capacity. *Jour. AWWA*, 75(8):406.

Roberts, P.V., and R.S. Summers. 1982. Performance of Granular Activated Carbon for Total Organic Carbon Removal. *Jour. AWWA*, 74(2):113.

Schimmoller, L.J. 1995. Performance of a Floc Blanket Reactor-Powdered Activated Carbon-Ultrafiltration System for Organics Removal. Masters thesis, University of Illinois.

Singer, P.C. 1994. Control of DBPs in Drinking Water. *Jour. of Env. Eng.*, 120(4):727.

Sketchell, J., H.G. Peterson, and N. Christofi. 1995. Disinfection By-Product Formation after Biologically Assisted GAC Treatment of Water Supplies with Different Bromide and DOC Content. *Water Res.*, 29(12):2635.

Snoeyink, V.L., H.T. Lai, and J.F.Y. Young. 1974. *Chemistry of Water Supply, Treatment and Distribution*. Rubin, A.J., ed. Ann Arbor, Mich.: Ann Arbor Science Publishers.

Snoeyink, V.L. 1990. Adsorption of Organic Compounds. In *Water Quality and Treatment*, 4th ed. New York: McGraw-Hill.

Sontheimer, H. 1983. Personal communication.

Sontheimer, H., and C. Hubele. 1987. The Use of Ozone and Granular Activated Carbon in Drinking Water Treatment. Huck, P.M., and P. Toft, eds. In *Treatment of Drinking Water for Organic Contaminants*. New York: Pergammon Press.

Summers, R.S., S.M. Hooper, G. Solarik, D.M. Owen, and S. Hong. 1995. Bench-Scale Evaluation of GAC for NOM Control. *Jour. AWWA*, 87(8):69.

Symons, J.M., S.W. Krasner, L.A. Simms, and M. Sclimenti. 1993. Measurement of THM and Precursor Concentrations Revisited: The Effect of Bromide Ion. *Jour. AWWA*, 85(1):51.

USEPA. 1992. Water Treatment Plant Simulation Program, Version 1.21 User's Manual. Washington, D.C.: Office of Groundwater and Drinking Water.

Ventresque, C., G. Bablon, and A. Jadas-Hecart. 1991. Ozone: A Means of Stimulating Biological Activated Carbon Reactors. *Ozone: Science and Engineering*, 13(1):91.

Wiesner, M.R., J.J. Rook, and F. Fiessinger. 1987. Optimizing the Placement of GAC Filtration Units. *Jour. AWWA*, 79(12):39.

CHAPTER • FOURTEEN

Application of Advanced Oxidation Processes for the Destruction of Disinfection By-Product Precursors

R. RHODES TRUSSELL

ISSAM NAJM

The term *advanced oxidation processes* (AOPs) was first used by Glaze, Kang, and Chapin (1987) and Aieta, Reagan, and Lang (1988) to describe a process that produces hydroxyl radicals for oxidation of organic and inorganic water impurities. Advanced oxidation processes can have multiple uses in water treatment. Examples include oxidation of synthetic organic chemicals (SOCs), color, taste-and-odor-causing compounds, sulfide, iron, and manganese. In addition, there is interest in evaluating their use for the destruction of precursors of halogenated disinfection by-products (DBPs) before the addition of chlorine.

This chapter reviews the various AOPs available for water treatment, and evaluates their ability to destroy natural organic matter (NOM) (as a measure of DBP precursor material) and their effect on the subsequent formation of DBPs.

CHEMISTRY OF ADVANCED OXIDATION REACTIONS

There have been a number of publications written about the chemistry of advanced oxidation reactions, and there is a technical journal dedicated to the subject (*Journal of Advanced Oxidation*). In this chapter, a total of five AOPs are discussed, including:

- ozone alone
- ozone with hydrogen peroxide addition, also known as the PEROXONE process (McGuire and Davis 1988)
- ozone with ultraviolet (UV) light
- UV light with hydrogen peroxide
- titanium dioxide (TiO_2)-catalyzed UV irradiation

It is important to note that the AOP concept was developed long after ozone had been in use in water treatment, and most early discussions do not include ozonation among the list of AOPs. In this chapter, the ozonation process itself is in the list of AOPs because recent studies have shown that, when applied for the purpose of oxidation, ozone is operated as an AOP in that the process takes advantage of the hydroxyl radical formed during the decomposition of ozone. More importantly, it is critical to understand the processes that inhibit or promote hydroxyl radical formation in the interaction of ozone and waters of various qualities in order to fully understand the appropriate action that can be taken to further promote the process.

Ozone added to water undergoes a series of complex chemical decomposition reactions in which various intermediate ions and radicals are formed and consumed. Figure 14-1 shows a schematic of the ozone reaction cycle in water (Staehelin, Buhler, and Hoigné 1984). The added ozone reacts with hydroxide ions (OH^-) present in the water to form the hydroperoxide radical $HO_2^{\cdot}$ and the superoxide ion O_2^-:

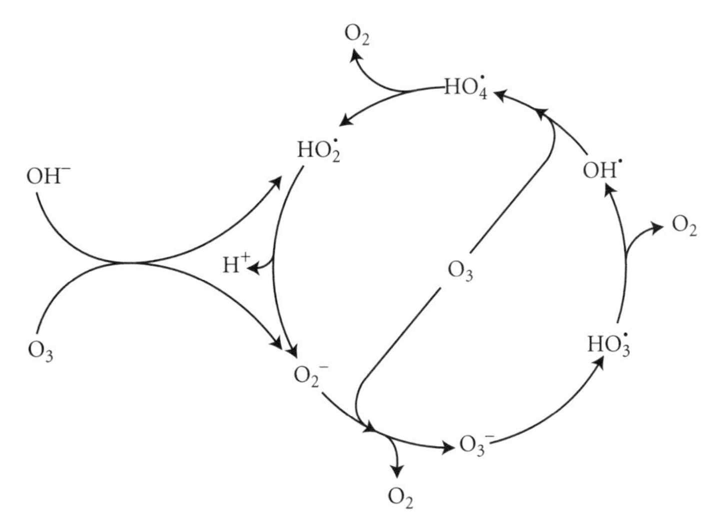

Source: After Staehelin, Buhler, and Hoigné (1984).

Figure 14-1 Reaction pathway of ozone in water

$$O_3 + OH^- \Rightarrow HO_2^{\cdot} + O_2^- \qquad (14\text{-}1)$$

These two reaction products then enter a continuous reaction cycle with the residual ozone present in the water (see Figure 14-1). One of the main intermediates of this cycle is the hydroxyl radical $OH^{\cdot}$, which is a highly reactive, nonselective oxidant that is key to the application of AOPs in drinking water treatment.

When hydrogen peroxide is added to ozone in water, it accelerates the ozone decomposition cycle by further enhancing the production of hydroxyl radicals. A schematic of the effect of hydrogen peroxide addition on the ozone reaction cycle is shown in Figure 14-2 (Staehlin, Buhler, and Hoigné 1984). Note that the basic ozone cycle remains unchanged. However, the hydrogen peroxide added dissociates first into the hydroperoxide ion HO_2^- and hydrogen ion H^+

$$H_2O_2 \Rightarrow HO_2^- + H^+ \qquad (14\text{-}2)$$

The hydroperoxide ion then reacts with molecular ozone to form the superoxide ion O_2^- and hydroxyl radical $OH^{\cdot}$:

$$O_3 + HO_2^- \Rightarrow OH^{\cdot} + O_2 + O_2^- \qquad (14\text{-}3)$$

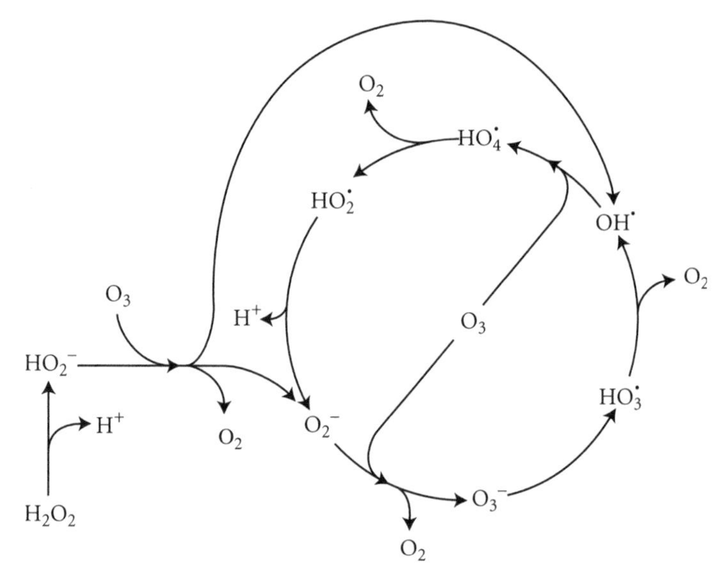

Source: After Staehelin, Buhler, and Hoigné (1984).

Figure 14-2 Reaction pathway of ozone with hydrogen peroxide in water

The superoxide ion and the hydroxyl radical then participate in the ozone reaction cycle as depicted in Figure 14-2. Under most conditions, reactions 14-2 and 14-3 are substantially faster than reaction 14-1. As a result, the addition of hydrogen peroxide to the ozonation process accelerates the production of hydroxyl radicals.

Another AOP that results in the production of hydroxyl radicals is UV irradiation of hydrogen peroxide. In essence, the energy from the UV light breaks the hydrogen peroxide molecule H_2O_2 into two hydroxyl radicals as follows (Volman 1949):

$$H_2O_2 \xrightarrow{h\nu} 2OH^{\cdot} \qquad (14\text{-}4)$$

It is noted that this reaction is relatively slow compared to the ozone-hydrogen peroxide reaction shown in Eq 14-2 (Guittonneau et al. 1988).

In addition, UV irradiation of a water containing ozone directly results in the formation of hydrogen peroxide (Peyton and Glaze 1983), which further reacts with the ozone in the water

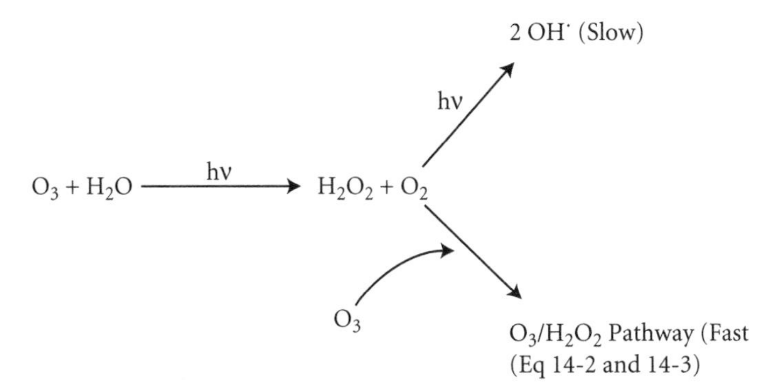

Figure 14-3 Reaction pathway of ozone and UV light in water

and the UV light according to the reactions depicted in Eq 14-2 and 14-3. The overall schematic of the ozone–UV reaction is shown in Figure 14-3. Considering the relatively slow rate of reaction for the H_2O_2–UV pathway compared to that of the H_2O_2–O_3 pathway, it is anticipated that hydroxyl radical formation is dominated by the H_2O_2–O_3 reaction pathway.

Finally, hydroxyl radicals can be produced by a catalytic reaction between UV light and water. Titanium dioxide (TiO_2) can serve as a catalyst for this reaction. The use of TiO_2-catalyzed UV processes has been successfully tested for the destruction of synthetic organic contaminants in water (Lichtin, DiMauro, and Svrluga 1989; Suri et al. 1993; Blake 1994). A reaction mechanism was proposed by Hand, Perram, and Crittenden (1995), a simplified schematic of which is shown in Figure 14-4. In essence, the UV energy excites the electrons in the metal from the valence band to the conduction band resulting in electron holes in the valence band and excess electrons in the conduction band. These holes and excess electrons then undergo reduction and oxidation reactions with oxygen and water (or hydroxide ions), respectively, at the surface of the metal. The reduction reaction involves the reaction of the excess electrons in the conduction band with dissolved oxygen in the water to form the short-lived super oxide ion O_2^-. The oxidation reaction is one in which water or hydroxide ions are oxidized to hydroxyl radicals.

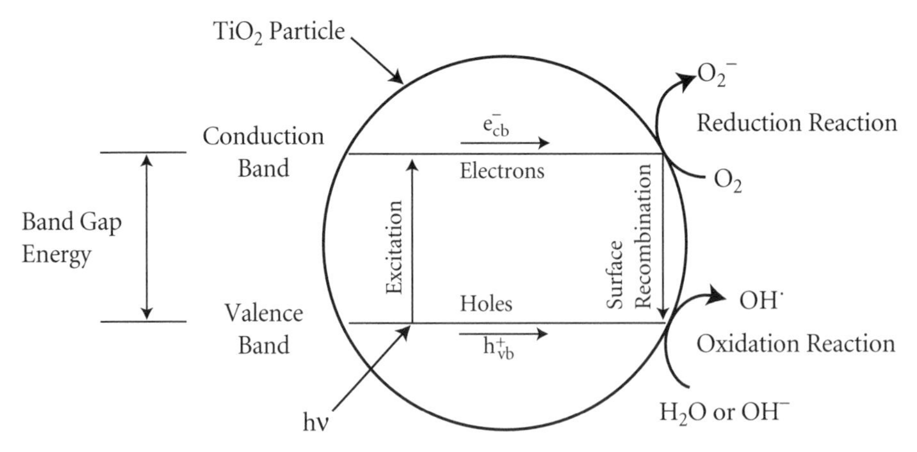

Source: After Hand, Perram, and Crittenden (1995).

Figure 14-4 Schematic reaction mechanism of TiO_2-catalyzed UV oxidation

RATES OF REACTIONS OF HYDROXYL RADICALS WITH WATER CONSTITUENTS

The kinetics of hydroxyl radical reactions with water constituents have been extensively evaluated. Since the focus of this chapter is on the application of AOPs for the destruction of NOM, the following discussion will concentrate on the three main constituents of natural waters that are known to dominate hydroxyl radical reactions. These include NOM, bicarbonate (HCO_3^-) ions, and carbonate (CO_3^{2-}) ions. In an ozone-based AOP process, all of these constituents will compete with the ozone-cycle reactions for the hydroxyl radicals.

For example, consider the O_3–H_2O_2 process in which ozone and hydrogen peroxide are added to natural waters. The resulting reactions are complex. However, for the purposes of this example and in order to illustrate the significance of water quality on the reaction of hydroxyl radicals with NOM, it is assumed that the five constituents that will compete for the hydroxyl radicals are the NOM, bicarbonate ions, and carbonate ions, as well as the ozone and hydrogen peroxide added. A list of the second-order reaction rate constants for hydroxyl radicals with each of these

Table 14-1 Second-order reaction rate constants for the reaction of hydroxyl radicals with water constituents in an O_3–H_2O_2 process

Reactants	k ($M^{-1}s^{-1}$)	Reference
NOM + $OH^{\cdot}$	1.9×10^8	Peyton (1996)
$O_3 + OH^{\cdot}$	20×10^8	Staehelin et al. (1984)
$CO_3^{2-} + OH^{\cdot}$	4.2×10^8	Staehelin and Hoigné (1985)
$H_2O_2 + OH^{\cdot}$	0.27×10^8	Christensen et al. (1982)
$HCO_3^- + OH^{\cdot}$	0.15×10^8	Staehelin and Hoigné (1985)

five constituents is presented in Table 14-1. These constants are based on the reaction rate expression

$$\frac{dC}{dt} = -kC[OH^{\cdot}] \tag{14-5}$$

Where:

C = reactant concentration (i.e., NOM, O_3, CO_3^{2-}, etc.), in mol (M)
k = second-order rate constant, in $M^{-1}s^{-1}$
$[OH^{\cdot}]$ = concentration of hydroxyl radical, in M

A first look at the second-order rate constants in Table 14-1 suggests that the ozone-hydroxyl radical reaction would completely dominate the reaction chemistry due to its high k value relative to the other constituents. However, as noted in the reaction shown in Eq 14-5, the actual rate of reaction of each constituent is calculated as the product of the rate constant with both the concentration of that constituent in solution and the concentration of the hydroxyl radicals. Therefore, in order to determine the half-life or rate of reaction of each constituent, the concentration of each reactant has to be known. For the purposes of this example, estimates of typical concentrations of NOM, O_3, CO_3^{2-}, H_2O_2, and HCO_3^- are made and listed in Table 14-2. A TOC concentration of 3 mg/L (0.25 mM) was used to represent the concentration of NOM. A 4-mM concentration of bicarbonate represents an alkalinity of 200 mg/L as $CaCO_3$, which at pH 8 also corresponds to a carbonate concentration of 0.02 mM. It is noted that the concentration of residual ozone in the process was assumed at 0.01 mM, which is equivalent to 0.5 mg/L of ozone.

Table 14-2 Properties of hydroxyl radicals reactions with various water constituents

Reactants	Typical Conc. (*m*M)	k ($M^{-1}s^{-1}$)	kC (s^{-1})	ε_i (*percent*)
$NOM + OH^{\cdot}$	0.25	1.9×10^8	0.48×10^5	35
$O_3 + OH^{\cdot}$	0.01	20×10^8	0.20×10^5	15
$CO_3^{2-} + OH^{\cdot}$	0.02	4.2×10^8	0.08×10^5	5.9
$H_2O_2 + OH^{\cdot}$	0.009	0.27×10^8	0.0024×10^5	0.18
$HCO_3^- + OH^{\cdot}$	4.0	0.15×10^8	0.60×10^5	44

Finally, the hydrogen peroxide concentration was assumed at 0.009 m*M*, which is equivalent to 0.3 mg/L.

Also listed in Table 14-2 are values of an efficiency factor representative of the relative fraction of hydroxyl radicals participating in each of the five simultaneous reactions. This factor was proposed by Hoigné (1982) to analyze the performance of a process with competing chemical reactions. The efficiency factor for each constituent i, ε_i, is calculated using Eq 14-6.

$$\varepsilon_i = \frac{k_i C_i}{\sum_{i=1}^{n} k_i C_i} \tag{14-6}$$

Where:

n = total number of competing reactions

and other variables are as defined previously.

The higher the value of ε_i, the more dominant constituent i is in the reaction pool. As such, the ε_i values calculated and listed in Table 14-2 clearly show that the NOM and the bicarbonate ion consume almost 80 percent of the hydroxyl radicals present in solution. Although the value of the rate constant for the reaction of ozone with hydroxyl radicals is approximately 10 times higher than that of NOM with hydroxyl radicals, only 15 percent of the available free radicals react with molecular ozone compared to 35 percent with the NOM present in the water. This is a result of the relative concentrations of the two constituents.

The efficiency factor can also be used to analyze the effect of alkalinity and TOC on the effectiveness of the AOP process for NOM oxidation (and thus destruction of DBP precursors). It is noted that the bicarbonate concentration assumed in Table 14-2 represents an alkalinity of 200 mg/L as $CaCO_3$. If the alkalinity of the water were 50 mg/L as $CaCO_3$, the value of ε for NOM would increase to 53 percent, compared to 35 percent at a carbonate alkalinity of 200 mg/L as $CaCO_3$.

The efficiency factor alone cannot be used to determine the actual amount of each constituent oxidized. In other words, the half-life of each reactant cannot be determined without obtaining information on the concentration of hydroxyl radicals in the water. This concentration is then used in the rate expression shown in Eq 14-5 (for $[OH^{\cdot}]$) in order to calculate the reaction rate of each constituent. Hong et al. (1996) developed a method for estimating the steady state concentration of hydroxyl radicals as a function of the steady state concentrations of ozone and hydrogen peroxide residuals in an O_3–H_2O_2 process. The authors assumed that under steady state conditions, the rate of formation of hydroxyl radicals is equal to their rate of consumption. Based on the authors' assumptions and their derivations of a steady state model, Eq 14-7 was developed to estimate the steady state concentration of hydroxyl radicals in an O_3–H_2O_2 process (all concentrations are expressed in mol/L).

$$[OH^{\cdot}]_{ss} = (10^{-4.85}[O_3][H_2O_2][H^+]^{-1}) \div \Big\{1.1 \times 10^8[O_3] + 2.7 \times 10^7[H_2O_2] + 1.9 \times 10^8[TOC] + 1.5 \times 10^7[HCO_3^-] + 4.2 \times 10^8[CO_3^{2-}]\Big\} \quad (14\text{-}7)$$

Assuming the following water quality: pH = 8.0; HCO_3^- = 4 mM; CO_3^{2-} = 0.02 mM, and total organic carbon (TOC) = 3 mg/L, Eq 14-7 is reduced to Eq 14-8:

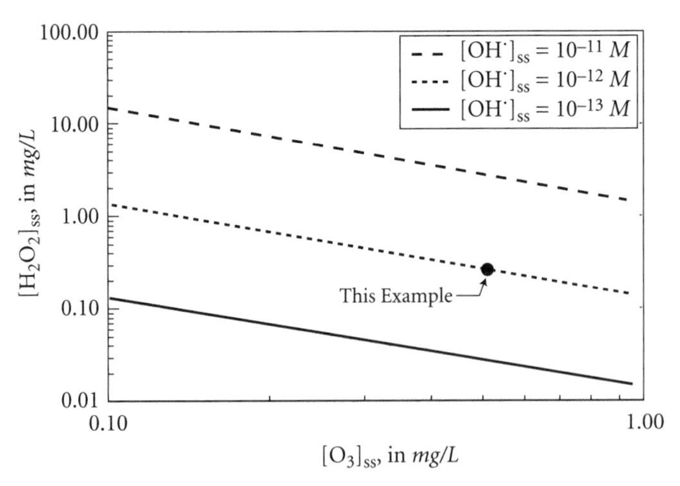

Source: After Hong et al. (1996).

Figure 14-5 Effect of steady state ozone residual and hydrogen peroxide on steady state hydroxyl radicals (pH = 8.0, alkalinity = 200 mg/L as $CaCO_3$, TOC = 3 mg/L)

$$[OH^{\cdot}]_{ss} = \frac{10^{3.15}[O_3][H_2O_2]}{1.1 \times 10^8[O_3] + 2.7 \times 10^7[H_2O_2] + 1.6 \times 10^5} \quad (14\text{-}8)$$

Figure 14-5, based on Eq 14-8, shows plots of the steady state concentration of hydroxyl radicals as a function of the steady state ozone and hydrogen peroxide concentrations in an O_3–H_2O_2 process. The plots are based on the water conditions noted above. The plots show that, in an O_3–H_2O_2 water treatment process with typical residual ozone and hydrogen peroxide concentrations, the concentration of hydroxyl radicals may range from a low of 10^{-13} *M* to a high of 10^{-11} *M*.

Assuming the same ozone and hydrogen peroxide residual concentrations used in the previous example (i.e., 0.5 mg/L ozone and 0.3 mg/L hydrogen peroxide), the hydroxyl radical concentration is estimated at 10^{-12} *M* using Eq 14-8. Now, the reaction rates for each constituent can be estimated (see Table 14-3). It is noted that the $[OH^{\cdot}]_{ss}$ assumption of Hong et al. (1996) represents a snapshot in time of any reactor configuration. Nevertheless, some

Table 14-3 Estimated reaction rates and half-lives of water constituents in an O_3–H_2O_2 process (assumed $[OH^{\cdot}] = 10^{-12}$ *M*; $[O_3] = 0.5$ mg/L; $[H_2O_2] = 0.3$ mg/L; pH = 8.0)

Constituent	Influent Conc.	k ($M^{-1}s^{-1}$)	Rate* (*mol/min*)	Half-Life (*min*)	Half-Life (*hr*)
NOM	3 mg/L as C	1.9×10^8	28×10^{-7}	60	1.0
CO_3^{2-}	0.02 m*M*	4.2×10^8	5.0×10^{-7}	28	0.46
HCO_3^-	4 m*M*	0.15×10^8	36×10^{-7}	770	13

*Rate = $k\,C\,[OH^{\cdot}]_{ss}$

useful perspective can be gained by comparing the half-lives ($\theta_{50\%}$) of the various water constituents in a closed-batch and plug-flow reactor (Table 14-3). The hydraulic retention time (HRT) values listed in Table 14-3 were calculated using the pseudo first-order reaction rate equation (Eq 14-9):

$$\frac{C}{C_o} = e^{-k[OH^{\cdot}]_{ss}\theta} = e^{-k'\theta} \qquad (14\text{-}9)$$

Where:

θ = hydraulic retention time in the reactor

k' = pseudo first-order reaction rate constant ($k' = k[OH^{\cdot}]$ = constant)

and other variables are as defined previously.

The half-life, $\theta_{50\%}$, is then calculated at $C/C_o = 0.5$

$$\theta_{50\%} = \frac{-Ln(0.5)}{k[OH^{\cdot}]} = \frac{0.693}{k'} \qquad (14\text{-}10)$$

The above analysis shows that the rate of NOM reaction with hydroxyl radicals has an unrealistically long half-life (approximately 1 hour). In an O_3–H_2O_2 process in which a 0.5 mg/L ozone concentration and 0.3 mg/L hydrogen peroxide concentration are maintained in contact with the same water assumed in Table 14-3 for a period of 10 minutes, only 10 percent destruction of the NOM would be achieved. This strongly suggests that the use of AOPs for the destruction of DBP precursor material may not be practically or economically feasible in large-scale water treatment applications.

CASE STUDIES OF ADVANCED OXIDATION FOR DBP PRECURSOR REMOVAL

There have been several studies conducted to evaluate the applicability of AOPs for the destruction of DBP precursor material. This section presents four case studies on ozonation, UV–peroxide, and TiO_2-catalyzed UV processes.

Ozonation

Ozonation is the most common AOP being considered and designed for full-scale drinking water treatment applications. A survey conducted by the International Ozone Association, Pan American Group in 1997 showed that more than 200 ozone plants were in operation in the United States as of June of that year. However, the vast majority of these plants are designed primarily for disinfection, taste-and-odor control, and/or color removal, and not specifically for NOM destruction. This is primarily due to the excessively high ozone doses and ozone contact times needed to achieve this objective, as well as concerns about the formation of undesirable organic and inorganic ozonation by-products. To illustrate this point, two examples of the effect of ozonation on the destruction of DBP precursor material will be considered.

Case 1: High-TOC water. This bench-scale ozone test was conducted by the authors using a high-TOC groundwater. Some of the water quality parameters are listed in Table 14-4. It is noted that the pH of the water during ozonation was 9.5. At this high pH, ozone decomposition and the generation of hydroxyl radicals should be rapid.

Table 14-4 Case 1: Values of select water quality parameters

Parameter	Unit	Value
TOC	mg/L	10.3
pH	—	9.5
Alkalinity	mg/L as $CaCO_3$	58
UV-254 absorbance	cm^{-1}	0.255
True color	PCU	17
Turbidity	NTU	0.4

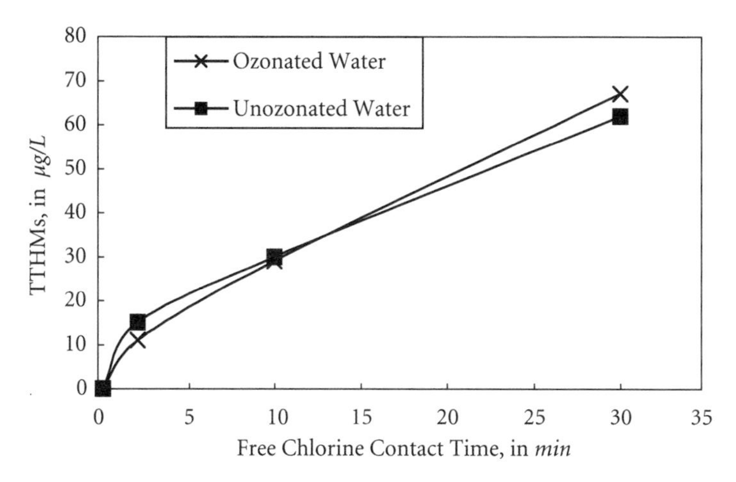

Figure 14-6 Effect of ozonation on the destruction of THM precursor material

The water was ozonated at a dose of 4.5 mg/L (O_3/TOC ratio = 0.44 mg/mg). The ozonated water and a portion of the unozonated water were then dosed with 6 mg/L chlorine (chlorine was added to the ozonated water sample after the ozone residual had completely dissipated). The chlorinated samples were incubated at 25°C and monitored for THM concentration at 2-, 10-, and 30-minute contact times. The results (Figure 14-6) show no effect of ozonation on the levels of THMs formed and, therefore, no apparent destruction of DBP precursor. A validation of this result is discussed later.

Case 2: Bromide-containing water. This second bench-scale ozonation test was conducted by the authors using a surface water sample. Values of some of the water quality parameters are listed in Table 14-5.

The water sample was ozonated with a dose of 2 mg/L (O_3/TOC ratio = 0.67 mg/mg). The ozonated and unozonated samples were then dosed with 3 mg/L chlorine and incubated in the dark at 20°C for 24 hours. (Similar to case 1, chlorine was added to the ozonated water sample after the ozone residual had dissipated.) At the end of the incubation period, the samples were analyzed for chlorine residual and TTHMs. The results of the analyses are listed in Table 14-6 and show that ozonation resulted

Table 14-5 Case 2: Values of select water quality parameters

Parameter	Unit	Value
TOC	mg/L	3.0
pH	—	7.5
Alkalinity	mg/L as $CaCO_3$	NA*
UV-254 absorbance	cm^{-1}	0.053
Bromide	mg/L	0.37
Turbidity	NTU	0.85

*NA: Not available.

Table 14-6 Case 2: Chlorination results from the ozonated and unozonated samples

Parameter	Unit	Ozonated	Unozonated
Ozone dose	mg/L	2	—
Chlorine dose	mg/L	3	3
Chlorine residual	mg/L	0.23	0.25
$CHCl_3$	µg/L	1.1	2.7
$CHCl_2Br$	µg/L	10	22
$CHClBr_2$	µg/L	37	48
$CHBr_3$	µg/L	46	33
Total THMs	**µg/L**	**94**	**105**

in a 10 percent destruction of THM precursor material. It should be noted that the TTHM results are partially influenced by the bromide concentration. As the hydroxyl radical destroys the NOM present in the water, the TOC/Br ratio decreases, resulting in an increase in the production of highly brominated THMs (i.e., bromoform) and a decrease in the highly chlorinated THMs (i.e., chloroform).

The reason why case 2 results showed a measurable destruction of DBP precursor material compared to that achieved in case 1 can best be illustrated by comparing the results to those reported by Riley, Mancy, and Boettner (1978). There, the researchers looked at the effect of ozone/TOC ratio on the destruction of chloroform precursors at various pH levels. Some of their results are reproduced in Figure 14-7. Superimposed on the Riley et al. curves are the results from the two case studies previously presented. Riley et al. (1978) showed that a substantial destruction of chloroform precursor material can be achieved at

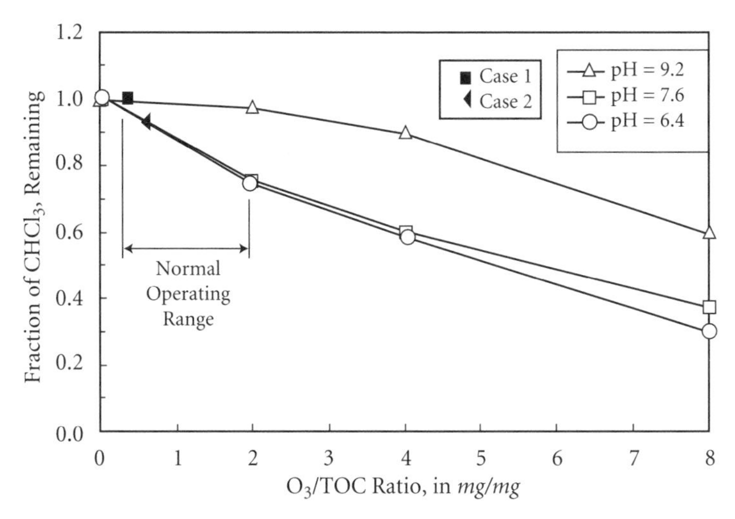

Source: Data from Riley, Mancy, and Boettner (1978).

Figure 14-7 Effect of ozone/TOC ratio and pH on the destruction of chloroform precursors

very high ozone/TOC ratios (in excess of 2 to 4 mg/mg, depending on the water pH value). However, practical applications of ozone rarely use ozone/TOC ratios greater than 1 mg/mg, and virtually never greater than 2 mg/mg. As such, the greatest destruction of chloroform precursor material possible under practical conditions is around 20 to 25 percent. In addition, the results of Riley et al. also showed that ozonation at a pH of 9.2 resulted in substantially less destruction of chloroform precursors compared to that at pH 7.6 and 6.4.

These results are not intuitive because one would expect higher production of hydroxyl radicals at higher pH values. However, the carbonate concentration also increases with increasing pH. Since the reaction rate constant of carbonate with hydroxyl radicals is almost 30 times that for the reaction with bicarbonate (see Table 14-3), the increase in the carbonate concentration results in an increase in the efficiency factor for carbonate, and a reduction in the efficiency factor for NOM (see Table 14-2). As such, an increase in the pH of the water results in a decrease in the efficiency of the ozonation process for the destruction of DBP precursors. It is noted that the results of the two case studies presented above compare well with those of Riley et al. (1978).

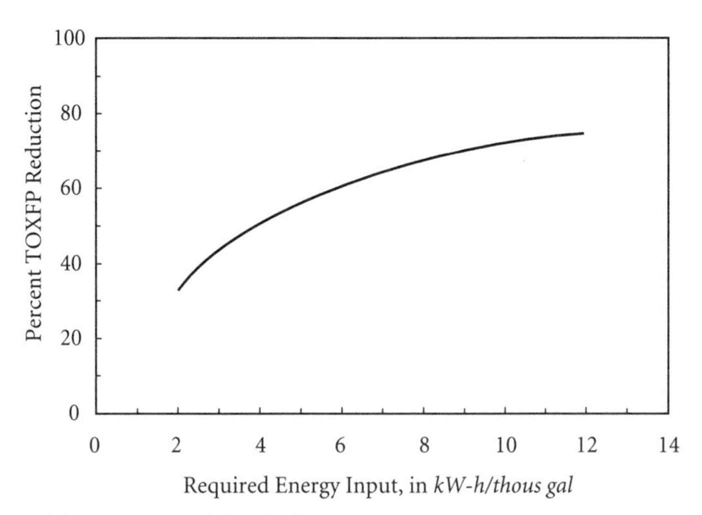

Source: After Symons and Worley (1995).

Figure 14-8 Energy requirement in a UV–H_2O_2 process for TOXFP reduction

UV Radiation With Hydrogen Peroxide Addition

A study was conducted at the University of Houston to investigate the applicability of the UV–H_2O_2 process for the destruction of DBP precursor material (Symons and Worley 1995). A mathematical model was also developed and used to simulate the performance of the process. Based on the results of the study, the authors presented an example of the applicability of this process for DBP precursor destruction in a 1-mgd water treatment plant. The model simulation showed that, for a natural water with a total organic halogen formation potential (TOXFP) of approximately 280 µg/L, the UV–H_2O_2 process can be used to achieve more than 70 percent reduction in the TOXFP of the water (the TOC of the water was measured at 4.3 mg/L). However, the energy required to achieve the precursor destruction was relatively high. The energy requirement is shown in Figure 14-8. In order to achieve 50 percent destruction of DBP precursors, approximately 4 kW-h are required per thous gal of water treated. The energy requirement increases to 10 kW-h/thous gal in order to achieve 70 percent destruction of the DBP precursors.

It should be noted that a hydrogen peroxide dose of approximately 65 mg/L was required. The energy values reported in Figure 14-8 are substantially less than those calculated by Symons and Worley (1995). The reason is that the authors conducted their testing in a single-stage continuous-flow stirred tank reactor (CSTR), but large-scale continuous-flow systems can be designed as three or five mixed reactors in series, thus increasing the process efficiency.

Titanium Dioxide-Catalyzed UV Radiation

The TiO_2-catalyzed UV process has been tested for the oxidation of SOCs, such as pesticides and/or solvents (Lichtin, DiMauro, and Svrluga 1989; Suri et al. 1993; Blake 1994), but little is known about its applicability for DBP precursor destruction in water treatment. Some of the more extensive work in this area has been conducted by researchers at the Michigan Technological University. In their work, the researchers screened 15 TiO_2 powders for their ability to destroy precursors of DBPs (Hand, Perram, and Crittenden 1993, 1995), and identified two promising powders, one of which is commercially available.

A settled, alum-treated water sample was tested in the laboratory. (The researchers noted that raw water samples could not be tested because of the fouling effect of the natural turbidity on the TiO_2 powders.) The water had a TOC of 5.9 mg/L, a pH of 6.4, an alkalinity of 130 mg/L as $CaCO_3$, and a THM formation potential (THMFP) of 290 μg/L (Hand, Perram, and Crittenden 1995). The researchers determined that the optimum TiO_2 powder dose was 1 g/L. The results of the study are shown in Figure 14-9 as a plot of the percent THMFP reduction as a function of energy input to the system. For example, approximately 1 kW-h/thous gal of water treated is required to achieve 45 percent destruction of THM precursors. The results suggest that approximately 3.5 kW-h/thous gal are required in order to achieve 80 percent destruction of THM precursors. It should be noted, however, that this process is not yet ready for large-scale implementation. One of the main unanswered questions at this time is how to recover and reuse the TiO_2 powder from the treated water.

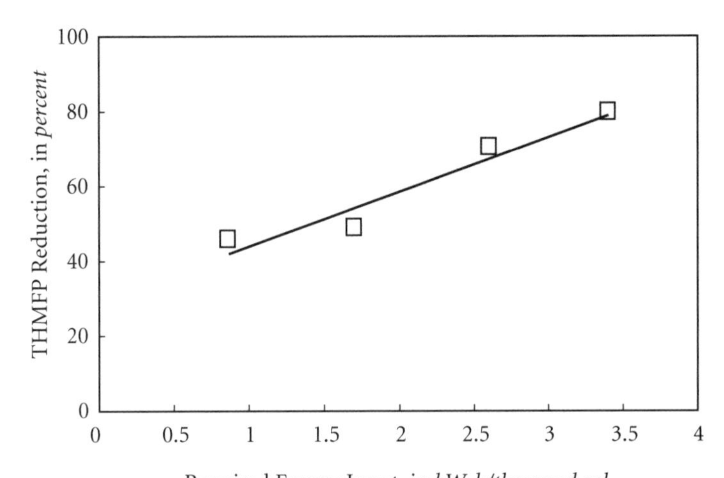

Source: Hand, Perram, and Crittenden (1995).

Figure 14-9 Energy requirement in a TiO_2-catalyzed UV process for THMFP reduction

SUMMARY AND CONCLUSIONS

There are several AOP processes applicable to water treatment, many of which are capable of destroying DBP precursors. The main AOP reaction mechanism is believed to be the production of highly reactive hydroxyl radicals, which are very strong oxidants. However, the cost of these processes when applied for this objective is quite high, which makes them unattractive, by themselves, as DBP precursor removal processes.

Nevertheless, the use of AOPs as part of a DBP control strategy in a water treatment plant is very applicable when combined with other objectives and other disinfectants. For example, AOPs can be used for the oxidation of taste-and-odor-causing compounds, or iron and manganese, which otherwise would require chlorine treatment. This strategy would result in lower use of chlorine, and thus lower DBP levels in the finished water.

REFERENCES

Aieta, E.M., K.M. Reagan, and J.S. Lang. 1988. Advanced Oxidation Processes for Treating Groundwater Contaminated With TCE and PCE: Pilot-Scale Evaluations. *Jour. AWWA*, 80(5):64.

Blake, D.M. 1994. Bibliography of Work on the Photocatalytic Removal of Hazardous Compounds From Water and Air. Golden, Colo.: National Renewable Energy Laboratory.

Christensen, H., K. Sehested, and H. Corfitzen. 1982. Reactions of Hydroxyl Radicals With Hydrogen Peroxide at Ambient and Elevated Temperatures, *Jour. Phys. Chem.*, 86(a):1588–1590.

Glaze, W.H., J.W. Kang, and D.H. Chapin. 1987. The Chemistry of Water Treatment Processes Involving Ozone, Hydrogen Peroxide, and Ultraviolet Radiation. *Ozone: Science and Engineering*, 9:335.

Guittonneau, S., et al. 1988. Comparative Study of the Photodegradation of Aromatic Compounds in Water by UV and H_2O_2/UV. *Environmental Technology Letters*, 9:1115.

Hand, D.W., J.C. Crittenden, and D.L. Perram. 1993. *Destruction of DBP Precursors Using Photoassisted Heterogeneous Catalytic Oxidation.* Denver, Colo.: American Water Works Association Research Foundation and American Water Works Association.

Hand, D.W., D.L. Perram, and J.C. Crittenden. 1995. Destruction of DBP Precursors With Catalytic Oxidation. *Jour. AWWA* (June 1995).

Hoigné, J. 1982. Mechanisms, Rates, and Selectivities of Oxidations of Organic Compounds Initiated by Ozonation in Water. In *Handbook of Ozone Technology and Applications.* Rice, R.G., and A. Netzer, eds. Ann Arbor, Mich.: Ann Arbor Science Publishers.

Hong, A., M.E. Zappi, C.H. Kuo, and D. Hill. 1996. Modeling Kinetics of Illuminated and Dark Advanced Oxidation Processes. *Jour. of Env. Eng., ASCE* (January 1996).

International Ozone Association, Pan American Group. 1997. Stamford, Conn.

Lichtin, N.N., T.M. DiMauro, and R.C. Svrluga. 1989. Catalytic Process for Degradation of Organic Materials in Aqueous and Organic Fluids to Produce Environmentally Compatible Products. US Patent 4,861,484.

McGuire, M.J., and M.K. Davis. 1988. Treating Water With Peroxone: A Revolution in the Making. *WATER/Engineering & Management* (May 1988).

Peyton, G., and W.H. Glaze. 1983. New Developments in the Ozone/UV Process. In *Proc. of the 6th Ozone World Congress*, Washington, D.C.

Peyton, G. 1996. Kinetic Modeling of Free-Radical Water Treatment Processes: Pitfalls, Practicality, and the Extension of the Hoigné/Bader/Staehelin Model. *Jour. Adv. Oxid. Technol.*, 1:2.

Riley, T.L., K.H. Mancy, and E.A. Boettner. 1978. The Effect of Preozonation on Chloroform Production in the Chlorine Disinfection Process. In *Water Chlorination: Environmental Impact and Health Effects*, Vol. 2. Jolley, R.L., H. Gorchev, and H.D. Hamilton, Jr., eds. Ann Arbor, Mich.: Ann Arbor Science Publishers.

Staehelin, J., R.E. Buhler, and J. Hoigné. 1984. Ozone Decomposition in Water Studied by Pulse Radiolysis. 2. OH and HO_4 as Chain Intermediates. *Jour. Phys. Chem.*, 88(24):5999.

Staehelin, J., and J. Hoigné. 1985. Decomposition of Ozone in Water in the Presence of Organic Solutes Acting as Promoters and Inhibitors of Radical Chain Reactions. *Environ. Sci. & Tech.*, 19:1206–1213.

Suri, R.P.S., et al. 1993. Heterogeneous Photocatalytic Oxidation of Hazardous Organic Contaminants in Water. *Jour. Water & Envir.*, 65(5):665.

Symons, J.M., and K.L. Worley. 1995. An Advanced Oxidation Process for DBP Control. *Jour. AWWA* (November 1995).

Volman, D.H. 1949. The Photochemical Decomposition of Hydrogen Peroxide. *Jour. Phys. Chem.*, 17:947.

CHAPTER • FIFTEEN

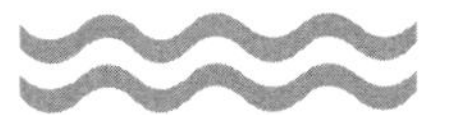

Control of Disinfection By-Products by Pressure-Driven Membrane Processes

JOSEPH G. JACANGELO

Membrane technologies are playing an increasingly important role in the treatment of water sources used for drinking purposes. The greater use of membranes results from more stringent water quality regulations, a decrease in adequate water resources, and an emphasis on water reuse. Additionally, the relative costs of membrane systems have decreased because of technological advancements, thus promoting their use as increasingly viable water treatment alternatives to traditional treatment methods. Whereas in the past, membrane processes were typically used for desalting purposes only, they are now being used for multiple purposes in the United States, including desalting, disinfection by-product (DBP) control, disinfection (pathogen removal), clarification, and removal of inorganic and synthetic organic chemicals (SOCs).

Table 15-1 Applications of pressure-driven membrane processes

Process	Application	Operating Pressure (*bar*)	Operating Pressure (*psi*)	MWCO (*daltons*)[a] or pore size (μm)[b]
Reverse osmosis	Desalting, SOC, IOC removal	10–100	150–1,500	<200[a]
Nanofiltration	Softening, NOM removal	5–9	75–130	200–1,000[a]
Ultrafiltration	Disinfection, particle removal	0.5–3	7–40	1,000–500,000[a]
Microfiltration	Disinfection, particle removal	0.5–3	7–40	0.05–5[b]

NOTES: SOC = synthetic organic chemical
IOC = inorganic chemical
NOM = natural organic matter
MWCO = molecular weight cutoff

This chapter provides an overview of pressure-driven membrane technology for DBP control. Descriptions of membrane processes, system configuration, and basic terminology are presented. The removal of natural organic manner (NOM) by various membrane processes and membrane fouling and control are also discussed.

CLASSIFICATION OF MEMBRANE PROCESSES

Membrane processes can be broadly categorized by their driving forces, i.e., pressure, electrical potential, concentration, and temperature. This chapter will focus primarily on the control of DBPs by those membrane processes that are pressure-driven. Table 15-1 summarizes the various pressure-driven membrane processes and their applications.

Reverse osmosis (RO) traditionally has been used to remove dissolved salts from brackish water and seawater, but can also be used to remove selected SOCs, inorganic chemicals (IOCs), and NOM. In North America there are 142 RO plants operating with a total capacity of 420,000 m^3/day (111 mgd) (Morin 1994). Most of these plants treat brackish groundwaters ranging in total dissolved solids (TDS) concentrations from approximately 1,000 to 8,000 mg/L.

Nanofiltration (NF), also called *membrane softening*, was initially designed for removal of ions contributing to hardness, i.e., calcium and magnesium (Culler and McClellan 1976; Conlon and McClellan 1989), but has been shown to be very effective for removal of NOM (Taylor, Thompson, and Carswell 1987; Taylor et al. 1989; Tan and Amy 1991; Taylor et al. 1992; Laîné et al. 1993; Allgeier and Summers 1995). Over 227,000 m^3/day (60 mgd) of capacity has been installed, primarily in Florida (Bergman 1996).

Ultrafiltration (UF) membranes cover a wide range of molecular-weight cutoffs (MWCOs) and pore sizes. Operational pressures range from 0.5 to 3 bar (7 to 40 psi), depending on the application. "Tight" UF membranes (MWCO 1,000 to 10,000 daltons) may be used to remove some portion of NOM from natural waters, while the objective of "loose" membranes (MWCO >50,000 daltons) is primarily for liquid–solid separation, i.e., particle and microbial removal. The pore size for the latter membranes is approximately 0.001 to 0.005 μm.

A major difference between microfiltration (MF) and loose UF is membrane pore size; that of MF (0.05 μm or greater) is approximately an order of magnitude greater than that of UF. The primary application for this membrane process is particle and microbial removal. Full-scale applications of MF and UF are now being realized. In 1986, there was less than 5,000 m^3/day (1.3 mgd) of installed capacity; by 1996, installed capacity had reached 186,000 m^3/day (49 mgd). Traditionally, most of the plants have been small (50 percent less than 3,800 m^3/day [1 mgd]). The first major MF plant, 18,900 m^3/day (5 mgd), was constructed in San Jose, Calif. (Yoo et al. 1995); however, there are now several drinking water plants in the United States either being planned or under design with capacities over 56,800 m^3/day (15 mgd).

MEMBRANE GEOMETRY AND COMPOSITION

For drinking water treatment, there are several commercially available membrane geometries, including spiral wound, tubular, hollow fiber, capillary, rotating disc, plate and frame, and cassette. In municipal-scale drinking water applications, spiral-wound and hollow-fiber geometries are the most commonly used. For RO and NF, spiral wound is the geometry most often used today. Spiral-wound membranes are manufactured from flat sheets wound around an inner collector tube that collects the water (permeate) that passes through the membrane (Taylor and Jacobs 1996). Feedwater passes through the open end of the spiral-wound element under pressure and either passes through the membrane or flows in a parallel fashion across the membrane and becomes more concentrated as it flows from element to element in the housing (pressure vessel). There are typically between four and eight spiral-wound elements per pressure vessel.

Hollow-fiber geometry is used more often for MF and UF applications because this geometry allows the membrane to be backwashed, a technique by which fouling due to particles and organic materials is controlled. There are two different flow regimes with a hollow-fiber geometry, *inside-out* and *outside-in*. Because the water is flowing through a concentric channel or lumen, the inside-out membrane allows good control over module hydrodynamics. An advantage of the outside-in membrane, as compared to the opposite flow regime, is that there is usually lower head loss through the module.

Membranes can be prepared from both organic (polymers) and inorganic (ceramics, metals, etc.) materials (Aptel and Buckley 1996). Numerous different types of polymer membranes exist, many varying widely in resistance to disinfectants, pH, temperature, and biological and mechanical degradation. Resistance to fouling is also, to a large extent, a function of membrane composition. Laîné and co-workers demonstrated that hydrophilic cellulosic-based membranes were substantially less prone to fouling by organic materials in a surface water as compared to relatively hydrophobic polyacrylonitrile and polysulfone membranes (Laîné et al. 1989). Because of the effect of hydrophobicity on membrane fouling by natural water constituents, many manufacturers are altering organic polymer functional groups to make previously hydrophobic membranes more hydrophilic in nature.

Inorganic membranes are attractive alternatives to polymeric membranes in that they typically have superior resistance to mechanical, chemical, thermal, and biological stresses. Vigorous chemical cleaning results in no loss of structural integrity, and as such, high transmembrane flux recovery after cleaning is achievable. However, these membranes can be brittle, and because their packing density is substantially lower than that of polymeric membranes and their material cost higher, inorganic membranes are not usually employed for large-scale water treatment applications.

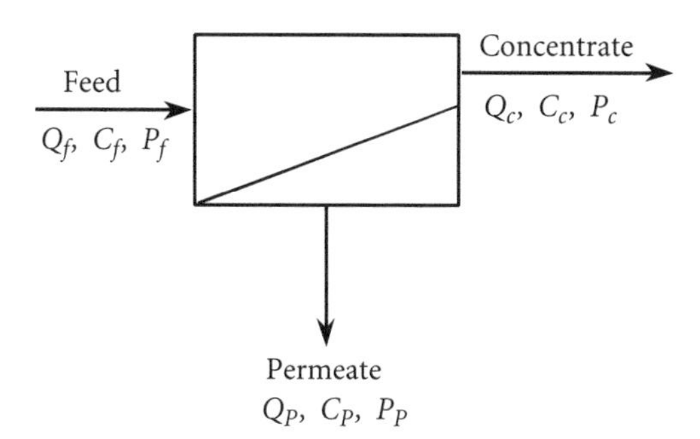

Figure 15-1 General membrane configuration showing various process streams

BASIC MEMBRANE TERMINOLOGY

Figure 15-1 shows a basic membrane process configuration, which is composed of the following three streams: feed, concentrate, and permeate. Each stream can be described in terms of flow (Q), concentration of solute (C), and pressure (P). Following are some basic terms used when discussing pressure-driven membranes.

Transmembrane Pressure

The primary driving force across the membrane is called the *transmembrane pressure*. When considering membrane processes for which osmotic pressure is not a factor (UF and MF), the transmembrane pressure can be described as:

$$\Delta P = \frac{P_f + P_c}{2} - P_p \tag{15-1}$$

Where:
- ΔP = transmembrane pressure, in bar or psi
- P_f = feed pressure, in bar or psi
- P_c = concentrate pressure, in bar or psi
- P_p = permeate pressure, in bar or psi

For RO and NF systems, it is important to consider the osmotic pressure gradient when calculating the net driving pressure, as in

$$\Delta P = \frac{P_f + P_c}{2} - P_p - \Delta\Pi \tag{15-2}$$

Where:
$\Delta\Pi$ = osmotic pressure gradient in bar or psi

Transmembrane Flux

This term, which quantifies the mass or volume transfer through a membrane surface, can be described by Darcy's law. For MF and UF systems

$$J = \frac{\Delta P}{\mu R_m} \tag{15-3}$$

Where:
J = transmembrane flux, in L/h/m^2 or gal/d/ft^2 (gfd)
ΔP = transmembrane pressure, in bar or psi
R_m = hydraulic resistance of the clean membrane, in m^{-1} or ft^{-1}

For NF and RO systems, Eq 15-3 can be modified to account for changes in osmotic pressure, as in

$$J = \frac{\Delta P - \sigma_k \Delta\Pi}{\mu R_m} \tag{15-4}$$

Where:
σ_k = an empirical constant
and other variables are as described previously.

Flux can be determined empirically as

$$J = \frac{Q_p}{A} \tag{15-5}$$

Where:
Q_p = permeate flow rate, in L/h or gal/d
A = membrane surface area, in m^2 or ft^2

Table 15-2 Typical operational characteristics for pressure-driven membrane processes

	Transmembrane Flux		Specific Flux		Recovery
Process	*(gfd)*	*(L/h/m²)*	*(gfd/psi)*	*(L/h/m²/bar)*	*(percent)*
Reverse osmosis	7–15	12–25	<0.2	<5	50–80
Nanofiltration	10–20	17–34	0.1–0.4	5–10	75–90
Ultrafiltration	40–100	68–170	2.5–7	60–250	90–98
Microfiltration	40–100	68–170	2.5–7	60–250	90–98

Table 15-2 shows typical transmembrane fluxes for the various membrane processes. As expected, the more permeable the membrane, the greater the flux. For RO, the fluxes range between 12 and 25 L/h/m^2 (7 and 15 gfd), whereas for MF and UF they usually range between 68 and 170 L/h/m^2 (40 and 100 gfd).

Specific Flux

The water volume transfer through the membrane per unit of membrane area and driving force is the specific flux as described by

$$J_s = \frac{J}{\Delta P} \qquad (15\text{-}6)$$

Where:

J_s = specific flux, in L/h/m^2/bar (gfd/psi)

and other variables are as described previously.

Dividing by the transmembrane pressure or net driving force is a method by which the transmembrane flux is normalized. Specific flux is a useful measure by which various membrane operations can be compared to each other. The specific fluxes for the various membrane processes are shown in Table 15-2. As noted, those for UF and MF can be over an order of magnitude greater than those for RO and NF.

Feedwater Recovery

The feedwater recovery of a system is described as the fraction of the feedwater produced as permeate

$$R = \frac{Q_p}{Q_f} \qquad (15\text{-}7)$$

Where:

R = feedwater recovery
Q_p = permeate flow, in L/h or gal/d
Q_f = feed flow in L/h or gal/d

Typical feedwater recoveries for the four systems are shown in Table 15-2. The highest recoveries (90 to 98 percent) are achieved by UF and MF, whereas typical NF recoveries range from 80 to 90 percent.

SYSTEM CONFIGURATION

Microfiltration and Ultrafiltration

Microfiltration and UF systems can be designed to operate in various different process configurations. Figure 15-2A shows a configuration in which the feedwater is pumped with a crossflow tangential to the membrane. The only pretreatment usually provided is a crude prescreening (usually to remove 50- to 300-μm particles). That water not passing through the membrane is recirculated as concentrate just ahead of the prescreen and blended with additional feedwater. In this mode, pressure on the concentrate side of the membrane is conserved and deducted from the total head requirements. A bleed stream may be used to control the concentration of solids in the recirculation loop. At various times, concentrate waste is discharged from the recirculation loop. A variation of this configuration includes water recirculated back to an open reservoir and blended with additional feedwater.

Microfiltration and UF systems may also be operated in a direct filtration configuration (Figure 15-2B), during which no crossflow is applied. This is often termed *dead-end filtration*. Prescreened water is applied directly onto the membrane. Since all of the prescreened feedwater passes through the membrane between backwashings, there is 100 percent recovery of this water. However, some raw water is normally used to flush the system. In the

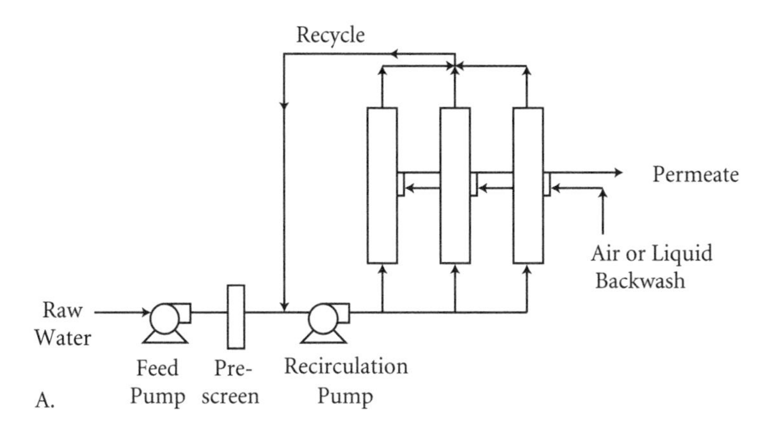

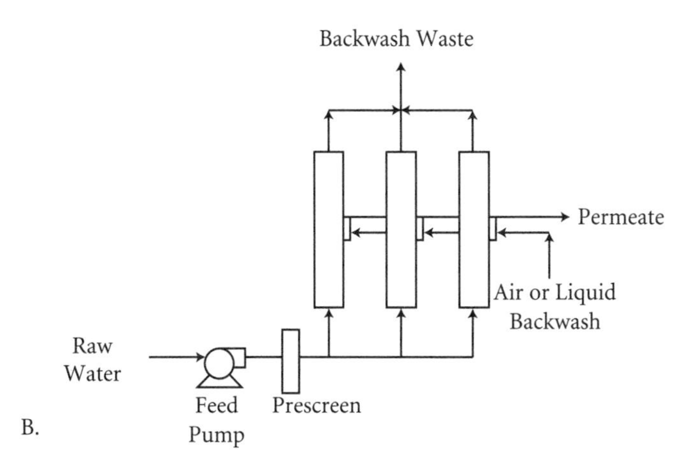

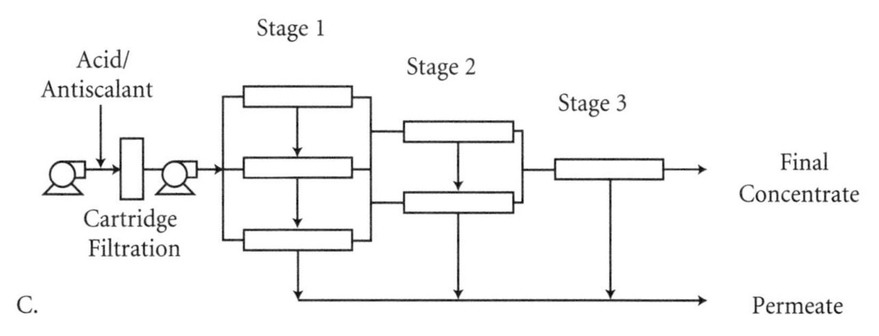

Figure 15-2 Pressure-driven membrane system configurations for (A) MF and UF with recycle, (B) direct flow MF and UF, and (C) RO and NF configured in a three-stage array

direct filtration mode, there is considerable energy savings, because no recirculation of the concentrate is required, and a capital savings, because no recirculation pumps or associated piping are necessary.

Reverse Osmosis and Nanofiltration

Reverse-osmosis and NF systems are usually configured in arrays comprised of stages. Figure 15-2C provides an example of a three-stage array. The first stage is composed of three pressure vessels (which usually contain four to eight membrane elements), the second stage has two pressure vessels, and the final stage, one. In full-scale plants, elements are usually 40 in. in length and 8 in. in diameter. Pilot modules can have 4- or 2.5-in. diameters. Permeate is collected from each pressure vessel. The concentrate from the first stage serves as the feed to the second stage; concentrate from the second stage serves as the feed to the third stage. Consequently, total system recovery is increased through each successive stage of the array. For many groundwater applications, the pretreatment required for RO or NF consists of the addition of acid or antiscalant and cartridge filtration. However, for surface waters, more extensive pretreatment is usually necessary. This may consist of conventional treatment (i.e., coagulation–clarification–porous-media filtration), MF, UF, slow sand filtration, ion exchange, or in some cases, GAC adsorption.

REMOVAL OF NATURAL ORGANIC MATTER

Historical Background on NOM Removal by Membranes

The first membranes, developed in the 1950s, were fabricated from cellulose acetate. Large-scale manufacturing of membranes began in the 1960s and the membranes were used primarily for desalting purposes (Jacangelo et al. 1989; Taylor and Jacobs 1996). With the regulation of THMs in the mid-1970s, there was an increasing interest in Florida to control the formation of DBPs resulting from the chlorination of groundwaters. The use of RO in Florida for control of DBPs began because of the high concentrations of total organic carbon (TOC) in groundwaters and because the technology could also remove high levels of salts from brackish waters. Granular activated carbon (GAC) was not an alternative treatment because of the high cost associated with frequent carbon regeneration required for effective use. In the 1980s, the concept of membrane softening or nanofiltration (NF) was developed (Conlon and McClellan 1989). The NF membranes were considered "loose" RO membranes, removing high levels of NOM and hardness, but only about half of the TDS. A major advantage of NF over RO was its lower operating pressure. The last part of that decade saw increasing research into the application of MF and UF for treatment of surface waters. These technologies were used primarily for particle and microbial removal. However, with the addition of an adsorbent (e.g., powdered activated carbon [PAC]) or coagulant ahead of the membrane process, increased rejection of NOM could be achieved. Microfiltration and UF plants with capacities between 20,000 and 95,000 m^3/day (5 and 25 mgd) are being constructed today.

Reverse Osmosis and Nanofiltration

Both RO and NF are used for removal of NOM. These membrane processes had been traditionally used for groundwaters containing moderate TDS concentrations, and high total hardness, color, and DBP precursors. However, as a result of the D/DBP Rule and the USEPA Information Collection Rule, several utilities have been evaluating NF membranes for surface water applications. The largest surface water NF plant, which will have a treatment capacity of 95,000 m^3/day (25 mgd) after construction, will be located in a suburb of Paris, France (Bourbigot, Cote, and Agbekodo 1993).

That NF is capable of effectively removing precursors of chlorinated organic compounds has been demonstrated by many investigators (Taylor, Thompson, and Carswell 1987; Conlon and McClellan 1989; Taylor et al. 1989, 1992; Reiss and Taylor 1991; Blau et al. 1992; Tan and Sudak 1992; Fu et al. 1993, 1994; Jacangelo et al. 1993, 1995; Laîné et al. 1993; Allgeier and Summers 1995; Chellam et al. 1997). It has also been shown to be a more effective DBP precursor removal process than conventional treatment. Figure 15-3 presents plots of the comparative removal of precursors of simulated distribution system (SDS) THMs and haloacetic acids (HAA6) from an eastern US surface water by MF,

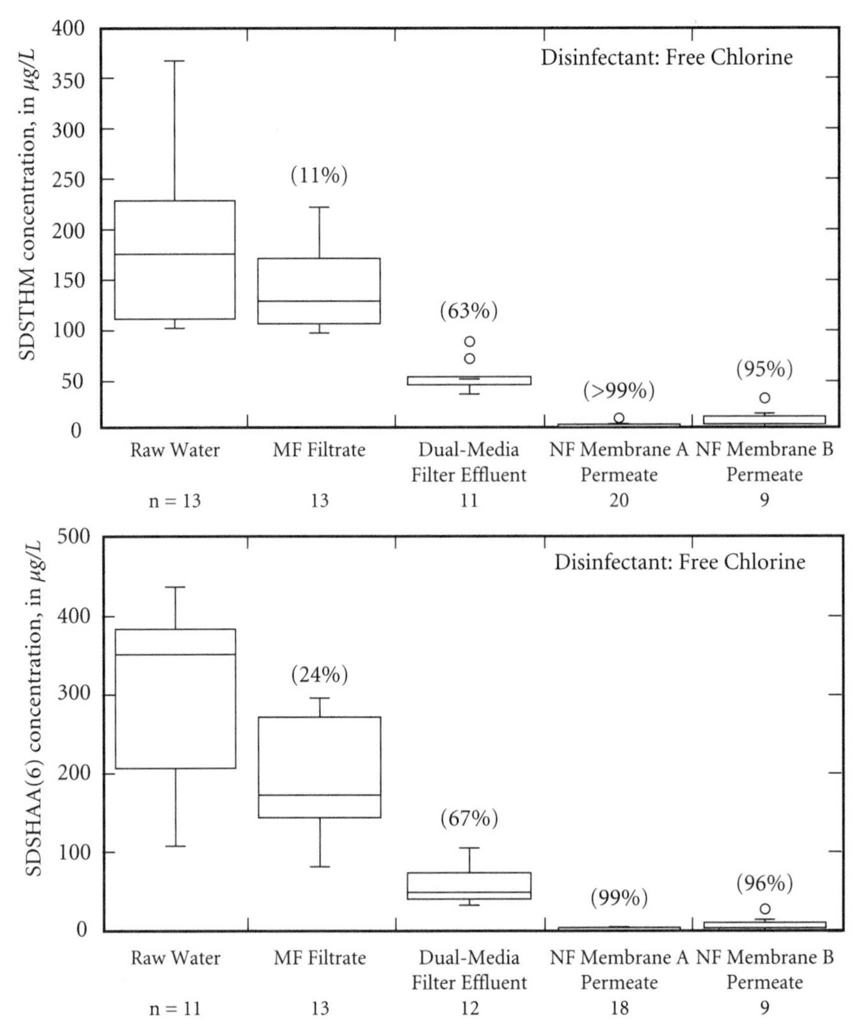

*n = number of observations.
Source: From Chellam et al. (1997).

Figure 15-3 Box and Whiskers plots showing removal of SDSTHM and SDSHAA(6) precursors by MF, conventional treatment, and NF (median percentage removals based on matched pair data are given in parentheses above each box)

Table 15-3 Removal of TOC and precursors of THMs and HAAs by nanofiltration

	Percent Removal		
Reference	TOC	THMs	HAAs
Taylor et al. (1987)	90	96	—
Blau et al. (1992)	97	96	97
Laîné et al. (1993)	54–87	31–95	—
Fu et al. (1994)	>92	>97	—
Allgeier and Summers (1995)	65–95	66–93	66–97
Chellam et al. (1997)	98	99	99

NF, and conventional treatment (Chellam et al. 1997). The data show that MF removed less than 25 percent of the DBP precursors. Conventional treatment removed a high percentage of the precursors (63 and 67 percent of the SDSTHM and SDSHAA6 precursors, respectively). However, when NF was used, both membranes evaluated removed an average of 95 percent or greater of the SDSTHM and SDSHAA6 precursors.

Table 15-3 shows that while high percentage removals of TOC, and THM and HAA precursors are achievable, they can vary from less than 60 to greater than 99 percent. The variation in removal of DBP precursors by NF depends on a number of factors, including the type of membrane used, the water quality of the source being treated, and the conditions under which the membrane system is operated. In general, membranes with MWCOs of less than 1,000 daltons are necessary to remove substantial levels of NOM (Fouroozi 1980; Taylor, Thompson, and Carswell 1987; Summers et al. 1994); a MWCO of less than 500 is usually necessary to reject greater than 90 percent of DBP precursors (Taylor and Jacobs 1996). Additionally, the type of membrane material will also influence its NOM rejection characteristics (McCarty and Aieta 1983).

Feedwater chemical characteristics can affect the removal of NOM and performance of the membrane. Low pH and high ionic strength decrease the rejection of NOM (Braghetta et al. 1997; Hong and Elimelech 1997). These conditions decrease the apparent molecular size of organic matter and its electrostatic repulsion from the membrane surface, thus causing an increased passage of

NOM into the permeate. Further, a decrease in permeate flux due to membrane fouling can result. The presence of divalent cations in feedwater also plays an important role in fouling by NOM (Lahoussine-Turcaud, Wiesner, and Bottero 1990; Jucker and Clark 1994; Hong and Elimelech 1997). By reducing the charge of carboxyl functional groups, cations such as magnesium and calcium can alter the electrostatic repulsion of humic macromolecules, resulting in an increase in deposition of these constituents on the membrane surface.

Reverse osmosis and NF are generally referred to as diffusion-controlled processes in relation to the mass transfer of ions across the membrane (Taylor and Jacobs 1996). However, to date, generalizations regarding the removal mechanisms of TOC or DBP precursors cannot be made. For example, work by Blau et al. (1992) and Laîné et al. (1993) showed that, for specific waters, varying the feedwater recovery or transmembrane pressure did not affect the permeate concentrations of organic DBP precursors. These results suggested that the observed rejection was by physical sieving and that the mass transfer of these materials was not diffusion-controlled. However, studying a variety of waters, Allgeier et al. (1995) used a bench-scale membrane test to show that TOC did vary with pressure and recovery, thus supporting a diffusion-controlled mechanism of rejection. Finally, Lozier and Carlson (1991, 1992) demonstrated, on a highly organic eastern US water, that TOC removal increased with increasing recovery when a cellulose acetate membrane was used but did not vary when a polyvinyl derivative membrane was evaluated. Thus, the mass transfer of NOM was dependent, at least in part, on the membrane surface characteristics.

Bromide has been shown to have a significant affect on the formation of DBPs after chlorination of membrane permeates (Laîné et al. 1993; Summers et al. 1994). Table 15-4 shows removal of bromide by NF in a variety of studies. In general, between 20 and 70 percent removal can be expected depending on the membrane employed and the conditions of operation. If a high percentage of bromide is not removed, a shift in DBP speciation can be observed after chlorination of the permeate as compared to chlorination of the feedwater. A higher percentage of brominated DBPs in each class of halogenated compounds, i.e.,

Table 15-4 Removal of bromide ion by nanofiltration

Reference	Feedwater Bromide Concentration (*mg/L*)	Bromide Removal (%)
Tan and Amy (1991)	0.06	11
Lainé et al. (1993)	0.40–0.60	17–35
Fu et al. (1994)	0.21	24–38
Allgeier and Summers (1995)	0.01–0.25	40–61
Chellam et al. (1997)	0.51	67

THMs or HAAs, have been identified under such conditions (Laîné et al. 1993). As the membrane MWCO decreases, an increase in TOC removal occurs. The resulting increase in the bromide-to-TOC ratio favors the formation of brominated DBPs after chlorination (Summers et al. 1994). However, if enough of the TOC is removed by the membrane, the absolute concentrations of the DBPs will be limited, regardless of such conditions as high bromide levels (Laîné et al. 1993).

Microfiltration and Ultrafiltration

Both MF and UF have been used primarily for removal of microorganisms and particles from surface waters (Jacangelo et al. 1989, 1991; Olivieri et al. 1991b; McGahey 1994; Yoo et al. 1995). Because of the large pore sizes (0.005 to 5 μm), the removal of NOM is substantially less than that observed with either NF or RO. Using a polypropylene 0.2-μm MF membrane, Olivieri et al. (1991a) demonstrated a 15 percent removal of TOC and total trihalomethane formation potential (THMFP) from a flowing stream. Wiesner, Schroelling, and Pickering (1991) reported approximately 30 percent removal of TOC using a 0.05-μm ceramic tubular membrane on blended river waters with an average TOC concentration of 8.2 mg/L. The THMFP was reduced by 10 to 20 percent by both the 0.05-μm and a 0.2-μm ceramic tubular membrane.

Feedwater pretreatment of UF and MF systems can be used to improve the level of removal of various natural water constituents, such as DBP precursors; it is also used to increase or maintain transmembrane flux and/or to retard fouling. The two most common types of pretreatment are coagulant and PAC addition. Most studies using coagulant addition have focused on the use of aluminum sulfate or ferric sulfate at concentrations ranging from 5 to 50 mg/L (Wiesner, Veerapaneni, and Brejchová 1992; Coffey et al. 1993). Metal coagulants are normally injected into the raw water through a feed line, after which the water then enters a reservoir for mixing and coagulation. Retention times in the reservoir range from 5 minutes to approximately one hour. Some investigators have injected the coagulant into a mixing pipe loop before addition to the feedwater of the MF membrane (Olivieri et al. 1991a). It appears that the coagulation of smaller particles into larger ones reduces the penetration of various materials, including colloidal and large organic macromolecules, into pores of the membrane. Coagulation may also increase the size of the particles comprising the cake layer on the membrane, increasing its permeability, and as a consequence, enhancing the feedwater flux through the membrane (Wiesner, Veerapaneni, and Brejchová 1992).

Olivieri et al. (1991a) demonstrated that THMFP removal from a surface water could be increased from 15 to 60 percent by the addition of 10 to 15 mg/L of ferric chloride as a pretreatment to MF. Similar removals were observed by Wiesner, Schroelling, and Pickering (1991) while working with tubular ceramic membranes. Coagulation improved the removal of TOC from 30 percent to approximately 60 percent using a 0.05-μm ceramic tubular membrane. Removal of THMFP also improved by approximately 30 percent using this pretreatment.

To enhance NOM removal by MF or UF, adsorbents, such as PAC or iron oxide particles (Chang and Benjamin 1996), have also been added to the feedwater for pretreatment. Table 15-5 summarizes the increased removal of NOM in various studies before and after an adsorbent was added as a pretreatment. The use of PAC for pretreatment for UF has been fairly well documented in the recent literature (Laîné, Clark, and Mallevialle 1990; Adham et al. 1991; Adham 1993; Jacangelo et al. 1995; Clark, Anselme, and Baudin 1996). This research has led to a 55,000-m^3/d (15-mgd) full-scale application at Vigneux-sur-Seine, France, a suburb of Paris. The plant uses an 8-mg/L PAC dose (Anselme et al. 1997). Less research has been conducted on PAC pretreatment of feedwaters before MF. Scanlan et al. (1997) applied PAC as a pretreatment for direct flow MF, but observed only 11 percent removal of the raw water TOC

Table 15-5 Removal of TOC by ultrafiltration and microfiltration with adsorbent pretreatment

Reference	Process	Influent TOC (*mg/L*)	% TOC Removal (UF or MF only)	Adsorbent Dose (*mg/L*)	% TOC Removal (UF or MF with adsorbent)
Laîné (1990)	UF + PAC	3	40	250	85
Adham et al. (1991)	UF + PAC	2.8–3.1	5–10	25	32–47
Jacangelo et al. (1995)	UF + PAC	1.1–11	12–14	10–200	17–82
Chang and Benjamin (1996)	UF + IOP	4.5	10–23	170	47–85
Scanlan et al. (1997)	MF + PAC	4.5	6.5	20	11

NOTES:
MF = microfiltration
UF = ultrafiltration
PAC = powdered activated carbon
IOP = iron oxide particles

(4.5 mg/L) at a dose of 20 mg/L and a contact time of 25 minutes. Longer contact times (up to 3 to 5 days) have been employed to use more of the adsorbent capacity of the PAC (Pirbazari, Badriyha, and Ravindran 1992). In this case, the feedwater flowed into a reactor to which a PAC slurry was added continuously. The water then passed across a membrane through which a fraction passed, the concentrate stream being recycled back to the reactor.

MEMBRANE FOULING AND CONTROL

A major obstacle to maintaining a constant operation of membranes is the degradation of transmembrane flux due to membrane fouling. Figure 15-4 conceptually describes membrane fouling for a typical nanofiltration membrane. Reduction of flux that can be restored by mechanical or chemical means is termed *reversible fouling.* Membranes that lose flux, which cannot be restored, are referred to as *irreversibly fouled membranes.* This is due to the deposition of materials on the membrane surface and/or in pores. Membrane fouling can be broadly described in four ways. Fouling due to *scaling* is caused by the precipitation of salts within the membrane element or module due to their increased

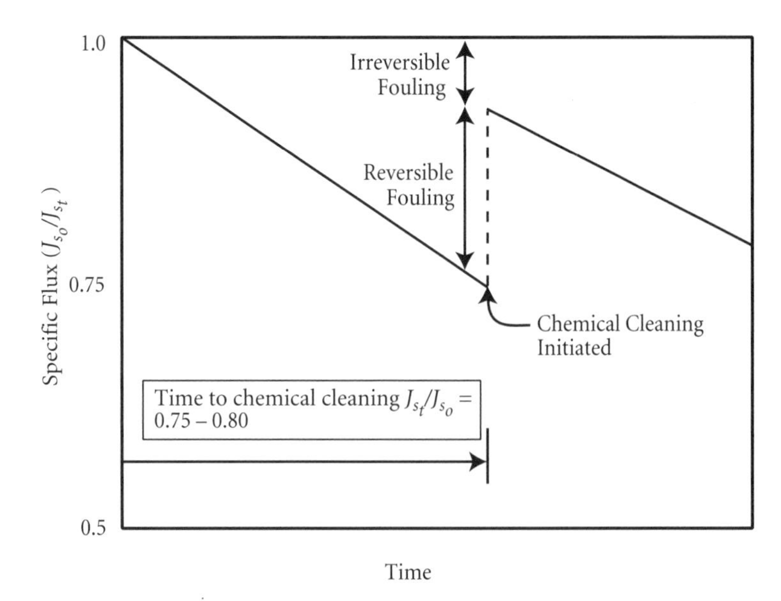

Figure 15-4 Conceptual plot of specific flux decline and initiation of chemical cleaning for a nanofiltration membrane

concentration during operation. Microfiltration and UF membranes are usually not subject to this type of fouling because they reject little or no salts. *Biofouling* causes a reduction of membrane productivity due to propagation of biological agents within or on the membrane surface. *Particle fouling* is due to plugging of the membrane surface, feed channel, or fiber inlets by particulate material. *Chemical fouling* results in a decrease in productivity due to the interaction of chemical solutes with the membrane surface layer.

Membrane system operational parameters influence the rate at which a membrane fouls. These parameters include permeate flux, feedwater recovery, membrane composition, and water chemistry (Wiesner and Aptel 1996; Braghetta, DiGiano, and Bell 1997; Chellam et al. 1997). After a membrane is fouled, it must be chemically cleaned. The variables considered in cleaning membranes include frequency of cleaning, duration of cleaning, chemicals and their concentrations, cleaning and rinse volumes, temperature of cleaning, recovery and reuse of cleaning chemicals, and neutralization and disposal of cleaning chemicals. A variety of different agents can

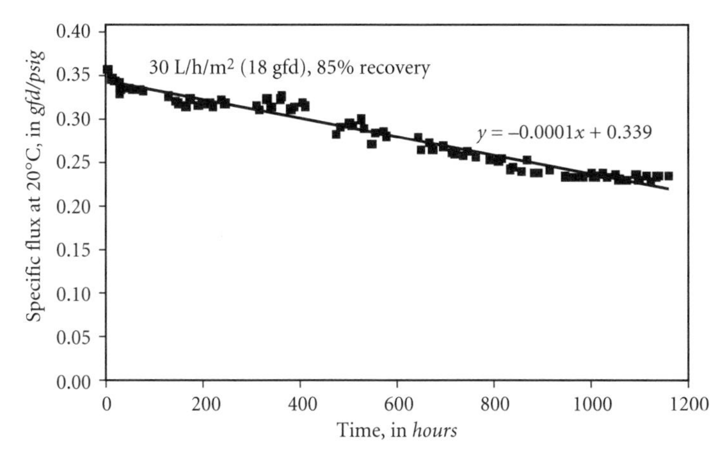

Source: From Chellam et al. (1996).

Figure 15-5 System specific flux profile for a two-stage NF pilot plant

be used for the chemical cleaning of membranes, including detergents, acids, bases, oxidizing agents, sequestering agents, and enzymes. Chlorine is often used in doses ranging from 2 to 50 mg/L with polysulfone or polysulfone-derivative materials, which are resistant to the compound.

One determinant in deciding to initiate chemical cleaning is the change in the ratio of the initial specific membrane flux (J_{s_0}) to the specific flux at some point during the operational run (J_{s_t}). For RO and NF, when $J_{s_t}/J_{s_0} = 0.85 - 0.90$, cleaning is usually performed. In comparison, the decline in specific flux for MF and UF is substantially greater before chemical cleaning is initiated ($J_{s_t}/J_{s_0} = 0.25 - 0.30$). Figure 15-5 shows an example of a specific flux profile for a sulfonated polyethersulfone NF membrane operated with MF as pretreatment (Chellam et al. 1996). The two-stage system was operated at a transmembrane flux of 30 L/h/m^2 (18 gfd) and 85 percent recovery. The observed fouling, which was linear, was determined to be approximately 10^{-4} gfd/psig/h. Using a 20-percent decline in specific flux as a criterion for initiation of chemical cleaning, the cleaning interval was approximately 25 days.

SUMMARY

As a result of the current and anticipated regulations on DBPs, interest in NOM removal has been increasing substantially in the US water supply community. This interest has brought about an increasing focus on employing membrane filtration for control of DBPs in finished waters. Among the DBP-control technologies, pressure-driven membrane processes have the greatest potential to achieve the highest sustained removal of NOM and DBP precursors. In the past, the costs of the process have been higher than those of coagulation and GAC adsorption; however, the rate of decrease in costs over the past five years has been greater than those associated with other treatments.

Membrane filtration for NOM removal had been limited traditionally to groundwaters because of the potential of surface waters to foul NF and RO membranes. However, with greater understanding of pretreatment techniques, surface water treatment using membranes is now being realized. In addition to NF and RO, the two lower-pressure processes, MF and UF, can also be used to control DBP formation by the addition of an adsorbent or coagulant for pretreatment.

ACKNOWLEDGMENT

The author thanks Shankararaman Chellam for reviewing this manuscript and Jennifer Day for helping with graphics.

REFERENCES

Adham, S. 1993. Evaluation of the Performance of Ultrafiltration With Powdered Activated Carbon Pretreatment for Organics Removal. PhD thesis, University of Illinois.

Adham, S.A., V.L. Snoeyink, M.M. Clark, and J.L. Bersillon. 1991. Predicting and Verifying Organics Removal by PAC in an Ultrafiltration System. *Jour. AWWA*, 83(12):68.

Allgeier, S., and S. Summers. 1995. Evaluating NF for DBP Control With the RBSMT. *Jour. AWWA*, 87(3):87.

Anselme, C., J.M. Laîné, I. Baudin, and M.R. Chevalier. 1997. Drinking Water Production By UF And PAC Adsorption: First Months of Operation for a Large Capacity Plant. In *Proc. of AWWA Membrane Technology Conference.* Denver, Colo.: American Water Works Association.

Aptel, P., and C.A. Buckley. 1996. Categories of Membrane Operations. In *Water Treatment Membrane Processes.* Mallevialle, J., P.E. Odendaal, and M.R. Wiesner, eds. New York: McGraw-Hill.

Bergman, R.A. 1996. Cost of Membrane Softening in Florida. *Jour. AWWA,* 88(5):32.

Blau, T.J., J.S. Taylor, K.E. Morris, and L.A. Mulford. 1992. DBP Control by Nanofiltration: Cost and Performance. *Jour. AWWA,* 84(12):104.

Bourbigot, M.M., P. Cote, and K. Agbekodo. 1993. Nanofiltration: An Advanced Process for the Production of High Quality Drinking Water. In *Proc. of AWWA Membrane Technology Conference.* Denver, Colo.: American Water Works Association.

Braghetta, A., F.A. DiGiano, and W.P. Ball. 1997. Nanofiltration of Natural Organic Matter: pH and Ionic Strength Effects. *Jour. of Env. Eng.,* 123(7):628.

Chang, Y., and M.M. Benjamin. 1996. Iron Oxide Adsorption and UF to Remove NOM and Control Fouling. *Jour. AWWA,* 88(12):74.

Chellam, S., J.G. Jacangelo, T. Bonacquisti, and B. Schauer. 1997. Effect of Pretreatment on Surface Water Nanofiltration. *Jour. AWWA,* 89(10):77.

Chellam, S., J.G. Jacangelo, T.P. Bonacquisti, and B.W. Long. 1996. Efficacy of Dual Membrane Systems for Removal of DBP Precursors From Surface Water. In *Proc. 1996 AWWA Annual Conference.* Denver, Colo.: American Water Works Association.

Clark, M.M., C. Anselme, and I. Baudin. 1996. Membrane-Powdered Activated Carbon Reactors. In *Water Treatment Membrane Processes.* Mallevialle, J., P.E. Odendaal, and M.R. Wiesner, eds. New York: McGraw-Hill.

Conlon, W.J., and S.A. McClellan. 1989. Membrane Softening: A Treatment Process Comes of Age. *Jour. AWWA,* 81(11):47.

Culler, P.L., and S.A. McClellan. 1976. A New Approach to Partial Demineralization. In *Proc. of FS/AWWA, FW and FPCA Joint Annual Conference.*

Fouroozi, J. 1980. Nominal Molecular Weight Distribution of Color, TOC, THM Precursors and Acid Strength in a Highly Organic Potable Water Source. Masters thesis. Orlando, Fla.: University of Central Florida.

Fu, P., D. Bedford, K. Thompson, C. Spangenberg, and S. Malloy. 1993. A Membrane Pilot Study on Highly Colored Groundwater in Southern California. In *Proc. of AWWA Membrane Technology Conference* (supplement). Denver, Colo.: American Water Works Association.

Fu, P., H. Ruiz, K. Thompson, and C. Spangenberg. 1994. Selecting Membranes for Removing NOM and DBP Precursors. *Jour. AWWA,* 86(12):55.

Fu, P., K. Thompson, and C. Spangenberg. 1994. A Pilot Study on Groundwater Natural Organics Removal by Low-Pressure Membranes. In *Proc. of AWWA Water Quality Technology Conference.* Denver, Colo.: American Water Works Association.

Hong, S., and M. Elimelech. 1997. Chemical and Physical Aspects of Natural Organic Matter (NOM) Fouling of Nanofiltration Membranes. *Jour. of Membrane Science*, 132:159.

Jacangelo, J.G., E.M. Aieta, K.E. Carns, E.W. Cummings, and J. Mallevialle. 1989. Assessing Hollow-Fiber Ultrafiltration for Particulate Removal. *Jour. AWWA*, 81(11):68.

Jacangelo, J.G., J. DeMarco, D.M. Owen, and S.J. Sandtke. 1995. Selected Processes for Removing NOM: An Overview. *Jour. AWWA*, 87(1):64.

Jacangelo, J.G., J.M. Laîné, K.E. Carns, E.W. Cummings, and J. Mallevialle. 1991. Low-Pressure Membrane Filtration for Removing *Giardia* and Microbial Indicators. *Jour. AWWA*, 83(9):97.

Jacangelo, J.G., J.M. Laîné, E.W. Cummings, and S.S. Adham. 1995. UF With Pretreatment for Removing DBP Precursors. *Jour. AWWA*, 87(3):100.

Jacangelo, J.G., J.M. Laîné, E.W. Cummings, A. Deutschmann, J. Mallevialle, and M.R. Wiesner. 1993. Evaluation of Ultrafiltration Membrane Pretreatment and Nanofiltration of Surface Waters. Denver, Colo.: American Water Works Association Research Foundation and American Water Works Association.

Jacangelo, J.G., N.L. Patania, and R.R. Trussell. 1989. Membranes in Water Treatment. *Civil Engineering*, 59(5):68.

Jucker, C., and M.M. Clark. 1994. Adsorption of Aquatic Humic Substances and Hydrophobic Ultrafiltration Membranes. *Jour. of Membrane Science*, 97:37.

Lahoussine-Turcaud, V., M.R. Wiesner, and J.Y. Bottero. 1990. Fouling in Tangential-Flow Ultrafiltration: The Effect of Colloid Size and Coagulation Pretreatment. *Jour. of Membrane Science*, 52:173.

Laîné, J.M., M.M. Clark, and J. Mallevialle. 1990. Ultrafiltration of Lake Water: Effects of Pretreatment on Organic Partitioning, THM Formation Potential, and Flux. *Jour. AWWA*, 82(12):82.

Laîné, J.M., J.P. Hagstrom, M.M. Clark, and J. Mallevialle. 1989. Effects of Ultrafiltration Membrane Composition. *Jour. AWWA*, 81(11):61.

Laîné, J.M., J.G. Jacangelo, K.E. Carns, E.W. Cummings, and J. Mallevialle. 1993. Influence of Bromide on Low-Pressure Membrane Filtration for Controlling DBPs in Surface Waters. *Jour. AWWA*, 85(6):87.

Lozier, J., and M. Carlson. 1991. Organics Removal From Eastern U.S. Surface Waters Using Ultra-Low Pressure Membranes. In *Proc. of AWWA Membrane Technology Conference*. Denver, Colo.: American Water Works Association.

Lozier, J., and M. Carlson. 1992. Nanofiltration Treatment of a Highly Organic Surface Water. In *Proc. 1992 AWWA Annual Conference*. Denver, Colo.: American Water Works Association.

McCarty, P.L., and E.M. Aieta. 1983. Chemical Indicators and Surrogate Parameters for Water Treatment. In *Proc. 1983 AWWA Annual Conference*. Denver, Colo.: American Water Works Association.

McGahey, C.L. 1994. Mechanisms of Virus Capture From Aqueous Suspension by a Polypropylene, Microporous, Hollow Fiber Membrane Filter. PhD dissertation. The Johns Hopkins University.

Morin, O.J. 1994. Membrane Plants in North America. *Jour. AWWA*, 86(12):42.

Olivieri, V.P., D.Y. Parker, Jr., G.A. Willinghan, and J.C. Vickers. 1991a. Continuous Microfiltration of Surface Water. In *Proc. of AWWA Membrane Technology Conference.* Denver, Colo.: American Water Works Association.

Olivieri, V.P., G.A. Willinghan, J.C. Vickers, and C. McGahey. 1991b. Continuous Microfiltration of Secondary Wastewater Effluent. In *Proc. of AWWA Membrane Technology Conference.* Denver, Colo.: American Water Works Association.

Pirbazari, M., B.N. Badriyha, and V. Ravindran. 1992. MF-PAC for Treating Waters Contaminated With Natural and Synthetic Organics. *Jour. AWWA*, 84(12):95.

Reiss, C.R., and J.S. Taylor. 1991. Membrane Pretreatment of a Surface Water. In *Proc. of AWWA Membrane Technology Conference.* Denver, Colo.: American Water Works Association.

Scanlan, P., B. Pohlman, S. Freeman, B. Spillman, and J. Mark. 1997. Membrane Filtration for the Removal of Color and TOC From Surface Water. In *Proc. of AWWA Membrane Technology Conference.* Denver, Colo.: American Water Works Association.

Summers, R.S., M.A. Benz, H.M. Shukairy, and L. Cummings. 1994. Effect of Separation Processes on the Formation of Brominated THMs. *Jour. AWWA*, 86(1):88.

Tan, L., and G.L. Amy. 1991. Comparing Ozonation and Membrane Separation for Color Removal and Disinfection By-product Control. *Jour. AWWA*, 83(5):74.

Tan, L., and R. Sudak. 1992. Color Removal From Groundwater by Membrane Processes. *Jour. AWWA*, 84(1):79.

Taylor, J.S., and E.P. Jacobs. 1996. Reverse Osmosis and Nanofiltration. *Water Treatment Membrane Processes.* Mallevialle, J., P.E. Odendaal, and M.R. Wiesner, eds. New York: McGraw-Hill.

Taylor, J.S., L.A. Mulford, S.J. Duranceau, and W.M. Barrett. 1989. Cost and Performance of a Membrane Pilot Plant. *Jour. AWWA*, 81(11):52.

Taylor, J.S., C.R. Reiss, K.E. Morris, P.S. Jones, D.K. Smith, T.L. Lyn, and S.J. Duranceau. 1992. Reduction of Disinfection By-product Precursors by Nanofiltration. Report to USEPA.

Taylor, J.S., D.M. Thompson, and J.K. Carswell. 1987. Applying Membrane Processes to Groundwater Sources for Trihalomethane Precursor Control. *Jour. AWWA*, 79(8):72.

Wiesner, M.R., and P. Aptel. 1996. Mass Transport and Permeate Flux and Fouling in Pressure-Driven Processes. *Water Treatment Membrane Processes.* Mallevialle, J., P.E. Odendaal, and M.R. Wiesner, eds. New York: McGraw-Hill.

Wiesner, M.R., D.C. Schroelling, and K. Pickering. 1991. Permeation Behavior and Filtrate Quality of Tubular Ceramic Membranes Used for Surface Water Treatment. In *Proc. of AWWA Membrane Technology Conference.* Denver, Colo.: American Water Works Association.

Wiesner, M.R., S. Veerapaneni, and Brejchová. 1992. Improvements in Membrane Microfiltration Using Coagulation Pretreatment. *Chemical Water and Wastewater Treatment II.* Klute, R., and H.H. Hahn, eds. New York: Springer-Verlag.

Yoo, R.S., D.R. Brown, R.J. Pardini, and G.D. Bentson. 1995. Microfiltration: A Case Study. *Jour. AWWA*, 87(3):38.

CHAPTER • SIXTEEN

Biofiltration for Removal of Natural Organic Matter

RAYMOND M. HOZALSKI
EDWARD J. BOUWER

INTRODUCTION

The benefits of using biological processes in drinking water treatment derive from the ability of bacteria to remove a portion of the natural organic matter (NOM) from drinking water by converting it into inorganic carbon (CO_2) and more biomass (cells). In drinking water treatment, the optimum location for biodegradation of NOM is in a rapid media filter. This is because (1) low organic loading coupled with high hydraulic loading does not favor a suspended growth process (i.e., washout), (2) a filter bed offers high specific surface area for attachment of bacteria and biofilm formation, and (3) costs are low due to retrofit of existing filtration systems. *Biologically active filters* or *biofilters* are the terms used to describe these dual-purpose porous media filters operated for removal of both particles and soluble biodegradable

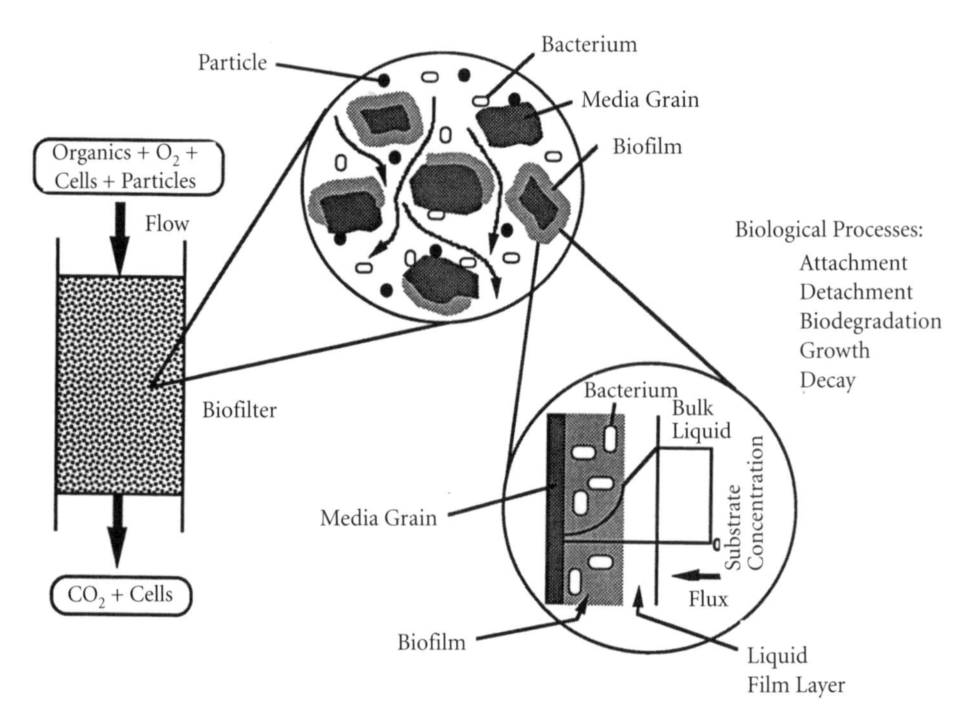

Figure 16-1 Schematic diagram illustrating the processes occurring in a drinking water biofilter

organic compounds. A schematic diagram of a biofilter is shown in Figure 16-1. The only significant difference between a biofilter and a standard rapid media filter is that, in a biofilter, bacteria are permitted or even encouraged to grow as biofilm on the surface of the filter-media grains.

Chemical coagulation and sedimentation can be an effective method for removing 50 percent or more of the NOM from water supplies (Sinsabaugh et al. 1986; Krasner and Amy 1995). However, simply reducing the bulk NOM level as measured by total organic carbon (TOC) concentration using traditional physicochemical processes may not be sufficient if some biodegradable organic carbon remains, as reflected in measurements such as assimilable organic carbon (AOC) or biodegradable dissolved organic carbon (BDOC). Assimilable organic carbon is a parameter used to monitor microbial regrowth potential in source and treated water supplies (van der Kooij, Visser, and Hijnen 1982). LeChevallier et al. (1991) suggested that AOC-P17 levels of 50 μg C/L or less are necessary to prevent bacterial regrowth and the occurrence of coliforms in distribution systems. However, a study in The Netherlands indicated that the AOC levels needed to be less than 10 μg C/L to achieve biological stability (i.e., no potential for regrowth) (van der Kooij and Hijnen 1990). Biofiltration offers the advantage of targeting those lower-molecular-weight organic compounds that are biodegradable but not readily removed by physicochemical processes. Ozonation and biofiltration, in combination with coagulation and settling, can be an effective method for removing organic carbon from water supplies to reduce the risk from disinfection by-products (DBPs) (LeChevallier et al. 1992; Symons 1994) and microbial regrowth (LeChevallier et al. 1992).

Biofiltration is considered a viable treatment alternative for meeting the requirements of the SDWA and the forthcoming D/DBP Rule for the following reasons: (1) potentially effective for removal of TOC, DBP precursors, and micropollutants (e.g., pesticides and taste-and-odor compounds) (Bouwer and Crowe 1988; LeChevallier et al. 1992; Symons 1994; Hozalski, Goel, and Bouwer 1995); (2) provides removal of biodegradable NOM to limit regrowth (LeChevallier et al. 1992; Servais, Billen, and Bouillet 1994; Hozalski, Goel, and Bouwer 1995); (3) increasing use of ozonation in practice, which tends to increase the AOC and BDOC concentrations; (4) relatively inexpensive to implement (use or retrofit existing filters); and (5) easy to operate and maintain.

Some potential drawbacks of biofiltration include (1) increased head-loss accumulation rate or reduced filter run times (Goldgrabe, Summers, and Miltner 1993; Ahmad and Amirtharajah 1995; Coffey et al. 1995), (2) increased bacteria levels in the filter effluent (Servais et al. 1992; Goldgrabe, Summers, and Miltner 1993), (3) increased particle concentration (but not necessarily turbidity) in the filter effluent (Goldgrabe, Summers, and Miltner 1993), and (4) aesthetic problems, such as biofilm and algae growth, on top of the filter media and other places in the treatment works (Coffey et al. 1995). Fortunately, many of these problems can be solved by the intermittent application of a disinfectant, such as chlorine, either upstream of the filter or in the backwash water (Coffey et al. 1995; Daniel and Teefy 1995). In addition, terminal disinfection provides an effective barrier for preventing the biomass produced in the biofilters from entering the distribution system (Servais et al. 1992).

HISTORY

Slow sand filters represent the first biological filters used in the treatment of drinking water. Their history dates back to the early 19th century (Collins et al. 1989). In these systems, most of the particle removal and biodegradation of dissolved organic compounds occurs in the top few centimetres of the filter in the region termed the *schmutzdecke layer*. These relatively low-tech systems can provide effective treatment with little or no chemical addition aside from terminal disinfection. However, the low loading rates (0.12 to 0.24 m/h; 0.05 to 0.1 gpm/ft^2) of slow sand filters have primarily limited their use in the United States to small systems where land is relatively inexpensive. Therefore, this chapter will primarily discuss biologically active filters operated at rapid filter loading rates (4.9 to 24.4 m/h; 2 to 10 gpm/ft^2).

One of the first documented examples of the use of biofiltration in a full-scale plant was the Mulheim process used to treat water from the Ruhr River in Germany (Sontheimer et al. 1978). This process consisted of preozonation, alum coagulation, flocculation, ozonation, double-layer filtration, granular activated carbon (GAC) filtration, and ground passage or slow sand filtration. Most of the biodegradation occurred in the GAC filters and in the ground passage or slow sand filtration systems, which were operated at loading rates of 8.8 gpm/ft^2 (22 m/h) and ≤0.25 gpm/ft^2 (0.625 m/h), respectively. The DOC concentration was reduced from 2.6 mg/L at the GAC filter influent to 0.9 mg/L after ground passage for a total removal of 65 percent across the biological treatment processes. The removal of DOC across the entire treatment system was 75 percent. Performance data for the first few months of operation indicated that most of the biodegradation was occurring during the ground passage step. However, it was observed that by increasing the empty-bed contact time (EBCT) in the GAC filters from 5 to 10 minutes, most of the DOC reduction then occurred in the GAC filters.

Before implementing ozonation–biofiltration in 1977, the plant was operated with breakpoint chlorination as a preoxidation step to reduce the relatively high ammonia concentrations in the Ruhr River water (4 to 6 mg/L N). However, the large chlorine doses required for the breakpoint chlorination (10 to 50 mg Cl_2/L) in combination with the DOC in the water (3.6 to 5 mg/L) resulted in the formation of large amounts of organochlorine compounds, including THMs. The conversion to ozonation–biofiltration eliminated the production of organochlorine compounds and resulted in significantly lower effluent DOC levels (from 1.8 to 0.9 mg/L). In addition, the complete removal of ammonia (conversion to nitrate) was achieved by biological nitrification. Overall, the drinking water quality was significantly improved without increasing treatment costs compared to the previous plant configuration (Heilker 1979). Additional benefits of the new process included better system stability, resulting in a 50 percent reduction in operator staff, and increased GAC filter runs (three to five times) (Heilker 1979).

The Mulheim process provided convincing evidence, almost 20 years ago, that biofiltration is an effective drinking water treatment process with many potential benefits. Work in other European countries, including France, England, and The Netherlands, also demonstrated the benefits of biofiltration and ozonation–biofiltration systems. Bouwer and Crowe (1988) reviewed the European experience with drinking water biofiltration and a summary of their findings for organic carbon removal is included as Table 16-1.

FACTORS THAT AFFECT BIOFILTRATION PERFORMANCE

The effectiveness of biofiltration can vary dramatically from one treatment plant to another (see Table 16-1) due to a number of factors, including (1) NOM source and characteristics, (2) preozonation dosage used (if any), (3) EBCT of the biofilter, (4) the method and frequency of backwashing, (5) filter media type (e.g., sand, anthracite, and GAC), and (6) water temperature. In the following sections, the effects of each of these factors on biofiltration performance are discussed in detail.

Table 16-1 Experience with biological processes in drinking water treatment for organic carbon removal

Process	Location	Influent Concentration	Reduction (Pseudo-Steady State)	Reference
Aerated Biological Filter				
Full-scale (upper layer of GAC)	Annet sur Marne, France	3.2 mg TOC/L	38%	Sibony (1982)
Fluidized-Bed				
Pilot-scale	Medmenham, Great Britain	3.7 mg TOC/L	5.4%	Short (1975)
Pilot-scale	Medmenham, Great Britain	2.8 mg BOD_5/L*	29%	Short (1975)
Pilot-scale	Colwick, Great Britain	10 mg TOC/L	10%	Foster (1972)
Rapid Sand Filtration				
Full-scale	Mulheim, Germany	2.9 mg DOC/L	10%†	Sontheimer et al. (1978)
Full-scale	The Netherlands	23–500 μg AOC/L	3–84%	van der Kooij (1984)
Slow Sand Filtration				
Full-scale	The Netherlands	NR	Δ3.1 mg TOC/L	KIWA (1983)
Full-scale	Bremen, Germany	NR	Δ0.9 mg TOC/L†	Eberhardt et al. (1977)
Full-scale	France	NR	<40%	Fiessinger (personal communication) (1987)
Microbially Active GAC				
Full-scale	Mulheim, Germany	1.8–2.6 mg DOC/L	75%†	Jekel (1977)
Pilot-scale	Choisy-le-Roi, France	2–4 mg TOC/L	30%†	Bourbigot (1981); Brunet et al. (1982)
Full-scale	Jefferson Parish, La.	NR	30–50%	AWWA Committee Report (1981)
Full-scale	Bremen, Germany	3.6–5 mg DOC/L	Δ1–1.7 mg DOC/L†	Sontheimer et al. (1978)
Full-scale (advanced wastewater treatment for groundwater recharge)	Orange County, Calif.	25 mg DOC/L	20%	McCarty et al. (1979)

Table 16-1 Experience with biological processes in drinking water treatment for organic carbon removal (continued)

Process	Location	Influent Concentration	Reduction (Pseudo-Steady State)	Reference
Soil–Aquifer Treatment				
Full-scale	Garonne River, France	2.3–2.9 mg TOC/L	60–70%	Bourbigot et al. (1985)
Full-scale	Karlsruhe, Germany	2.5–4.5 mg DOC/L	40–65%†	Nagel et al. (1982)
Full-scale (bank filtration)	Rhine River, Germany	5–7 mg DOC/L	75%†	Sontheimer (1980)
Full-scale (dune infiltration)	Castricum, The Netherlands	13–17 μg EOCl/L‡	84–87%	Piet and Zoeteman (1985)
Full-scale (dune infiltration)	Amsterdam, The Netherlands	19 μg/L Σ organic nitrogen compounds	92–94%	Piet and Zoeteman (1985)

NR = not reported.
*BOD_5 = 5-day biochemical oxygen demand.
†Employed preozonation.
‡EOCl = extractable organic chlorine.
Source: Bouwer and Crowe (1988).

Natural Organic Matter Source and Characteristics

Humic substances comprise a large portion of NOM in water supplies and the molecular weights of these organic materials can vary widely. The distribution of molecular weight can significantly affect the biodegradability of the NOM (Bouwer, Hozalski, and Goel 1995; Goel, Hozalski, and Bouwer 1995) as well as the potential for removal of the organics by physicochemical processes in a water treatment plant (Sinsabaugh et al. 1986; Amy et al. 1992; Bouwer, Hozalski, and Goel 1995; Hozalski, Goel, and Bouwer 1995). Lower-molecular-weight (MW) compounds should be more biodegradable because they are more easily transported across the cell membrane and attacked by metabolic enzymes. Before higher MW compounds are biodegradable, they generally have to be broken down into lower-MW compounds outside the cell by enzyme catalyzed hydrolysis reactions, which usually occur at relatively slow rates (Manem 1988). Therefore, it is expected that NOM sources with a greater percentage of their DOC in the lower-MW fractions are more amenable to biodegradation.

The source of the NOM in water supplies greatly influences the composition of the organic matter and distribution of functional groups. One source of organic matter in freshwater supplies is algal and cyanobacterial biomass, which is mostly aliphatic and totally devoid of lignins. This material is low in phenolic and aromatic constituents (Rashid 1985). In contrast, organic matter originating from soils is derived from terrestrial vegetation that has a high lignin content. Lignin contains a high fraction of aromatic compounds, consequently NOM of soil origin tends to have a higher aromatic fraction than NOM created in the water by algae and cyanobacteria. Soil-derived humic materials tend to have a strong aromatic core with aliphatic side chains. The molecules contain more polycyclic and polyphenolic constituents. The location and age of the soil-derived NOM also influences the organic matter composition. The NOM formed by leaching of peat (i.e., of soil origin) in surface waters is predominantly moderate-to-low-MW material that comprises fulvic acids. This material is relatively soluble and does not tend to adsorb to soil solids with the same propensity as humic acids. In aged groundwater, any organic matter that can sorb has usually already been removed during the long contact time with aquifer solids. Thus, the organic materials that remain in solution are the less adsorbable fractions.

A study was performed at Johns Hopkins University (JHU) to examine how differences in NOM source and characteristics relate to their biodegradability and influence biofiltration effectiveness. Source waters that contained either cyanobacterial products, peat fulvic acid, humic acid, or groundwater NOM were used to systematically study the biodegradation of the NOM in order to establish relationships between NOM composition and biodegradation performance (Bouwer, Hozalski, and Goel 1995; Goel, Hozalski, and Bouwer 1995). Results of batch biodegradation experiments with the raw NOM sources are shown in Table 16-2 along with the values of two parameters that were determined for each source, the ultraviolet (UV) absorbance at 254 nm to TOC ratio (UV/TOC) and the percentage of DOC smaller than an MW cutoff of 1K or 1,000 daltons (Goel, Hozalski, and Bouwer 1995). The maximum amount of TOC biodegraded in the batch

Table 16-2 Effect of NOM composition on TOC removal by biodegradation in batch experiments

NOM Source	UV/TOC* (*L/mg*)	DOC <1K (*percent*)	TOC Biodegraded (*percent*)
Ald HA†	0.090	9.7	0
FGW‡	0.039	46.6	21
DSW§	0.031	22.2	22
Ana Ex**	0.020	60.3	37

*UV absorbance measured at 254 nm using a light-path length of 1 cm.

†Ald HA = Aldrich Humic Acid, Aldrich Chemical Company, Milwaukee, Wisc.

‡FGW = Florida groundwater, from Barefoot Bay community water supply, Brevard County, Fla.

§DSW = Dismal swamp water, from Great Dismal Swamp in southeastern Virginia.

**Ana Ex = *Anabaena* exudates, soluble exudates from lab cultures of *Anabaena sp.*

cultures was inversely proportional to UV/TOC and proportional to the amount of low-MW NOM. The UV/TOC ratio provides a measure of the aromaticity of NOM, so that sources with high UV/TOC are considered to have a preponderance of unsaturated carbon bonds, including relatively recalcitrant aromatic structures. This study represents one of the first attempts at correlating NOM biodegradability with two relatively easy-to-measure physical or chemical parameters (UV/TOC and percent DOC <1K). More work is needed in this area to further test the correlations observed in this study and to test correlations with other parameters in order to develop an approach for predicting NOM biodegradability.

Ozone Dose

The biodegradability of raw or nonozonated NOM in water supplies varies considerably and is generally poor because this material often consists of the remnants of natural decomposition processes. Ozonation has been shown to increase the biodegradability of humic substances (Kuhn et al. 1978; Huck, Fedorak, and Anderson 1991; LeChevallier et al. 1992; Goel, Hozalski, and Bouwer 1995) via formation of hydroxyl, carbonyl, and carboxyl groups, increased polarity and hydrophilicity that decreases adsorbability, loss of double bonds and aromaticity, and a shift in the molecular weight distribution, resulting in an increased percentage of lower MW compounds (Anderson, Johnson, and Christman 1985; Glaze et al. 1989). The products of ozonation of organic compounds are usually more polar chemicals that are known to be readily biodegradable, such as aldehydes, oxalic acid, and pyruvic acid (Hoigné 1988).

There are many examples in the literature that demonstrate increased NOM biodegradability in response to ozonation. Preozonation enhanced biological growth in GAC adsorbers in a study performed in the Federal Republic of Germany (Kuhn et al. 1978). Ozonation of a raw water in The Netherlands increased the AOC by 10 to 20 times (van der Kooij, Visser, and Hijnen 1984). Bradford, Palmer, and Olson (1994) observed a 2- to 10-fold increase in AOC after ozone treatment of some Southern California groundwaters.

Although ozone has been shown to enhance the biodegradability of NOM, the increase in biodegradability for a given ozone dose can vary significantly depending on the source and characteristics of the NOM. In addition to looking at the effect of NOM source on biodegradability of raw or unozonated NOM, another goal of the work performed at JHU's laboratory was to examine the effects of ozone on biodegradability using the four different NOM sources described earlier (Goel, Hozalski, and Bouwer 1995) and shown in Table 16-2. A significant enhancement in biodegradability and total TOC removal for low to moderate ozone doses (<2 mg O_3/mg TOC) was observed for NOM sources with a large percentage of high-MW NOM (i.e., low percentage DOC <1K) and with moderate to high UV/TOC as shown in Table 16-3. The sources consisting primarily of low-MW NOM with low to moderate UV/TOC ratio did not show a significant improvement in TOC removal from the application of low to moderate ozone doses. Sources consisting of relatively high MW material with a high UV/TOC ratio would tend to be more recalcitrant without ozonation and would be expected to have more reactive sites for oxidation by ozone or other radical species that are products of ozone decomposition.

Many utilities in the United States already use ozone, and increased use of ozone as an alternative to chlorine is likely to occur in the United States when the D/DBP Rule is promulgated. Given the significant enhancement of NOM biodegradability resulting from ozonation of water supplies, the use of biofiltration should increase proportionally in order to provide a means

Table 16-3 Effect of ozonation on TOC removal by chemical oxidation and subsequent biodegradation in batch experiments

NOM Source*	Ozone Dose, (*mg O_3/mg TOC*)	TOC Removal (*percent*)		
		O_3	Biodegradation	Total
DSW	0	NA†	22	22
DSW	1.6	22	27	49
DSW	7.3	26	31	57
Ald HA	0	NA	0	0
Ald HA	1.3	6	23	29
Ald HA	1.9	24	38	62
Ald HA	2.8	27	41	68
Ana Ex	0	NA	37	37
Ana Ex	0	NA	34	34
Ana Ex	1.3	6	25	31
Ana Ex	4.3	25	50	75
FGW	0	NA	21	21
FGW	1.9	7	18	25
FGW	2.7	12	18	30
FGW	4.1	12	31	43

*Refer to Table 16-2 footnotes.
†NA = not applicable.

of controlling these biodegradable materials prior to passage of the treated water into the distribution system.

Empty-Bed Contact Time

Recent biofiltration research has focused on the effect of operational parameters, especially EBCT, on the removal of organic carbon. Empty-bed contact time is defined as the empty-bed volume of the filter divided by the volumetric flow rate through the filter. The hydraulic residence time in the filter is equivalent to the EBCT multiplied by the bed porosity. Many researchers have observed that removal of organic carbon as measured by TOC, DOC, BDOC, AOC, and/or THMFP increases with increasing EBCT. LeChevallier et al. (1992) evaluated the removal of TOC and AOC in GAC-sand filters as a function of EBCT following ozonation of a water source at an applied dosage of 0.5 to 1.0 mg O_3/mg TOC. The researchers reported increases in TOC removal from 29 percent at 5 minutes of EBCT, to 33 percent at 10 minutes of EBCT, to 51.2 percent at 20 minutes of EBCT. LeChevallier and co-workers also observed increases in AOC removal over the same EBCT range. Sontheimer and Hubele (1987) observed an increase in DOC removal from 27 to 41 percent as residence time within the biologically active GAC filters was increased from 5 to 20 minutes. Removal of AOC appears to occur more rapidly than the removal of the biodegradable fraction of the TOC. Prevost et al. (1992) found that 2 minutes of contact time was sufficient to remove 62 to 90 percent of the AOC in biologically active filters, but 10 to 20 minutes of EBCT was required to remove 90 percent of the BDOC. LeChevallier et al. (1992) concluded from their results that a 15- to 20-minute EBCT may be needed for TOC removal, but this EBCT range may be excessive for treatment of AOC.

Conversely, some results in the literature suggested that EBCT did not effect organic carbon removal. Price et al. (1993) observed similar removals of AOC and THMFP in GAC-sand filters operating at 1, 3, and 5 gpm/ft^2 (2.4, 7.3, and 12 m/h), which covered an EBCT range of 4.5 to 22.5 minutes. Krasner, Sclimenti, and Coffey (1993) evaluated the effect of EBCT on TOC and AOC removal in biologically active anthracite and GAC filters after ozonation of California State Project water at an applied dosage of 0.5 mg O_3/mg TOC. The researchers concluded that although biological activity was established sooner on the anthracite filters operated with an EBCT of 4.2 minutes, the filters with lower EBCTs of 1.4 and 2.1 minutes were eventually able to achieve comparable removals. Similar results were observed for the GAC filters operated at EBCTs of 1.4 and 4.2 minutes. Results from the JHU's laboratory using bench-scale sand biofilters also indicated that NOM removal (as measured by TOC concentration) was independent of biofilter EBCT (Hozalski, Goel, and Bouwer 1995) as shown in Figure 16-2. Similar results were obtained for three other biofiltration experiments that examined different NOM sources and ozone doses (Hozalski, Goel, and Bouwer 1995).

The lack of an effect of EBCT (4 to 20 minutes) in these biofiltration experiments may have resulted from any one or a combination of the following factors: (1) relatively high ozone dosages were used (2 to 4 mg O_3/mg TOC), which led to the formation of a large fraction of easily biodegradable material; (2) the biofilters were not

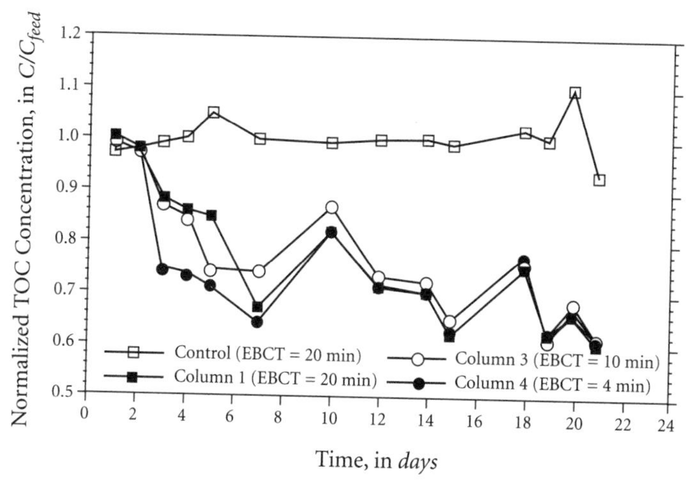

Figure 16-2 Effect of EBCT on normalized effluent TOC concentration over time in a biofiltration experiment with ozonated *Anabaena* exudates (2.2 ± 0.6 mg O_3/mg TOC) as the carbon source

backwashed so that large biomass populations were able to accumulate on the filter media; and (3) the low hydraulic loading rates used in the experiments resulted in low shear stress within the filter bed, which could also contribute to a high biomass accumulation (Hozalski, Goel, and Bouwer 1995). In summary, the literature and studies at JHU indicate that effective removal of BDOC from drinking water can be achieved in biofilters operated at typical rapid media filter EBCTs in the range of 2 to 20 minutes.

Backwashing

The goal of backwashing biofilters is to remove biomass and nonbiological particles to restore head-loss capacity and permit desired filter run times while retaining enough biomass to prevent the breakthrough of biodegradable organic matter. The effects of backwashing on biofiltration performance have not been studied extensively and are not well known. There is some evidence that biofilter performance is not impaired if backwashing is carried out with nonchlorinated water (Ahmad and Amirtharajah 1995; Miltner, Summers, and Wang 1995; Hozalski and Bouwer 1998). Hozalski and Bouwer (1998) examined the effect of backwashing with water alone (i.e., no chlorine) on removal of TOC in sand biofilters fed a solution of Dismal Swamp NOM (primarily composed of fulvic acids) ozonated at a dose of 0.58 ± 0.12 mg O_3/mg TOC. Although an increase in normalized effluent TOC concentration was observed following the second backwash after 96 hours of operation, subsequent backwashes did not have a significant effect on the removal of TOC in the sand biofilters. Despite the removal of substantial quantities of biomass during each backwash (typically greater than 10^7 colony-forming units (CFU) per cm^3 of filter-bed volume), the residual biomass in the biofilter was able to maintain effective treatment in the subsequent filter cycle.

The use of chlorinated backwash water, which is common in water treatment practice, can have detrimental effects on biofiltration performance. Ahmad and Amirtharajah (1995) observed an increasing rate of head-loss buildup in successive filter runs when GAC biofilters were backwashed with water only; however, for filters backwashed with chlorinated water, there was a decreasing rate of head-loss development in successive filter runs. This implies that repeated backwashing with chlorinated water is eliminating biomass without sufficient recovery or replacement of biomass between backwashes. Miltner, Summers, and Wang (1995) observed lower concentrations of attached biomass in dual-media (anthracite–sand) biofilters backwashed with chlorinated water (1 mg/L free chlorine) compared to those backwashed with water only. A short-term loss of control of ozonation by-products (formaldehyde, acetaldehyde, methyl glyoxal, and AOC-NOX) was observed in the chlorine backwash biofilters. The biofilters recovered to steady state removal values for ozonation by-products within 12 hours after the chlorinated backwash; however, the steady state removal values (samples taken after the 12-hour recovery period) for ozonation by-products, TOC, and THMFP were lower for the chlorine backwashed biofilters as compared to the biofilters backwashed with nonchlorinated water (Miltner, Summers, and Wang 1995).

Ahmad and Amirtharajah (1995) also evaluated the effect of bed expansion on removal of biomass by water backwash. Total biomass removal as measured by plate counts and adenosine triphosphate (ATP) analysis increased steadily for bed expansions from 10 to 60 percent, where the maximum removal was observed, and decreased for higher bed expansions. A bed expansion of 60 percent for the GAC filter media used in their study corresponded to an expanded bed porosity of 0.69. Theoretically, the fluid shear stress is at a maximum for expanded bed porosities of 0.68 to 0.71 (Amirtharajah 1978). Filters in US water treatment practice are typically designed and operated for bed expansions ranging from 15 to 30 percent (Cleasby 1990), making it physically impractical to achieve the maximum shear stress.

Because insufficient removal of biomass (and nonbiological particles) may be a problem for nonchlorinated-water backwashing at low bed expansions (15 to 30 percent), the use of enhanced backwashing methods such as air scour may be necessary. Ahmad and Amirtharajah (1995) evaluated the effects of air scour backwash under collapse-pulsing conditions on head-loss development in biofilters. The total headloss across biofilters backwashed with air scour was lower than for biofilters backwashed with water only. In addition, there was no significant change in the rate of head loss development for the air-scour backwashed biofilters in successive filter runs. As mentioned, water-backwashed biofilters exhibited an increasing rate of head-loss development and chlorinated water backwashed biofilters exhibited a decreasing rate of head-loss development in successive filter runs (Ahmad and Amirtharajah 1995). Therefore, the authors concluded that chlorinated backwash was detrimental to biological filtration and that biological filters need air scour. Unfortunately, Ahmad and Amirtharajah (1995) did not report results concerning the effects of the different backwash methods on subsequent removal of BDOC.

Filter Media

Another factor that can affect NOM removal is the type of filter media in the biofilter. This is because the filter media provides the surface for microbial colonization and, in the case of sorptive filter media like GAC, can have other benefits, such as sorption and release of substrates and protection from toxic chemicals or chlorine. LeChevallier et al. (1992) reported that GAC performed better than mixed media (e.g., anthracite, sand, and garnet) for treatment of AOC and TOC. However, in this study, the improved performance of the GAC biofilters may have been exaggerated by the use of new or nonexhausted GAC and by the intermittent pilot-plant operation resulting from weekend shutdown. Krasner, Sclimenti, and Coffey (1993) observed higher removals of the relatively recalcitrant ozone by-product glyoxal in GAC filters than in similarly designed and operated sand or anthracite filters. However, the removals of AOC and formaldehyde in the GAC and anthracite biofilters were similar after 56 days of operation. Wang, Summers, and Miltner (1995) observed higher removals of TOC, AOC, and DBP precursors in coal-based GAC biofilters compared to sand and anthracite-sand biofilters. Phospholipid analysis of the biomass on the filter media indicated that GAC was able to support three to eight times more biomass (nmol lipid-P/g media) than either sand or anthracite-sand (Wang, Summers, and Miltner 1995). The researchers attributed the higher biomass levels observed on the GAC to the larger surface area and better surface texture of the GAC. However, normalizing attached biomass levels on the basis of sample weight (instead of volume) may not yield a fair comparison due to the density differences of the media. Wang and co-workers also observed differences in biofiltration performance among several types of GAC. Enhanced TOC, AOC-NOX, THMFP, and TOXFP removal performance was observed for coal-based GAC biofilters compared to wood-based GAC biofilters despite comparable biomass levels. The authors speculated that the difference in performance of the GAC filters may have been due to differences in the ability of the various GACs to adsorb and release substrates.

Critical examination of the earlier studies mentioned in combination with some recent research (e.g., Coffey et al. 1995; Daniel and Teefy 1995) leads to the conclusion that the performance of GAC biofilters and anthracite biofilters is comparable under pseudo-steady-state conditions. However, under extreme conditions, i.e., during biofilter startup, low temperature, or intermittent (or constant) exposure to chlorine, GAC biofilters typically outperform anthracite biofilters. (The effect of temperature on biofiltration performance is discussed in the next section.) Finally, there is also some evidence to suggest that sand biofilters can perform effectively with respect to removal of

BDOC (Coffey et al. 1995; Hozalski, Goel, and Bouwer 1995; Hozalski and Bouwer 1998). Therefore, it appears that GAC filter media is not a requirement for good biofiltration performance in all cases, but when extreme conditions are expected, GAC is most favorable for ensuring adequate system stability.

Temperature

Biological processes are typically temperature sensitive, with performance in terms of microbial degradation rate or substrate removal increasing with increasing temperature until the microbial inactivation temperature is attained. Over the range of temperatures encountered in water treatment, microbial reaction rates tend to double for every 10°C increase in temperature (Metcalf and Eddy 1991). Essentially all studies that have examined the effect of temperature on biofiltration performance have observed increases in carbon removal (as measured by TOC, DOC, AOC, BDOC, or trace constituents) with increasing temperature. In addition, in studies that examined the effect of temperature and filter media (i.e., GAC versus anthracite or sand), the performance of GAC filters was much less sensitive to temperature change than similarly operated anthracite or sand filters.

Servais et al. (1992) reported that the EBCT needed to be doubled when the temperature decreased from 20 to 8°C in order to maintain the 20°C BDOC removal efficiency in GAC biofilters despite the fact that biomass levels were similar at both temperatures. An increase in AOC removal with increasing temperature (9.5 to 16.5°C) was observed in both dual-media (anthracite and sand) and GAC-sand biofilters in an East Bay Municipal Utility District (EBMUD) (Calif.) pilot study (Price et al. 1993). However, the dual-media biofilters were much more temperature sensitive than the GAC-sand filters. The removal of aldehydes in full-scale anthracite filters (loading rate = 4 gpm/ft^2) at the Fremont, Calif., water treatment plant was consistently greater than 90 percent when the raw water temperature was above 17.5°C (Daniel and Teefy 1995). However, the removal dropped to approximately 30 percent for several samples when the temperature approached 10°C. Aldehyde removal in a similarly operated GAC filter was not significantly affected by the same temperature fluctuation. In studies at the Metropolitan Water District of Southern California's Ozonation Demonstration Plant, similar removal of glyoxal as a function of EBCT (0.5 to 2.1 minutes) was observed for GAC and anthracite filters when the water temperature was between 20 and 25°C (Coffey et al. 1995). The glyoxal removal in both filters was reduced for a water temperature of 10°C, but the decrease in removal for the anthracite filter was much more substantial. For example, both biofilters removed approximately 85 percent of the glyoxal in the filter influent when the temperature was in the range of 20 to 25°C, but, at 10°C, the removals of glyoxal for the GAC and anthracite filters were 70 and 45 percent, respectively. Therefore, GAC filters are recommended for maintaining effective removal of BDOC in drinking water biofilters operated under low temperature conditions (10°C or less).

SUMMARY

Biofiltration is an emerging water treatment technology that can aid utilities in meeting the requirements of the SDWA and the proposed D/DBP Rule. In reviewing the history and current literature concerning biofiltration, it is apparent that this promising technology has an important place in water treatment. In the future, improvements in biofiltration can be expected in the areas of backwashing practices and technologies, designs to reduce the effect of system perturbations, and process control technologies in general.

ACKNOWLEDGMENTS

The authors thank the American Water Works Association Research Foundation (AWWARF) and American Water Works Association for financial support of the research performed in our laboratory. In addition, the authors acknowledge the technical contributions of Sudha Goel who was a co-research assistant (along with the first author) in the AWWARF study performed in our laboratory (*Removal of Natural Organic Matter in Biofilters*). Finally, the authors thank Amit Amirtharajah, Brad Coffey, and Peter Huck for their contributions to a literature review concerning biofiltration, which provided a basis for some of the material discussed in this chapter.

REFERENCES

Ahmad, R., and A. Amirtharajah. 1995. Detachment of Biological and Non-Biological Particles from Biological Filters During Backwashing. In *Proc. 1995 AWWA Annual Conference*. Denver, Colo.: American Water Works Association.

Amirtharajah, A. 1978. Optimum Backwashing of Sand Filters. *Jour. of Env. Eng., ASCE*, 104(EE5):917.

Amy, G.L., R.A. Sierka, J. Bedessem, D. Price, and L. Tan. 1992. Molecular Size Distributions of Dissolved Organic Matter. *Jour. AWWA*, 84(6):67–75.

Anderson, L.J., J.D. Johnson, and R.F. Christman. 1985. The Reaction of Ozone With Isolated Aquatic Fulvic Acid. *Organic Geochemistry*, 8(1):65–69.

Bouwer, E.J., and P. Crowe. 1988. Assessment of Biological Processes in Drinking Water Treatment. *Jour. AWWA*, 80(9):82.

Bouwer, E.J., R.M. Hozalski, and S.B. Goel. 1995. *Removal of Natural Organic Matter in Biofilters*. Denver, Colo.: American Water Works Association.

Bradford, S.M., C.J. Palmer, and B.H. Olson. 1994. Assimilable Organic Carbon Concentrations in Southern California Surface and Groundwater. *Water Res.*, 28(2):427–435.

Cleasby, J.L. 1990. Filtration. *Water Quality and Treatment*. Pontius, F.W., ed. New York: McGraw-Hill.

Coffey, B.M., S.W. Krasner, M.J. Sclimenti, P.A. Hacker, and J.T. Gramith. 1995. A Comparison of Biologically Active Filters for the Removal of Ozone By-Products, Turbidity, and Particles. In *Proc. 1989 AWWA Water Quality Technology Conference*. Denver, Colo.: American Water Works Association.

Collins, M.R., T.T. Eighmy, J.M. Fenstermacher, and S.K. Spanos. 1989. *Modifications to the Slow Sand Filtration Process for Improved Removals of Trihalomethane Precursors*. Denver, Colo.: American Water Works Association Research Foundation.

Daniel, P., and S. Teefy. 1995. Biological Filtration: Media, Quality, Operations, and Cost. In *Proc. 1995 AWWA Annual Conference*. Denver, Colo.: American Water Works Association.

Glaze, W.H., M. Koga, D. Cancilla, K. Wang, M.J. McGuire, S. Liang, M.K. Davis, C.H. Tate, and E.M. Aieta. 1989. Evaluation of Ozonation By-Products from Two California Surface Waters. *Jour. AWWA*, 81(8):66–73.

Goel, S., R.M. Hozalski, and E.J. Bouwer. 1995. Biodegradation of NOM: Effect of NOM Source and Ozone Dose. *Jour. AWWA*, 87(1):90.

Goldgrabe, J.C., R.S. Summers, and R.J. Miltner. 1993. Particle Removal and Head Loss Development in Biological Filters. *Jour. AWWA*, 85(12):94.

Heilker, E. 1979. The Mulheim Process for Treating Ruhr River Water. *Jour. AWWA*, 71(11):623–627.

Hoigné, J. 1988. The Chemistry of Ozone in Water. *Process Technologies for Water Treatment*. Plenum Publishing Corporation.

Hozalski, R.M., and E.J. Bouwer. 1998. Deposition and Retention of Bacteria in Backwashed Filters. *Jour. AWWA*, 90(1):71–85.

Hozalski, R.M., S. Goel, and E.J. Bouwer. 1995. TOC Removal in Biological Filters. *Jour. AWWA*, 87(12):40–54.

Huck, P.M., P.M. Fedorak, and W.B. Anderson. 1991. Formation and Removal of Assimilable Organic Carbon During Biological Treatment. *Jour. AWWA*, 83(12):69–80.

Krasner, S.W., and G.L. Amy. 1995. Jar Test Evaluations of Enhanced Coagulation. *Jour. AWWA*, 85(5):62–71.

Krasner, S.W., M.J. Sclimenti, and B.M. Coffey. 1993. Testing Biologically Active Filters for Removing Aldehydes Formed During Ozonation. *Jour. AWWA*, 85(5):62.

Kuhn, W., H. Sontheimer, L. Steiglitz, D. Maier, and R. Kurz. 1978. Use of Ozone and Chlorine in Water Utilities in the Federal Republic of Germany. *Jour. AWWA*, 70(6):326–331.

LeChevallier, M.W., W.C. Becker, P. Schorr, and R.G. Lee. 1992. Evaluating the Performance of Biologically Active Rapid Sand Filters. *Jour. AWWA*, 84(4):136–146.

Manem, J. 1988. Interactions Between Heterotrophic and Autotrophic Bacteria in Fixed-Film Biological Processes Used in Drinking Water Treatment. Urbana-Champaign, Ill.: University of Illinois.

Metcalf, and I. Eddy. 1991. *Wastewater Engineering—Treatment, Disposal, and Reuse.* New York: McGraw-Hill.

Miltner, R.J., R.S. Summers, and J.Z. Wang. 1995. Biofiltration Performance: Part 2, Effect of Backwashing. *Jour. AWWA*, 87(12):64–70.

Prevost, M., J. Coallier, J. Mailly, R. Desjardins, and D. Duchesne. 1992. Comparison of Biodegradable Organic Carbon (BOC) Techniques for Process Control. *Jour. Water Supply Technology-Aqua*, 41(3):141.

Price, M.L., R.W. Bailey, A.K. Enos, M. Hook, and S.W. Hermanowicz. 1993. Evaluation of Ozone/Biological Treatment for Disinfection Byproducts Control and Biologically Stable Water. *Ozone: Science and Engineering*, 15:95–130.

Rashid, M.A. 1985. *Geochemistry of Marine Humic Compounds.* New York: Springer-Verlag.

Servais, P., G. Billen, and P. Bouillot. 1994. Biological Colonization of Granular Activated Carbon Filters in Drinking Water Treatment. *Jour. of Env. Eng.*, 120(4):888–899.

Servais, P., G. Billen, P. Bouillot, and M. Benezet. 1992. A Pilot Study of Biological GAC Filtration in Drinking-Water Treatment. *Journal Water SRT-Aqua*, 41(3):163–168.

Sinsabaugh, R.L., R.C. Hoehn, W.R. Knocke, and A.E. Linkins. 1986. Removal of Dissolved Organic Carbon with Iron Sulfate. *Jour. AWWA*, 78(5):74–82.

Sontheimer, H., E. Heilker, M.R. Jekel, H. Nolte, and F.H. Vollmer. 1978. The Mulheim Process. *Jour. AWWA*, 70(7):393–396.

Sontheimer, H., and C. Hubele. 1987. The Use of Ozone and Granular Activated Carbon in Drinking Water Treatment. *Treatment of Drinking Water for Organic Contaminants.* Huck, P.M., and P. Toft, eds. New York: Pergammon Press.

Symons, J.M. 1994. Precursor Control in Waters Containing Bromide. *Jour. AWWA*, 86(6):48.

van der Kooij, D., ed. 1984. *The Growth of Bacteria on Organic Compounds in Drinking Water.* Rijswijk, The Netherlands: KIWA.

van der Kooij, D., and W.A.M. Hijnen. 1990. Criteria for Defining the Biological Stability of Drinking Water as Determined With AOC Measurements. In *Proc. 1990 AWWA Water Quality Technology Conference.* Denver, Colo.: American Water Works Association.

van der Kooij, D., A. Visser, and W.A.M. Hijnen. 1982. Determining the Concentration of Easily Assimilable Organic Carbon in Drinking Water. *Jour. AWWA*, 74(10):540–545.

Wang, J.Z., R.S. Summers, and R.J. Miltner. 1995. Biofiltration Performance: Part 1, Relationship to Biomass. *Jour. AWWA*, 87(12):55–63.

CHAPTER • SEVENTEEN

Anion Exchange for the Removal of Total Organic Carbon and Disinfection By-Product Precursors

DENNIS CLIFFORD
PHILIP KIM
PAUL FU

Natural organic matter (NOM) in water contains significant amounts of high-molecular-weight, soluble and colloidal humic and fulvic acid anions, which are often associated with the soluble and colloidal iron, manganese, and silica in the water. Since the mid-1950s, NOM has been known to be a fouling agent for the strong-base anion (SBA) resins used in water treatment.

In spite of known resin fouling problems, macroporous weak-base anion (WBA) resins were used in the 1960s to remove color from river water, and macroporous strong-base resins were used in the 1970s to successfully treat highly colored groundwater. Also in the 1970s, polyacrylic strong base resins were developed, which were less prone to irreversible fouling compared with the standard polystyrene resins in universal use. Following the discovery of the formation of trihalomethanes (THMs) and other disinfection by-products (DBPs) in water in the mid-1970s, various strong- and weak-base anion exchange resins were found to be capable of removing DBP precursors from water. Experimental use of resins for total organic carbon (TOC) removal continued through the 1980s and early 1990s and this recent work has clarified some of the mechanisms of NOM removal by strong-base resins.

This chapter summarizes what is known about anion exchange resins for TOC, DBP precursors, and color removal, and provides some rules of thumb about (1) choosing between weak and strong, macroporous and gel, and polystyrene and polyacrylic resins, (2) the influence of pH and background ions, and (3) the potential for combinations of granular activated carbon (GAC) and ion exchange (IX) for removing TOC.

CHEMISTRY OF NOM AND RESINS

In slightly acidic, neutral, and alkaline waters, the acidic functional groups on NOM are negatively charged. Thus, TOC molecules are naturally attracted to anion exchange resins, which contain positively charged amine functional groups attached to a polystyrene or polyacrylic polymer matrix.

Weak-base anion resins contain weakly basic—primary, secondary, or tertiary amine—functional (exchange) groups, which are positively charged (protonated) only in acidic solution. The structures of two WBA resins that have been used for TOC removal from water are shown in Figure 17-1. These resins function as anion exchangers only when the solution pH $\leq$6. Above pH 6, the amine functional groups are neutral and do not exchange ions. These WBA resins can, however, function as adsorbents for TOC at pHs $\geq$6.

Strong-base anion resins contain quaternary amine functional groups typically attached to a polystyrene or polyacrylic matrix, as shown in Figure 17-2. The more common macroporous polystyrene resin is MP 500, whereas IRA 458 is a microporous polyacrylic resin, which is often used because of its lower organic fouling potential. Both resins have quaternary amine functional groups,

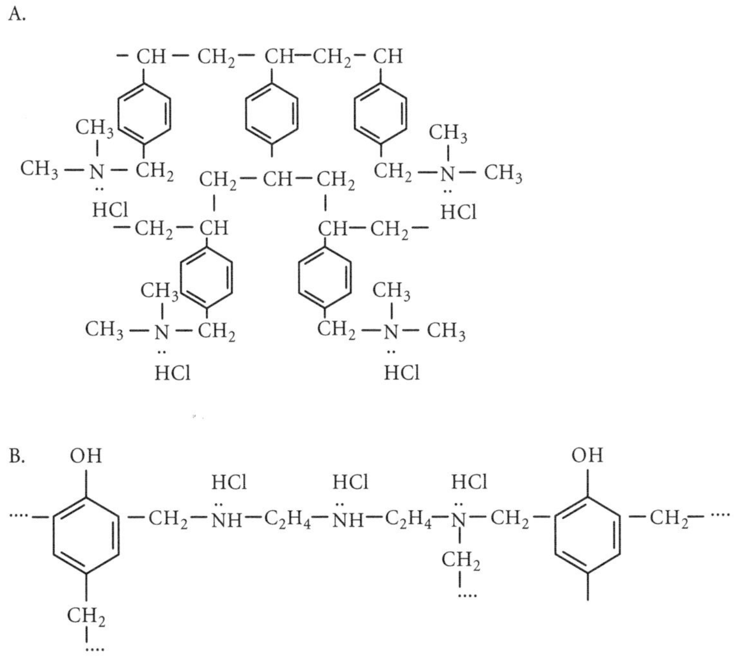

A.

$-CH-CH_2-CH-CH_2-CH$

CH_3 CH_3

CH_3-N-CH_2 $CH_2-CH-CH_2$ CH_2-N-CH_3

HCl HCl

$-CH_2-CH$ $CH-CH_2-$

CH_3 CH_3

CH_3-N-CH_2 CH_2-N-CH_3

HCl HCl

B. OH OH

HCl HCl HCl

$....-$ $-CH_2-NH-C_2H_4-NH-C_2H_4-N-CH_2-$ $-CH_2-....$

CH_2

CH_2

Source: A. Duolite Data Leaflet 2 (1974); B. Duolite Tech Sheet 130 (1968).

Figure 17-1 A. Structure of ES-368 polystyrene tertiary-amine WBA resin and B. A-7 phenol formaldehyde WBA resin

which are ionized (positively charged) and function as anion exchangers throughout the 3 to 13 pH range. Regarding porosity, macroporous resins have measurable BET surface area (measured by N_2 adsorption), whereas microporous (or gel) resins have no measurable BET surface area. In aqueous solution, both types of resins have apparent porosity because they are swollen with water and readily allow hydrated ions to enter the hydrated polymer.

RESIN FOULING

Typical SBA resins are well-hydrated, positively charged organic polymers with at least some aromatic character. Natural organic matter in water is composed of large, slow-diffusing, negatively charged molecules with aromatic character. Thus, it is easy to see why NOM is strongly attracted to SBA resins. In fact, NOM is

A. $-CH-CH_2-$

CH_3

$CH_2-N^+-CH_3$ Cl^-

$-CH-$

CH_3

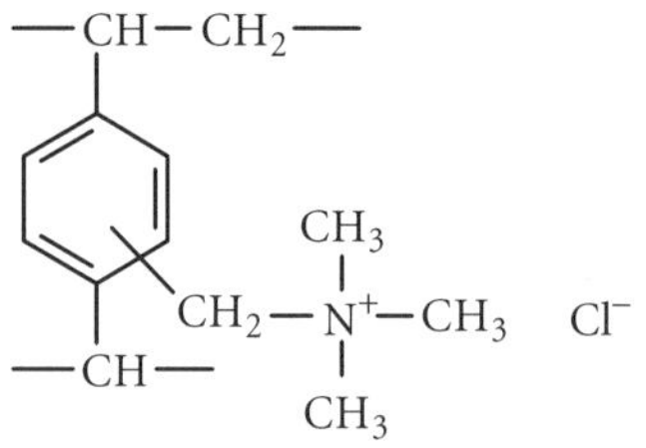

B. $-CH-CH_2-CH-CH_2-$

$C=O$

CH_3

$NH-CH_2-CH_2-CH_2-N^+-CH_3$ Cl^-

CH_3

$-CH-$

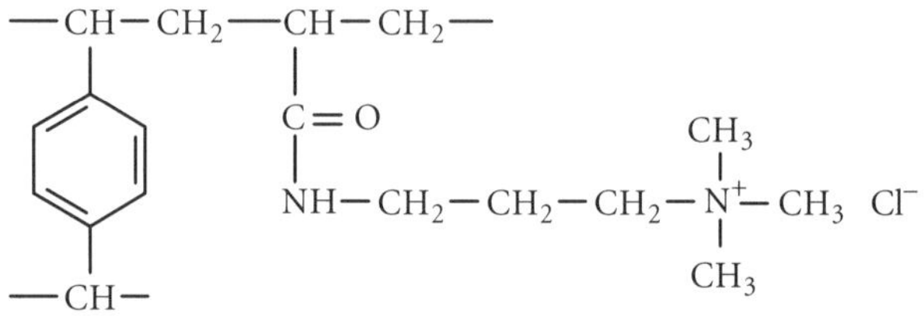

Source: A. Miles, Inc. (1992); B. Kunin (1983).

Figure 17-2 A. Structure of MP 500 polystyrene type 1 SBA resin and B. IR-458 polyacrylic SBA resin

sometimes irreversibly sorbed onto SBA resins because the NOM molecules become deeply embedded in the polymer beads and complexed with the aromatic rings on the polymer chain. Thus, NOM has been recognized for more than 40 years as a common foulant of the anion exchangers used in demineralization processes (McGarvey and Reents 1954; Frisch and Kunin 1957; Wilson 1959). Such fouling causes product water quality deterioration, resin darkening, brown-colored regenerant, exchange capacity loss, and increased frequency of regeneration. Early on it was found that occasionally regenerating the fouled SBA resin with a mixture of NaCl (1.7 *M*) and NaOH (0.5 *M*) would reverse most of the fouling and lead to acceptable run lengths and resin life (Wilson 1959).

Because NOM and color-causing organics were viewed as resin foulants, the use of SBA resins for color and TOC removal was curtailed for some years. The specter of organic fouling that haunted SBA resins did not, however, plague WBA resins as they were used in some early studies to remove color from river water.

One reason for choosing WBA resins to trap organics was based on some early experiments by Abrams and Breslin (1965).

Table 17-1 Adsorption and elution* of humates by anion resins

Resin	Matrix	Porosity	Relative Sorption	Relative Elution
WBA	Phenolic	Macro	100	100
WBA	Epoxy-Amine	Gel	62	62
WBA	Polystyrene	Macro	39	43
SBA, type 1	Polystyrene	Gel	64	2
SBA, type 2	Polystyrene	Gel	49	6
SBA, type 1	Polystyrene	Macro	49	<1

*NaOH elution (Abrams and Breslin 1965).

Their experiments showed that a macroporous phenolic WBA resin was better than any other WBA or SBA resin tested. Table 17-1 from their early work compares the performance of the phenolic resin A-7 (see Figure 17-1B) to polystyrene resins (similar to MP 500 in Figure 17-2A) and epoxy-amine resins (not shown). For humate adsorption and elution, the WBA resins were better than SBA resins, and the polystyrene resins were clearly inferior to the phenolic and epoxy-amine resins.

WBA RESINS FOR COLOR REMOVAL

The earliest work on color removal by resins was a two-year pilot study of Merrimack River water by WBA resin reported by Coogan et al. (1968). Their 0.5 to 1 gal/min pilot plant treated slightly acidic river water containing 25 to 100 color units by passing it through a diatomaceous-earth filter and then through A-7 resin with empty-bed contact times (EBCTs) in the range of 3.75 to 7.5 min. Color removal results are shown in Figure 17-3, a typical effluent history of one of their runs. There is a gradual breakthrough of organics beginning right from the start of the run. In this regard, the effluent history is like that of GAC, but GAC would normally have a much longer EBCT. It is noteworthy that there are no concentration peaks in the effluent history. (Later, it will be shown that such peaks are common in TOC breakthrough curves from SBA resins in NOM removal applications.)

Although A-7 is an anion exchanger at low pH, it had virtually no ion exchange capacity at the pH of the influent water (6.4

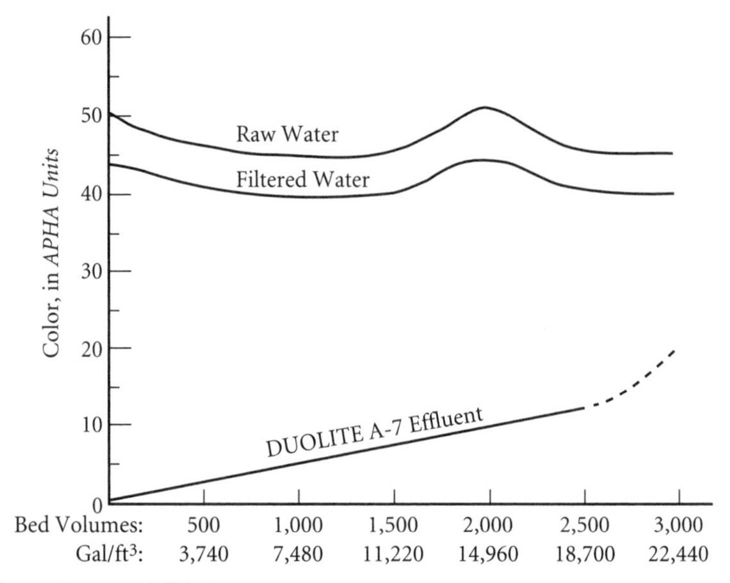

Source: Coogan et al. (1968).

Figure 17-3 Performance of A-7 WBA resin on Merrimack River water of average color (feed pH 6.4 to 6.9, EBCT 7.5 min)

to 6.9). Thus, the removal of color was by adsorption of anion–cation pairs, e.g., calcium fulvate, rather than by ion exchange. The main findings of the study were as follows:

- With APHA color values of 50 to 100 in the raw water, diatomaceous-earth prefiltration to remove about 10 percent of the presumably larger color bodies was required ahead of the resin bed.
- EBCTs in the 3.75 to 7.5 min range were acceptable, with the 3.75 min time giving only a small decrease in run length, which was 2,000 to 2,500 bed volumes (BVs) for 7.5 min.
- Typically, 80 to 100 percent color removal was attained at the start of a run, decreasing to 50 to 75 percent at the end.
- NaOH at a level of 4 lb/ft^3 (1.6 eq/L) was adequate to regenerate the resin during more than 200 exhaustion–regeneration cycles in which the resin capacity for TOC remained relatively constant.

- Iron removals were about 25 percent across the diatomaceous-earth filter and 95 percent through the resin bed, which led to the suggestion that soluble iron was being removed by complexation with the adsorbed color bodies.

This study and others reported by Abrams led researchers to try A-7 and other weak-base resins for adsorption rather than ion exchange of TOC (Abrams 1982).

Rook and Evans (1979) tested Duolite ES 368, a polystyrene WBA resin, for reduction of THMFP at the Rotterdam Water Works. Their reason for testing the resin was that they considered GAC to be slow and poor for adsorption of fulvic acids. This study differed from the earlier Merrimac River study in that the resin was installed after full-scale coagulation, flocculation, sedimentation, and dual-media filtration. At this point, color was only 5 (platinum cobalt units); TOC was 3.5 mg/L, typical of treated surface waters; and the pH in and out of the resin was about 7.5. Under these conditions, the TOC removal mechanism was adsorption of fulvic acid ion pairs rather than ion exchange, because this WBA resin has no ion exchange capacity at pH 7.5. Another unique aspect of Rook and Evans' work was the use of a two-step (0.6 eq HCl, then 0.6 eq NaOH/L resin) regeneration. Others had used only NaOH for WBA resins or much more NaOH for SBA resins in a one-step process. Their process used a total of 1.2 eq HCl + NaOH/L resin, whereas the typical SBA processes used five times that amount of regenerant. Although their process did a good to fair job of removing color (76 percent), THMFP (65 percent), ultraviolet (UV) absorbance (46 percent), and TOC (37 percent) during a 1,800-BV run, they were disappointed because there was no consistent correlation between THMFP removal and the removal of one of the easier-to-measure surrogates—TOC, UV, and color.

STRONG-BASE ANION RESINS FOR COLOR AND TOC REMOVAL

The successful use of SBA resins to treat groundwater for organics removal was reported by Kolle (1984) at the Führberg Water Works in Hannover, Germany, where the influent groundwater contained troublesome amounts of Fe, Mn, ammonia, TOC, and

TOCl. After aeration, coagulation with iron (III), neutralization, permanganate and chlorine oxidation, filtration and breakpoint chlorination, the effluent still contained 6 mg/L TOC and was loaded with TOCl. Bacterial growth in the distribution system was a big problem with a bacterial doubling time of only 5 hours. They tested a macroporous polystyrene SBA resin that eliminated much of the TOC, which led to a near-complete elimination of the chlorine ahead of the IX plant, which was the last step in the process.

The 2,500-m^3/d (15.8-mgd) full-scale process used a macroporous type 1 polystyrene SBA resin (Lewatit MP 500 A) in four single beds operated in parallel with an EBCT = 1.2 min. Regeneration was accomplished with a two-BV mixture of 1.7 *N* (9.3 percent) NaCl and 0.5 *N* (3 percent) NaOH. A striking feature of the process was the reuse of regenerant seven times with the NaCl and NaOH concentrations adjusted after each use. The TOC was not removed from the recycled regenerant, and it built up to 25,000 mg/L prior to disposal, which was triggered when the sulfate concentration increased to the chloride level. This reuse method allowed for a 20,000:1 ratio of product water to spent regenerant.

Typical breakthrough curves for DOC, chloride, and sulfate from the early portion of the run in the Hannover process are shown in Figure 17-4. After sulfate breakthrough at 250 BV, the DOC removal remained at about 50 percent until 5,000 BV—the point of run termination. Although far from complete, the 50 percent TOC removal was sufficient to increase the bacterial doubling time in the product water from 5 to 12 hours, and to drastically reduce the chlorine consumption with its attendant organic halide formation problems. The overall treatment efficiency was reduced by only 10 percent because of fouling and resin losses over a period of two years of full-scale operation.

More recently, Baker, Lavinder, and Fu (1995) reported on a 40-gpm (150-L/min) pilot study using Thermax A72-MP—a polystyrene type 1 macroporous SBA resin—for TOC and color removal from lime-softened water in Broward County, Fla. The water had to be treated because the former method of chlorinating the softener effluent to remove color was no longer acceptable due to excessive THM formation. The addition of Fe(III) during lime softening improved color removal, but resulted in a hard-to-dewater sludge. Because the softening process had just

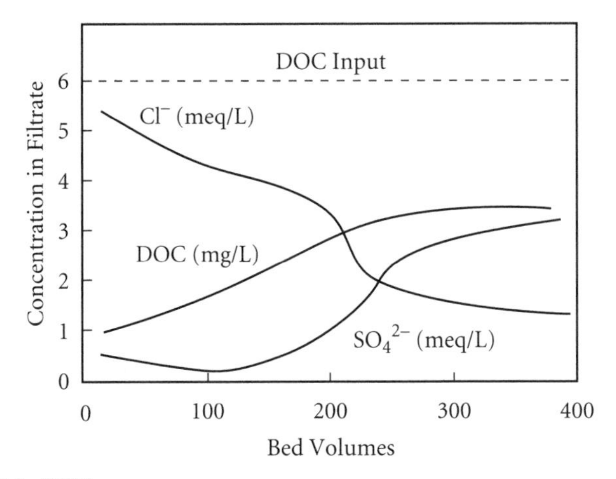

Source: Kolle (1984).

Figure 17-4 Breakthrough curves for polystyrene SBA resin (MP 500 A)

been refurbished, it was not economically feasible to replace it with nanofiltration softening membranes. Ozonation was also considered but was unacceptable because it effectively reduced color but not TOC and trihalomethane formation potential (THMFP). The softened water fed to the anion resin exhibited the following characteristics: TDS (248 mg/L), TOC (10.2 mg/L), color (17 pcu), and sulfate (27 mg/L). The resin EBCT was 2.7 min and run length was set to 3,400 BV based on previous work with similar waters. A unique feature of the process was that NaOH was not a component of the regenerant solution; only NaCl at a level of 10 lb/ft^3 (2.7 eq/L) resin was used. The spent regenerant was not recycled, but was disposed of in the sanitary sewer after dilution with rinse water and backwash water.

The pilot plant was operated for 30 exhaustion–regeneration cycles (103,000 BV) without any apparent resin fouling or deterioration in process performance. A typical effluent history for TOC, THMFP, and color is presented in Figure 17-5. Both THMFP and TOC exhibited similar gradual breakthroughs, while color removal was consistently greater than 85 percent throughout the run. On average, the resin removed 95 percent of the TOC in the >10K MW fraction; 65 percent of the 1 to 10K fraction, but only 36 percent of the <1K fraction. The authors concluded that

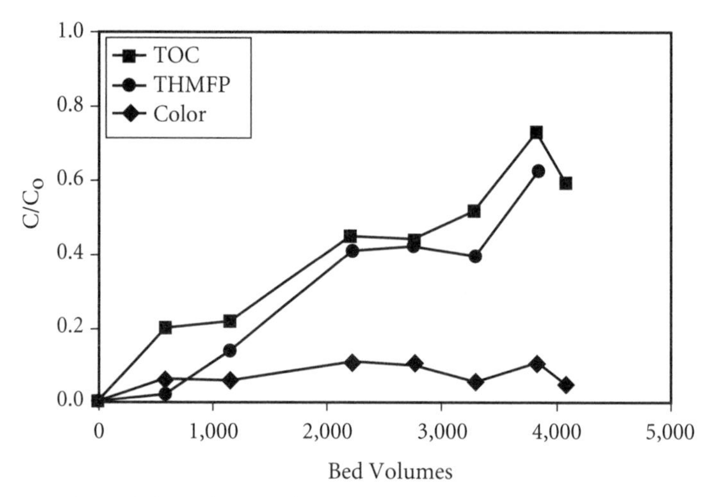

Source: Baker, Lavinder, and Fu (1995).

Figure 17-5 Performance of polystyrene SBA resin on softened Broward County, Fla., water, showing TOC, color, and THMFP breakthrough curves for cycle 17

anion exchange was effective for color and NOM removal from this south Florida lime-softened groundwater.

FUNDAMENTALS OF TOC REMOVAL BY ION EXCHANGE

Ion Exchange Versus Adsorption for TOC Uptake

Before the work of Fu and Symons (1990), it was believed that NOM was adsorbed onto resins rather than ion exchanged. To shed light on the NOM uptake process, Fu and Symons studied two macroporous resins, five gel resins, and two macroporous polymeric adsorbents to remove TOC from <1K, 1 to 5K, 5 to 10K, and >10K MW fractions of Lake Houston surface water. The percentages of TOC in the raw water fractions were 52, 9, 28, and 11, respectively.

Their ion exchange isotherm experiments indicated that each milliequivalent of TOC in the 1 to 5K fraction sorbed was balanced by an equal amount of chloride ion released (see Figure 17-6). Overall, their results showed that the sorption mechanism for TOC removal was ion exchange with only a small percentage of the TOC in the <1K and >10K fractions being removed by

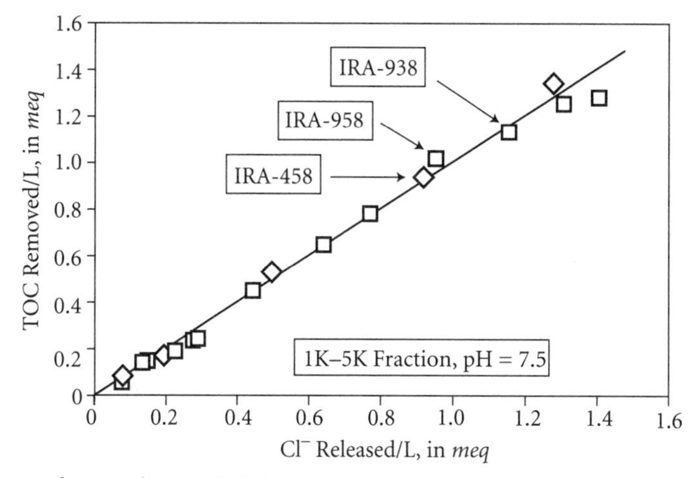

Source: After Fu and Symons (1990).

Figure 17-6 Proof of ion-exchange mechanism

adsorption. This is in contrast to the previously described color removal results of Coogan et al. (1968) who observed that fulvic acid anions (presumably paired with mobile cations, such as Ca^{2+} and Na^{+}) were removed primarily by adsorption onto the phenolic weak base resin, Duolite A-7.

Comparison of SBA Resins

Using equilibrium isotherm tests, Fu and Symons found that the resin matrix and porosity exhibited significant influences on TOC removal. Figure 17-7 compares the sorption isotherms for six SBA resins tested using the 1K to 5K MW TOC fraction as the sorbate. Most apparent in Figure 17-7 is the superior performance of the polyacrylic resins IRA 958 and 458. Even the microporous polyacrylic IRA 458 resin with no measurable BET surface area outperformed the macroporous polystyrene resins (IRA 938, IRA 904, and XE 510) with BET surface areas of 7, 60, and 400 m2/g, respectively. Typically, for TOC MW ≥1K, the hydrophilic polyacrylic resins were preferred over the relatively hydrophobic polystyrene resins. Generally, the researchers found that the ability of a resin to sorb NOM was positively correlated with its tendency to swell in water, i.e., its hydrophilic nature. Thus, their superior NOM adsorption performance coupled with

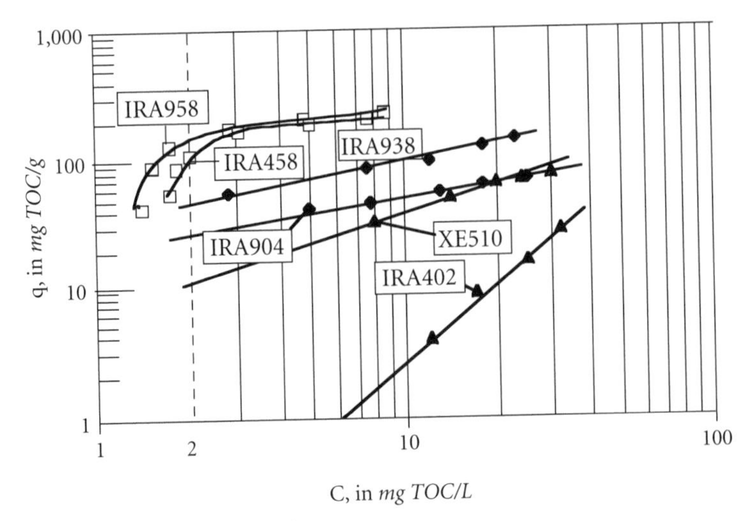

Source: After Fu and Symons (1990).

Figure 17-7 Isotherms for 1K to 5K fraction

their well-known organic fouling resistance (Baker, Davies, and Yarnell 1977) made the polyacrylic resins (IRA 458 and 958) logical choices for further research. The authors did not suggest, however, that polyacrylic resins were always to be preferred to polystyrene resins for NOM removal. Rather, they pointed out, as others have, that the choice among resin matrices will depend on the NOM characteristics of the water to be treated.

Resin porosity also influenced NOM adsorption. For all MW fractions >1K, macroporous resins performed better than the gel types, i.e., the greater the resin porosity, the better the TOC removal. At <1K MW, the resin porosity did not influence TOC removal, which was, however, influenced by the resin matrix. In this low MW range, the hydrophobic polystyrene resins performed somewhat better than the hydrophilic polyacrylic resins.

Multicomponent Column Behavior

Using the best resin from the isotherm tests, a macroporous acrylic resin (IRA 958), Fu and Symons (1990) studied the column chromatographic behavior of the <1K and 5 to 10K fractions. With only 6 mg/L sulfate and 18 mg/L chloride in the dialyzed feed water, the run length to sulfate breakthrough was

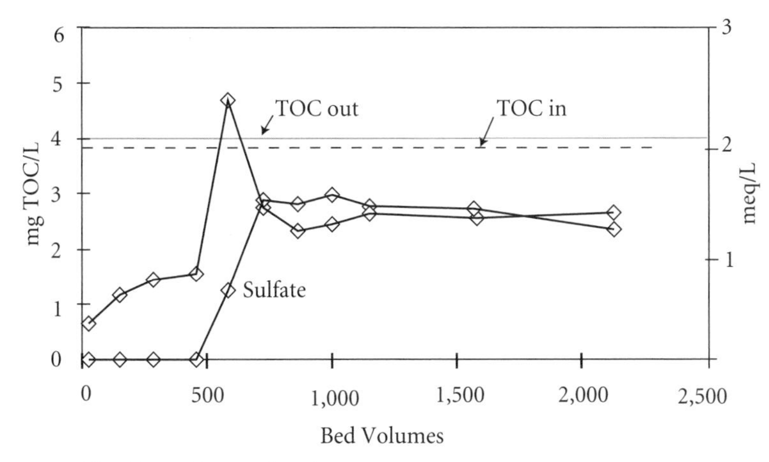

Source: After Kim and Symons (1991).

Figure 17-8 IRA-958 effluent history, treated Houston (Texas) water with Virgin IRA 958 resin, pH at 6.8, and EBCT at 10 min

exceptionally long, 3,000 BV, compared with typical drinking waters. Before sulfate breakthrough, TOC removal was about 75 percent. The 25 percent TOC not removed was considered to be nonionic. At sulfate breakthrough, a TOC peak appeared, which was the ionic TOC less-preferred than sulfate being driven off the column by the advancing sulfate wave front. The TOC more-preferred than sulfate was still being removed after sulfate breakthrough; at this point the TOC removal was about 50 to 60 percent until 9,000 BV. After that, TOC removal dropped gradually to zero at about 12,000 BV. These column results were similar to those of Kolle (1984) in that TOC removal started out at 70 to 80 percent and dropped to a long period (3,000 to 9,000 BV) of about 50 percent removal following sulfate breakthrough.

Total Organic Carbon and THMFP Behavior During a Pilot Study

Kim and Symons (1991) continued the work described previously by operating small columns of IRA 958 acrylic SBA resin at a Houston, Texas, surface water treatment plant. Following conventional treatment, the water fed to their experimental minicolumns was cartridge filtered (1 and 0.45 μm), dechlorinated (Na_2SO_3), and pH adjusted (to 6.8). Figure 17-8 illustrates virgin resin breakthrough curves for sulfate and TOC (which was DOC for this coagulated, filtered water).

As previously described, background sulfate concentration determined the performance of the system. When sulfate broke through at about 500 BV, it caused the TOC that was less preferred than sulfate to peak at about 170 percent of the influent TOC concentration. In plots of the effluent histories of the various TOC fractions, all peaked at essentially the same time, indicating that all fractions, except <0.5K, contained some TOC less preferred than sulfate. The TOC effluent histories also indicated that, generally, the higher the MW the better was the TOC removal before and after the sulfate breakthrough. Running beyond sulfate breakthrough was not very effective with this coagulated water because only about 25 percent overall TOC removal was achieved after 1,300 BV.

The Kim and Symons (1991) TOC regeneration study determined that the most effective TOC regenerant was a mixture of 2*N* NaCl with 0.5*N* NaOH, similar to the 1.7*N* NaCl plus 0.5*N* NaOH mixture used by Kolle (1984) in their full-scale studies. This mixture was used in the regenerant reuse study, which showed that the IRA 958 resin could be exhausted and regenerated nine times with the same regenerant. The regenerant had been compensated with salt and caustic after each regeneration to maintain their original concentration. Although the TOC elution during the eighth reuse cycle was only 13 percent, the overall performance of the resin for TOC removal up to sulfate breakthrough was essentially unchanged. Kim and Symons suggested that the TOC not eluted during regeneration was building up in the deeper pores of the resin where it did not interfere with the reversible ion exchange, which appeared to be taking place closer to the resin surface. More study is needed about the detrimental effects of TOC buildup after many exhaustion–regeneration cycles.

Figure 17-9 is the effluent history of THMFP for the IRA 958 column that had been exhausted eight times and regenerated with recycled regenerant. The THMFP breakthrough was similar to TOC breakthrough; THMFP peaked just before sulfate breakthrough and its level was >100 µg/L after only 200 BV. This showed the limited usefulness of ion-exchange alone for THMFP control.

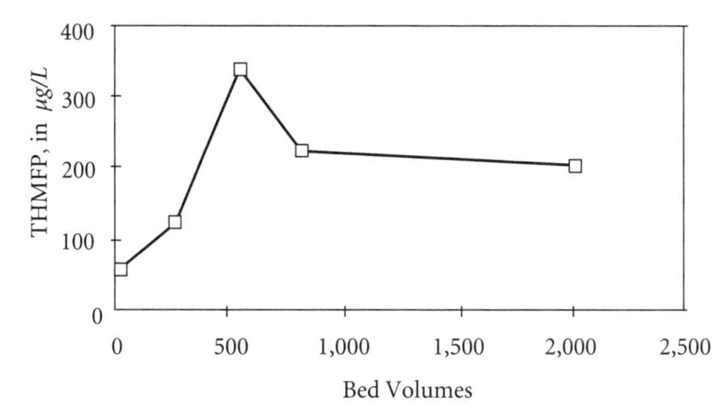

Source: After Kim and Symons (1991).

Figure 17-9 THMFP effluent history for IRA-958 effluent, eight times regenerated

ION EXCHANGE VERSUS GRANULAR ACTIVATED CARBON AND ION EXCHANGE PLUS GRANULAR ACTIVATED CARBON

Previous investigators had looked at GAC versus anion exchange for control of color, TOC, and DBP precursors in drinking water. For example, Medlar, Turner, and Keene (1981) studied the potential use of a polystyrene macroporous SBA resin (IRA 904) for control of THMFP, TOC, and color on a raw river water. A summary of their findings is presented in Table 17-2. The THMFP, TOC, and color removals were nearly identical for this raw river water with influent TOC of 2 to 10 mg/L and color 10 to 60 pcu. However, based on a run length of 6,300 BV and EBCT of 8 min, the estimated cost of GAC with thermal regeneration was more than double that of the resin with salt (NaCl) regeneration.

Boening, Beckman, and Snoeyink (1980) compared six anion resins (WBA and SBA) with five GACs for the removal of soil- and leaf-generated fulvic acids and soil and commercial humic acids. They usually found that the resins were equal to or better than the carbons for adsorbing the humic and fulvic acids. One resin, the strong-base polystyrene macroporous resin IRA 904, was much better than all the GACs, and all the other resins tested including IRA 458, the strong-base polyacrylic gel that performed

Table 17-2 GAC versus SBA resin comparison

Parameter	GAC	Resin
THMFP removal	68%	70%
TOC removal	54%	54%
Color removal	67%	67%
Regeneration	Thermal	NaCl
Run length, BV	6,300	6,300
10-year cost, $ mil	4.5	1.9

NOTE: Resin: IRA 904, macroporous, polystyrene; GAC: Filtrasorb 300.
Source: Medlar, Turner, and Keene (1981).

much better than IRA 904 in the experiments of Fu and Symons (1990). Curiously, pH 5.5 was better than 7.0 or 9.5 for adsorption of the organics onto the IRA 904 resin. This suggests that adsorption, in addition to ion exchange, was a major factor in removing the organics. To choose a resin, they recommended pilot tests with the actual water to be treated.

When Kim and Symons (1991) tested the sequence of ion exchange (IRA 958) followed by GAC (Calgon F-400) for THMFP control, they found it to be far superior to GAC alone or anion exchange alone. When regenerating the resin at sulfate breakthrough and reusing the regenerant, the IX–GAC process consistently achieved 90 percent TOC and 95 percent THMFP removal to 5,000 BV, whereas GAC alone achieved only 75 percent THMFP removal after about 2,000 BV. Granular activated carbon plus ion exchange is an excellent method for controlling DBPs, but the costs may be prohibitive.

CONCLUSIONS

Resins

Compared with GAC, WBA or SBA resins have greater sorption capacity for NOM, remove NOM faster, and are easier to regenerate. When operated at pH 6.5 to 8.5, the WBA resins adsorb the NOM, whereas SBA resins operate by the mechanism of ion exchange. The choice between polystyrene and polyacrylic resins will depend on the water supply being treated. Generally polyacrylic resins will be easier to regenerate, but often the polystyrene resins will have greater capacity for NOM. Pilot tests with several exhaustion–regeneration cycles will typically be required. Macroporous resins should be chosen over microporous resins for the removal of NOM because of its typically high MW, but there are exceptions where microporous polyacrylic resins may be preferred.

Organics

Each source water will have a different concentration and distribution of TOC components, and each MW fraction will have a different affinity for the chosen IX resin. Generally, the <0.5K MW fraction will not be removed by resins while the >10K fraction will be effectively removed.

Run Length

For color removal by WBA resins in the 6 to 9 pH range, run lengths typically are in the range of 1,500 to 5,000 BV. Because NOM removal is by adsorption rather than ion exchange, WBA run lengths are not significantly influenced by sulfate concentration, and organic (color or TOC) peaking is not obvious in the effluent history.

For TOC removal by SBA resins, run lengths of 200 to 700 BV can be expected before sulfate breakthrough. Total organic carbon peaking is generally observed immediately prior to sulfate breakthrough because the less preferred NOM is displaced by sulfate. After sulfate breakthrough, one can typically expect 50 percent TOC removal for several thousand more BV.

Regeneration

Weak-base resins in NOM removal service are easily regenerated with NaOH in a single-step regeneration or by lesser amounts of NaOH and HCl in a two-step regeneration. No work has been reported on the reuse of WBA regenerants.

Strong-base resins used for color and NOM removal can be regenerated by a mixture of NaCl (1.7 *M*) and NaOH (0.5 *M*), or by NaCl (2 *M*) alone. Several studies have shown that the spent regenerant can be directly reused 7 to 10 times before disposal. Because they are inherently more hydrophilic, polyacrylic SBA resins are more easily regenerated than polystyrene resins.

THMFP Removal

Various studies have demonstrated that it is not possible to reliably predict THMFP removal based on surrogates, such as color or TOC removal. The relationship depends on the source water and the prior treatment. For some waters, TOC removal will track THMFP removal, and 50 percent reduction may be sufficient to meet the MCL. For waters for which high levels of THMFP removal are required, a combination process consisting of strong-base IX followed by GAC adsorption has been demonstrated to remove more than 95 percent of the THMFP.

REFERENCES

Abrams, I.M. 1982. Organic Fouling of Ion Exchange Resins. In *Physicochemical Methods for Water and Wastewater Treatment.* Pawlowski, L., ed. Amsterdam: Elsevier Publishing Company.

Abrams, I.M., and R.P. Breslin. 1965. Recent Studies on the Removal of Organics from Water. In *Proc. 26th International Water Conference of the Engineers Society of Western Pennsylvania.* Pittsburgh, Pa.

Baker, B., V.R. Davies, and P.A. Yarnell. 1977. Use of Acrylic Strong Base Anion Resin in Treatment of Organic Bearing Waters. In *Proc. 38th Annual Int'l. Water Conf.,* Pittsburgh, Pa.

Baker, J., S. Lavinder, and P. L.-K. Fu. 1995. Removal of Natural Organic Matter with Anion Exchange Resins. In *Proc. 1996 AWWA Annual Conference.* Denver, Colo.: American Water Works Association.

Boening, P.H., D.D. Beckmann, and V.L. Snoeyink. 1980. Activated Carbon Versus Resin Adsorption of Humic Substances. *Jour. AWWA,* 73(1):54.

Coogan, G.J., G.R. Bell, D.A. Jayes, and I.M. Abrams. 1968. A New Method for Color Removal From Surface Water. *Jour. New England Water Works Assoc.* (March 1968).

Duolite Data Leaflet No. 2. 1974. Philadelphia, Pa.: Rohm and Haas Company.

Duolite Tech Sheet 130. 1968. Philadelphia, Pa.: Rohm and Haas Company.

Frisch, N.W., and R. Kunin. 1957. *Ind. Eng. Chem.,* 49:1365.

Fu, P. L.-K., and J.M. Symons. 1990. Removing Aquatic Organic Substances by Anion Exchange Resins. *Jour. AWWA,* 82(10):70.

Kim, P. H.-S., and J.M. Symons. 1991. Using Anion Exchange Resins to Remove THM Precursors. *Jour. AWWA,* 83(12):61.

Kolle, W. 1984. Humic Acid Removal with Macroreticular Ion Exchange Resins at Hannover. In *NATO/CCMS Adsorption Techniques in Drinking Water Treatment,* CCMS 112, USEPA 570/9-84~005. Washington, D.C.: US Environmental Protection Agency.

Kunin, R. 1983. *Amber Hi-Lites,* No. 173. Philadelphia, Pa.: Rohm and Haas Company.

McGarvey, F.X., and A.C. Reents. 1954. Get Rid of Fouling in Ion Exchangers. (Sept. 1954). *Chem. Eng.,* 61:205.

Medlar, S., S. Turner, and W. Keene. 1981. Economic Comparison of Granular Activated Carbon and Anion Exchange Resin for Trihalomethane Precursor Removal. Presented at the 181st National ACS Meeting, Atlanta, Ga.

Miles, Inc. 1992. Lewatit Ion Exchange Resins for Treatment of Industrial and Surface Waters. Pittsburgh, Pa.: Miles, Inc.

Rook, J.J., and S. Evans. 1979. Removal of Trihalomethane Precursors from Surface Waters Using Weak Base Resins. *Jour. AWWA,* pp. 520–524.

Wilson, A.L. 1959. Organic Fouling of Strongly Basic Anion-Exchange Resins. *Jour. Appl. Chem.,* 9:352.

CHAPTER • EIGHTEEN

Treatment Costs for Disinfection By-Product Control

DOUGLAS M. OWEN

Cost is one of the major factors affecting the use of any technology to meet regulatory requirements. Although treatment requirements are in many instances regulatory-driven, i.e., mandated by USEPA or state agencies, cost is a significant factor in the selection and implementation of a successful technological solution. In this chapter, the factors affecting technology implementation and the costs associated with technologies to control DBPs are presented.

CONSIDERATIONS IN IMPLEMENTING TECHNOLOGIES TO CONTROL DISINFECTION BY-PRODUCTS

Previous chapters in this book have discussed chemistry, health effects, regulations, and treatment alternatives for disinfection by-products (DBPs). In many cases, there is an array of treatment alternatives that can be used to meet a specific DBP water quality goal. Given this matrix of solutions, other criteria may be used to select the preferred alternative. Questions relating to these criteria include

1. How well can the treatment alternative fit into the existing process train? Is it compatible with existing treatment processes?
2. Does the treatment alternative meet other water quality objectives? Can multiple customer water quality requirements, beyond DBPs alone, be met through the technology?
3. What does the technology cost?
4. Is the technology capital- or operationally intensive? Can it be implemented within the fiscal constraints (i.e., rate structure and bonding capacity) of the facility?
5. Is the treatment technology compatible with existing operational paradigms? Does the utility need to investigate a change in operational concepts?

These types of questions must be answered in order to select a technology that can be implemented successfully. Without exception, however, a major component is cost.

FACTORS AFFECTING COST ESTIMATES

Design Criteria and System-Specific Requirements

In the following discussion, costs will be presented for various technologies to control DBPs. It is important to note that there are a broad range of factors that will affect these costs, including:

1. The type and efficiency of existing treatment processes and units.
2. The process design criteria required to meet the water quality objectives.
3. The existing facility layout and the ability to incorporate the new or modified technology in the existing treatment facilities.
4. The capital cost of the technology and the ability of the system to float bonds or find other financial mechanisms

to finance construction. In particular, interest rates, which are critical in calculating the unit cost of the technology as the capital is amortized over a period of time, are strongly dependent on the size of the facility and the bonding capacity of the purveyor.

5. The ability to finance the operational costs through existing or modified rate structures.

In the following discussion, cost estimates will be provided for DBP treatment alternatives. The assumptions associated with the cost estimates will be explicitly stated, including the design criteria, interest rates for amortizing capital, and other important factors affecting the costs presented. It is important to recognize that design criteria will vary depending on the efficiency of existing treatment. For example, when enhancing coagulation by increasing coagulant dosages to reduce DBP precursors, the extent to which the dosage must increase depends on such factors, as (1) the source water quality, including precursor concentration, alkalinity, and pH; (2) the existing chemical dosage; (3) the dosage required to meet the water quality objective; (4) the existing configuration of chemical feed facilities; and (5) existing and future sludge-handling requirements. Therefore, the design criteria on which the costs are based in this chapter may not be specific to a given facility. These issues will be discussed for each of the DBP control technologies.

Cost Basis

Costs are presented in terms of capital, operations and maintenance (O&M), and on a unit basis (¢/thous gal produced) for a range of system sizes. As would be expected, while capital and O&M costs increase with system size, unit costs decrease as system size increases. This occurs because of economies of scale for many technologies, particularly those that are capital-intensive.

Capital costs are based on the design flow for the facility, which is often the maximum daily flow. The O&M cost, on the other hand, is based on the annual average flow. An interest rate of 7 percent and a 20-year capital recovery period is used in amortizing capital on an

Table 18-1 Assumptions for selected construction cost components

Cost Factor	Percent of Capital Cost
Engineering	15
Site Work, Interface Piping	15
Subsurface Considerations	10
Standby Power	5

annual basis. This annual capital recovery cost, when added to annual O&M costs, results in a total annual cost for a given technology. This total annual cost, divided by the average annual production, yields a unit cost (¢/thous gal produced).

It is recognized that the interest rate factor depends on the prevailing inflation rate and borrowing costs, as well as system size. Small systems often cannot obtain interest rates as low as larger systems. Nevertheless, a 7 percent value is typically used in the industry for comparing the unit costs of different treatment alternatives and is not varied by system size in this chapter.

Similarly, a design period of 20 years is also dependent on the technology. Facilities that are concrete- and piping-intensive may have expected lives much longer than 20 years, while mechanically-intensive facilities may have an expected useful life less than 20 years. Again, a 20-year period is typical when amortizing capital to develop unit costs. Since both capital and O&M costs are provided for the various technologies, any interest rate or design period can be applied to annualize capital costs.

Cost estimates for each process include cost allowance factors. Table 18-1 summarizes example percentages applied to capital costs to account for engineering, site work, subsurface considerations, and standby power. Unit costs for land and energy, and unit chemical costs, are presented in Tables 18-2 and 18-3, respectively. Other assumptions can be found in greater detail in other documents (e.g., USEPA 1998a).

It also must be stressed that the costs presented here are central-tendency costs. That is, they are costs associated with a typical system across the country. There are a wide variety of geographic factors, in

Table 18-2 Unit costs for energy and land

Parameter	Unit Cost
Land Cost, $ per acre	20,000
Electricity, $ per kW·h	0.08
Labor, $ per hour	30
Diesel Fuel, $ per gal	1.25
Natural Gas, $ per ft^3	0.006
Building Energy Use, kW·h/ft^2/yr	102.6

addition to those previously described, which will affect the individual cost for any particular system. Some of these include prevailing labor rates and material costs, cost of land, energy costs, and other factors that are geographically specific. Therefore, it is not recommended that the costs shown here be used to develop a budget estimate for a specific system. Rather, these costs can be used to understand the conceptual cost of certain strategies and relative differences in the costs for different technologies to control DBPs, as applied using criteria similar to those used to develop the estimates.

Finally, the cost estimates provided are based on "greenfield" construction, i.e., it assumes that the processes will be constructed on open, available land, as opposed to retrofitting existing unit processes. Because retrofitting is such a site-specific process, it is not possible to provide general estimates for retrofit construction. It must be recognized that these central-tendency costs need to be modified to account for retrofitting existing facilities. Highly mechanical equipment modifications (e.g., ammonia or chlorine dioxide addition) may have lower retrofit costs, while facilities-intensive processes (e.g., granular activated carbon [GAC]) may have higher retrofit costs.

System Size

Costs obviously vary with system size. System sizes used by USEPA, as shown in Table 18-4, are also used for presenting costs in this chapter.

Table 18-3 Unit costs for chemicals

Chemical	Units	Cost
Granular Activated Carbon	$/lb	1.0
Powdered Activated Carbon	$/lb	0.95
Alum, Dry Stock	$/ton	300
Alum, Liquid Stock	$/ton	230
Ammonia, Anhydrous	$/ton	300
Ammonia Aqueous	$/ton	320
Carbon Dioxide, Liquid	$/ton	340
Carbonic Acid	$/ton	400
Chlorine, 1-ton cylinder	$/ton	350
Chlorine, 150-lb cylinder	$/ton	400
Diatomaceous Earth	$/ton	650
Ferric Chloride	$/ton	350
Ferric Sulfate	$/ton	350
Ferrous Sulfate	$/ton	350
Hexametaphosphate	$/ton	1,280
Lime, Hydrated	$/ton	110
Lime, Quicklime	$/ton	95
Polymer	$/lb	2.25
Potassium Permanganate	$/ton	2,700
Soda Ash	$/ton	400
Sodium Chloride	$/ton	99
Sodium Hydroxide, 50% Solution	$/ton	371
Sodium Hypochlorite, 12% Solution	$/ton	1,000
Sulfuric Acid	$/ton	116
Hydrogen Peroxide	$/gal	1.25
Liquid Oxygen	$/ton	350
Sulfur Dioxide	$/ton	350
Surfactant, 5% Solution	$/gal	0.15
Chlorine, Bulk	$/ton	280
Sodium Chlorite	$/ton	400
Phosphoric Acid	$/ton	300

Table 18-4 USEPA flow categories

Median Population Served	Average Flow (*mgd*)	Design Capacity (*mgd*)
57	0.0056	0.024
225	0.024	0.087
750	0.086	0.27
5,500	0.70	1.8
15,000	2.1	4.8
35,000	5.0	11
60,000	8.8	18
88,000	13	26
175,000	27	51
730,000	120	210
1,550,000	270	430

Source: USEPA (1998a).

COSTS FOR TECHNOLOGIES TO CONTROL DBPs

Introduction

Technologies to control DBPs can be divided into two broad categories: (1) alternative disinfection technologies to reduce the formation of specific DBPs of concern and (2) technologies to remove the precursor material that forms DBPs when disinfectant is added. In general, the costs of the alternative disinfectants are less than those for the precursor removal technologies, particularly those precursor removal technologies that are highly efficient and can remove 50 percent or more of the precursor material. Consequently, systems will often investigate the use of alternative disinfectants first. Unfortunately, all disinfectants form some type of DBP and therefore precursor removal may have to be used to some extent to meet water quality goals. In the following discussion, the costs for alternative disinfectants are discussed, followed by precursor removal technologies. These technologies are presented in the order of increasing cost, based on typical applications.

Costs for Alternative Disinfectants

Free chlorine has been the disinfectant of choice in the water industry since the early 1900s. In general, chlorine is inexpensive, easy to apply, and efficient in inactivating pathogens. In efforts to control DBPs, other chemical disinfection technologies have been pursued, including chloramines, chlorine dioxide, and ozone. The disinfecting capacity increases in that order (i.e., chloramines least and ozone most), as does cost. Capital and operating costs for these disinfectants are discussed below, based on specific design criteria.

Chloramines. As described in earlier chapters, chloramines are formed by adding ammonia to free chlorine. Chloramine costs presented here are based on a dosage of 4.0 mg/L as chlorine, because this is a typical design dosage for secondary disinfection in systems using chloramines. A 4:1 chlorine-to-ammonia mass ratio was assumed, which allows for a small amount of excess free ammonia. The use of ammonium sulfate and aqua ammonia was assumed for systems smaller and larger than 1 mgd, respectively.

Costs include pumps, piping, injectors, and all other equipment necessary for chemical feed and storage. It was assumed that chemical feed systems are enclosed and chlorine storage is contained with appropriate scrubbing facilities. Ammonia also is housed and has appropriate containment facilities. Storage volumes were based on 10-day storage at design flow.

The capital, O&M, and unit costs to implement chloramines for various system sizes are summarized in Table 18-5.

Chlorine dioxide. Chlorine dioxide has widespread use in Europe and has been used more specifically in North America for controlling taste and odors. Recent improvements in chlorine dioxide technology have focused on maximizing the purity of chlorine dioxide produced to minimize the production of inorganic by-products (e.g., chlorite and chlorate), which may have adverse health effects.

Chlorine dioxide is consumed rapidly through oxidation–reduction reactions and therefore is principally used as a primary disinfectant (i.e., not to maintain a residual in the distribution system). Dosages vary significantly based on source water quality and overall disinfection requirements. Therefore, costs are presented for two dosages for comparison—0.5 mg/L and 1.0 mg/L.

Table 18-5 Estimated upgrade costs for chloramines as secondary disinfectant

Design Capacity (*mgd*)	Upgrade Capital Cost (*$M*)	Upgrade O&M Cost (*¢/thous gal*)	Unit Upgrade Cost at 7% (*¢/thous gal*)
0.024	0.011	21	71
0.087	0.012	5.5	19
0.27	0.015	1.9	6.3
1.8	0.04	1.4	3.0
4.8	0.07	0.70	1.5
11	0.11	0.49	1.1
18	0.16	0.40	0.87
26	0.21	0.037	0.79
51	0.28	0.33	0.60
210	0.17	0.29	0.39
430	0.85	0.26	0.34

In the costs presented here, chlorine dioxide is produced with an automatic generator using chlorine and sodium chlorite. The amount of chlorine consumed per chlorine dioxide generated on a mass basis is 0.5:1. Similarly, the amount of sodium chlorite used to produce chlorine dioxide is 1.35:1 on a mass basis. It was assumed that chlorine dioxide generation equipment would be located in a building. Sodium chlorite is also housed and chlorine storage is contained with appropriate scrubbing equipment.

The capital, O&M, and unit costs for chlorine dioxide are summarized in Table 18-6. Equipment was amortized at 5 years and building/containment facilities were amortized over a 20-year period.

Ozone. Ozone disinfection is a technology implemented extensively throughout Europe and has gained acceptance in the United States primarily since the 1980s. Ozone is a powerful disinfectant and decomposes rapidly. Therefore, it only is considered as a primary disinfectant and is not used to maintain a residual in the distribution system.

With the possible exception of chlorine dioxide, ozone is the only disinfectant that can be used to disinfect particularly resistant pathogen cysts, such as *Cryptosporidium*. Consequently, ozone dosages will vary substantially depending on the disinfectant

Table 18-6 Estimated installation and upgrade costs for chlorine dioxide as a primary disinfectant

	ClO_2 Dose = 0.5 mg/L			ClO_2 Dose = 1.0 mg/L		
Design Flow (*mgd*)	Upgrade Capital Cost (*$M*)	Upgrade O&M Cost (*¢/thous gal*)	Unit Upgrade Cost at 7% (*¢/thous gal*)*	Upgrade Capital Cost (*$M*)	Upgrade O&M Cost (*¢/thous gal*)	Unit Upgrade Cost at 7% (*¢/thous gal*)*
0.024	0.10	1,548	2,010	0.10	1,552	2,014
0.087	0.10	364	472	0.10	367	475
0.27	0.10	107	137	0.10	110	140
1.8	0.22	15	23	0.22	17	25
4.8	0.33	6.3	10	0.33	8.6	13
11	0.33	3.9	5.7	0.33	6.0	7.7
18	0.68	3.3	5.0	0.68	5.3	7.0
26	0.68	2.9	4.4	0.68	4.8	6.4
51	0.76	2.3	2.7	0.76	4.1	4.7
210	0.77	1.8	2.2	0.77	3.5	4.2
430	0.83	1.6	2.1	0.83	3.1	3.1

*Equipment capital amortized over 5 years; building capital amortized over 20 years.

objective; much lower dosages are required for inactivation of *Giardia* compared to *Cryptosporidium*. Further, because ozone reacts quickly with a variety of compounds in water, the dosage will be strongly dependent on ozone-demanding substances and temperature. Therefore, when reviewing the costs, it is important to evaluate the design criteria closely.

Costs were prepared for ozone dosages of 2, 5, and 7 mg/L at a design temperature of 5°C. It was assumed that pure oxygen was the feed gas for systems less than 1 mgd, and air preparation equipment was used for systems larger than 1 mgd. In both cases, a medium-frequency generator was assumed. The system-specific energy consumption is 10 kwh/lb of ozone produced. It also was assumed that a separate, baffled concrete contact chamber would be required for ozone disinfection. A 10-minute contact time was used in the estimates.

Table 18-7 summarizes the capital, O&M, and unit costs for a variety of system sizes.

Table 18-7 Estimated installation and upgrade costs for ozone as a primary disinfectant

	Ozone Dose = 2 mg/L			Ozone Dose = 5 mg/L			Ozone Dose = 7 mg/L		
Design Capacity (*mgd*)	Upgrade Capital Cost* (*$M*)	Upgrade O&M Cost (*¢/thous gal*)	Unit Upgrade Cost at 7% (*¢/thous gal*)†	Upgrade Capital Cost* (*$M*)	Upgrade O&M Cost (*¢/thous gal*)	Unit Upgrade Cost at 7% (*¢/thous gal*)†	Upgrade Capital Cost* (*$M*)	Upgrade O&M Cost (*¢/thous gal*)	Unit Upgrade Cost at 7% (*¢/thous gal*)†
0.024	0.22	161	1,177	0.23	322	1,384	0.24	644	1,752
0.086	0.24	38	297	0.28	75	377	0.30	150	473
0.27	0.29	10	197	0.40	21	145	0.17	42	183
1.8	0.89	1.8	35	1.3	4.4	52	1.5	6.3	62
4.8	1.5	1.8	21	1.9	4.4	28	2.0	6.3	31
11	1.9	1.8	12	2.6	4.4	17	2.8	6.3	21
18	2.4	1.8	8.9	3.0	4.4	13	3.7	6.3	17
26	2.6	1.8	7.2	3.9	4.4	12	4.8	6.3	15
51	3.8	1.8	5.6	6.2	4.4	10	7.4	6.3	13
210	9.2	1.8	4.0	18	4.4	7.8	24	6.3	11
430	16.5	1.8	3.6	35	4.4	7.4	47	6.3	11

*Costs include clearwell with 60 minutes contact time.

†Equipment capital amortized over 5 years; building capital amortized over 20 years.

Costs for DBP Precursor Removal Technologies

Disinfection by-product precursor removal technologies fall into the following three broad categories: (1) enhancing the removal of precursors through coagulation or precipitative softening (also used to meet a variety of other treatment goals), (2) GAC adsorption, and (3) membrane separation. These latter technologies are more expensive and typically are more efficient for removal of DBP precursors than enhanced coagulation, but may or may not meet other water quality objectives required to provide the desired water quality to the customer.

Enhanced coagulation. The term *enhanced coagulation* was coined during development of the D/DBP Rule (1992 to 1993). It should not be confused with *optimized coagulation*, which is the use of the coagulation process to meet multiple water quality objectives (e.g., removal of particles and NOM). Enhanced coagulation is a strict regulatory definition that addresses the use of the coagulation process to remove DBP precursor compounds (i.e., NOM). The fundamental premise is that the higher the coagulant dosage, and the lower the coagulation pH, the greater the extent of precursor material that can be removed. While this concept is adequate, it is important to note that there are a wide variety of site-specific conditions that affect the success of enhanced coagulation. The US Environmental Protection Agency (USEPA) has developed a guidance manual on the topic of enhanced coagulation (USEPA 1998b), and American Water Works Association (AWWA) has sponsored a project evaluating the secondary effects (e.g., effect on operations, sludge production, particle removal, and disinfection) of enhanced coagulation (AWWA 1998). Therefore, the implementation of this technology, and its subsequent effects, are extremely site-specific. Consequently, it is very important that the costs be reviewed relative to the design criteria in this section, so that cost estimates are not misapplied to a given utility.

There are many methods to improve removals in coagulation, such as changing the coagulant type (e.g., from alum to ferric chloride), increasing coagulant dosages, and lowering the pH of coagulation. In the effort for developing compliance forecasts during the regulatory negotiation, only increased coagulant dosage was evaluated. This was done because increasing the dosage is the method of operation a utility will typically consider first when attempting to improve organic removal by coagulation. Other alternatives will be evaluated as different constraints (e.g., sludge disposal options) become limiting.

A 40-mg/L increase in alum dosage is used here for cost purposes; this is likely a conservative case. Many recent studies (e.g., AWWA 1997) indicate that lower incremental dosages are required by most utilities, or that utilities can use smaller dosage increases together with reducing pH at lower total cost than those associated with increasing the dosage by 40 mg/L. Therefore, the costs presented here provide a reasonable upper estimate for enhanced coagulation. A caustic dosage of 15 mg/L also was assumed to maintain pH.

The cost estimates include all chemical pumping, piping, and storage facilities. A 10-day storage requirement at maximum daily flow was used. A building was provided to house chemical pumps, electrical equipment, and chemical storage with containment. Finally, sludge disposal was based on sludge drying lagoons. It is well understood that many facilities may be able to use significantly less

Table 18-8 Estimated upgrade costs for increasing coagulant dosage*

Design Capacity (*mgd*)	Capital Cost (*$M*)	O&M Cost (*¢/thous gal*)	Unit Cost at 7% (*¢/thous gal*)
0.024	0.001	10	15
0.087	0.003	10	13
0.27	0.006	10	12
1.8	0.08	6.3	9.1
4.8	0.14	5.9	7.7
11	0.18	5.8	7.0
18	0.26	5.7	6.8
26	0.36	5.7	6.7
51	0.64	5.7	6.6
210	1.4	5.6	6.3
430	2.7	5.6	6.3

*Alum dosage increase of 40 mg/L.

expensive disposal methods (e.g., sanitary sewer) or may require more expensive disposal options (mechanical dewatering and land disposal). It was not possible to determine a pooled cost for these alternatives based on any understanding of existing practice, however, because the practices may change as sludge quantities increase. During a peer-reviewed evaluation of the overall disposal costs during the regulatory negotiation, it was found that the sludge lagoon estimate provided a reasonable central tendency for water systems.

The costs presented in Table 18-8 have been shown to lie within a reasonable envelope of expected costs based on several separate utility estimates (e.g., Cheng et al. 1995; Chowdhury et al. 1995; Crozes et al. 1995). Therefore, although the dosage may be overestimated for many facilities, other factors (e.g., sludge disposal and lower coagulant dosages with additional pH adjustment) tend to compensate for the differences. Again, these costs may be considered a practical maximum.

Enhanced precipitative softening. The definition of enhanced precipitative softening, even from a regulatory perspective, has been even more elusive than enhanced coagulation. Softening processes have a number of unique configurations primarily designed at removing hardness. Not all softening processes are designed for the same finished water hardness, thus the costs vary widely.

Softening typically falls into two broad categories: calcium removal softening and "high lime" softening. In the latter, a pH above 10.5 is used and magnesium is precipitated. Sodium carbonate (soda ash) may be required to provide the needed carbonate for this process. Chemical dosages and associated treatment costs are typically much higher for high lime softening.

Costs for enhanced precipitative softening are not presented here. Preliminary research (Shorney and Randtke 1994) indicates that the costs of enhanced precipitative softening can range between one and five times the cost of enhanced coagulation. Higher costs are associated with systems needing to change from calcium softening to high lime softening.

Granular activated carbon. Granular activated carbon historically has been used in drinking water treatment to remove specific organic compounds associated with synthetic or taste-and-odor-causing organic chemicals. Recently, however, there has been considerable research into the use of GAC for the removal of DBP precursors (e.g., Summers et al. 1995). The extent to which GAC can be implemented successfully depends on the adsorbability of the compounds in the source water, the finished water quality goal, and the trade-off between capital facility investment and reactivation operations.

Granular activated carbon has a fixed capacity for removing organic material, including DBP precursors. Therefore, the overall cost depends on the amount of GAC used in the contactor and the frequency at which the exhausted GAC is reactivated. That is, GAC contactors can be increased in size to reduce the frequency at which reactivation occurs, or the contactors can be reduced in size and the reactivation frequency increased. The total GAC mass that must be reactivated does not change. The frequency of emptying GAC contactors and reactivating is a cost function that must be evaluated for each system and depends on the organic content of the applied water and the treatment goal.

In the costs presented here, it is assumed that GAC will be used in a postfilter adsorption mode. That is, a separate contactor containing GAC will be provided after filtration. Another mode of operation not evaluated here, filter adsorption, is one in which the GAC is placed in the filter bed and serves the dual functions of particle removal and adsorption/biodegradation of organic compounds (see chapter 13). The GAC in filter adsorbers is typically replaced only every two to three years, and therefore adsorptive capacity is completely exhausted within a few months, after which the GAC operates in a biological mode for any organic removal.

A general rule of thumb is that postfilter GAC contactors should not be emptied and reactivated at a frequency greater than once every 60 days. Conversely, a practical maximum for GAC reactivation used for DBP control is reactivation once every 180 days (AWWARF and USEPA 1989). During the regulatory negotiation held for the D/DBP Rule during 1992 to 1993, two GAC designs were defined. One consisted of a 10-minute empty-bed contact time (EBCT) with a 180-day reactivation frequency and the other a 20-minute EBCT with a 60-day reactivation frequency. These two conditions represented what was considered to be the least stringent and most stringent GAC designs for DBP precursor removal. It was assumed that carbon was replaced and reactivated off-site by a contractor for systems with design flows less than 5 mgd, and on-site reactivation with multiple hearth furnace technology for design flows greater than 5 mgd. Capital costs include concrete contactors and virgin carbon storage and reactivated carbon storage vessels, as well as carbon transfer systems, gas-fired regeneration furnaces, and ancillary regeneration equipment. The costs associated with these systems are summarized in Table 18-9.

Membrane treatment. Membranes can be subdivided into two broad categories—membrane filtration and nanofiltration–reverse osmosis. The former can remove microbial contaminants, including cysts, bacteria, and in some cases, viruses, but remove very little DBP precursor material. Accordingly, these are not discussed further here. Nanofiltration (NF) and reverse osmosis (RO) are membrane processes that require higher transmembrane pressures (e.g., 80 to 250 psi, depending on the ionic strength of the source water), and can remove between 95 and 99 percent of DBP precursors (Allgeier and Summers 1995).

The cost of NF membranes depends on the source water quality (e.g., ionic strength and pH), the type of membrane (e.g., cellulose

Table 18-9 Estimated upgrade costs for installation of GAC adsorption

	EBCT = 10 min Reactivation Frequency = 180 days			EBCT = 20 min Reactivation Frequency = 60 days		
Design Capacity (*mgd*)	Capital Cost (*$M*)	O&M Cost (*¢/thous gal*)	Total Cost at 7% (*¢/thous gal*)	Capital Cost (*$M*)	O&M Cost (*¢/thous gal*)	Total Cost at 7% (*¢/thous gal*)
0.024	0.12	32	600	0.23	295	1,374
0.087	0.20	15	227	0.38	155	566
0.27	0.32	11	107	0.71	113	328
1.8	0.95	14	49	2.5	84	175
4.8	2.1	12	38	5.9	66	138
11	3.9	9.4	30	9.8	46	97
18	6.2	9.2	27	16	45	91
26	8.9	4.7	22	20	43	82
51	13	3.7	17	34	23	56
210	42	2.9	12	110	18	41
430	75	2.7	10	180	16	34

acetate or thin film composites), and the associated feed pressure. Capital costs were separated into membrane costs and facility costs.

Capital and O&M cost estimates for large systems, i.e., more than 1 mgd, were developed from NF cost data presented in a NF plant survey conducted by Bergman (1996). Capital and O&M costs presented in that report were derived from cost data submitted by existing plants. The Bergman survey contained capital cost and O&M cost data obtained between 1988 and 1996. It is recognized that spiral-wound membrane modules, which include the majority of NF membranes, have decreased in cost significantly in recent years. Based on vendor information, costs for spiral-wound membrane modules have been reduced by approximately 50 percent over the past five years. For this reason, costs for membrane modules presented by Bergman (1996) were reduced by 50 percent. The reduced-cost items include new membrane capital costs and O&M membrane replacement costs.

Additionally, the costs given in the Bergman survey consist solely of NF costs from Florida plants. The source waters treated by these plants are relatively warm, resulting in higher membrane

Table 18-10 Estimated upgrade costs for installation of nanofiltration

Design Capacity (*mgd*)	Capital Cost (*$M*)	O&M Cost (*¢/thous gal*)	Unit Cost at 7% (*¢/thous gal*)
0.024	0.15	308	473
0.087	0.38	213	326
0.27	0.84	154	234
1.8	3.2	90	135
4.8	6.3	68	102
18	17	50	75
26	25	50	75
51	49	50	75
210	202	50	75
430	413	50	75

fluxes than would be expected for lower temperature waters of comparable quality. As a result, the costs presented by Bergman (1996) may not be representative of costs for all areas of the country. For this reason, the costs were adjusted to equivalent costs for 20°C water. Using the Bergman survey data, the average facility cost was found to be between two and three times the cost for membrane equipment.

For small systems, membrane module capital costs were based on vendor quotations. Facility costs were estimated by multiplying the membrane capital costs by three, the factor calculated by analyzing the cost data for large plants (Bergman 1996). Operation and maintenance costs for small systems were estimated by extrapolating the large system O&M cost curve below 1 mgd. It should be noted that in the Bergman survey the lowest flow was 1 mgd.

As temperature decreases, membrane flux will decrease, and therefore more membrane area will be required to meet a given capacity and the corresponding costs will increase. The extent to which these costs change depend both on the change in flux for a given membrane with temperature and the extent to which the water demand changes at low- and high-temperature conditions. Capital and O&M costs for nanofiltration are given in Table 18-10.

SUMMARY

Costs for alternative disinfectant technologies, as well as costs for technologies to remove the precursors for DBPs, have been presented. Unit costs (¢/thous gal produced) were calculated based on amortized capital costs, summed with annual O&M costs based on annual average production. There are many factors that affect the costs presented here, many of which are system-specific. Therefore, the costs presented in this chapter are intended to provide relative costs for implementing the various technologies, based on the design criteria presented and various system sizes. The costs are highly dependent on the design criteria and may be system-specific.

Figures 18-1 and 18-2 illustrate the unit costs for alternative disinfectants and DBP precursor removal technologies, respectively, for a range of system sizes. These costs illustrate the relative differences among these technologies. In general, the costs for alternative disinfectants are less than those for precursor removal, except at the smallest system sizes. Chloramine addition is by far the least expensive alternative disinfectant, followed by chlorine dioxide and ozone. Chloramines, however, provide less disinfecting capacity than chlorine, and much less than chlorine dioxide and ozone. The general tendency is the stronger the disinfectant, the more expensive the technology.

Similarly, increasing the coagulant dosage, even when taking into account sludge disposal costs, is much less expensive than GAC or membrane treatment. Again, this least-cost pattern follows the extent to which DBP precursor removal is achieved. That is, technologies that remove more precursor material have higher costs. Because of the modular nature of nanofiltration membranes, economies of scale tend to flatten out between design flows of 5 and 20 mgd. Beyond that, the unit cost for NF does not change significantly.

As expected, small-system unit costs are higher than large-system unit costs. For very small systems, modular systems that are simple to operate (e.g., membranes) can be significantly less expensive than any technology other than chloramines and increasing coagulant dosage. As such, a technology that has the greatest potential to improve water quality in terms of DBPs is

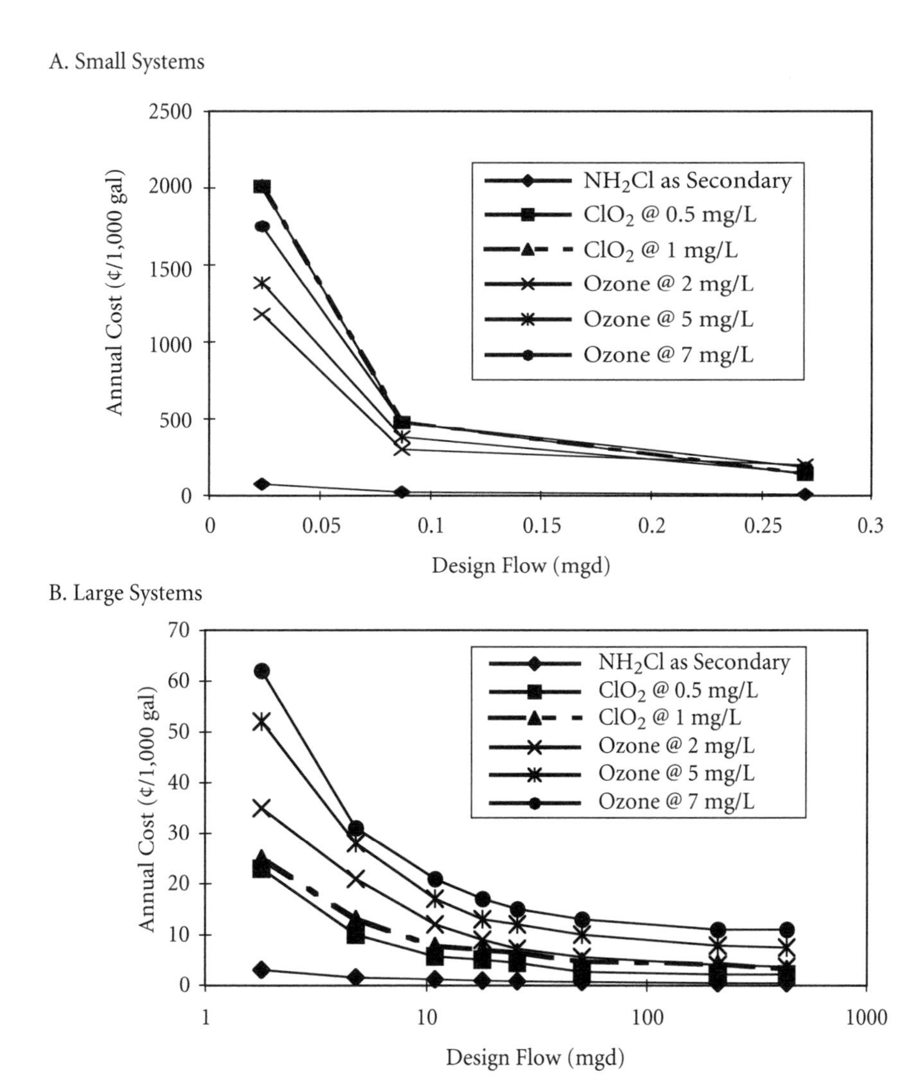

Figure 18-1 Upgrade costs for alternative disinfection

not necessarily the most expensive technology for small systems. Again, this assumes that sufficient pretreatment is available for the membrane system.

When comparing the costs, it is important to note that the effectiveness of each of these technologies is not equivalent. Rather, Figures 18-1 and 18-2 illustrate a comparison of costs (least expensive to most expensive) that utilities may consider

A. Small Systems

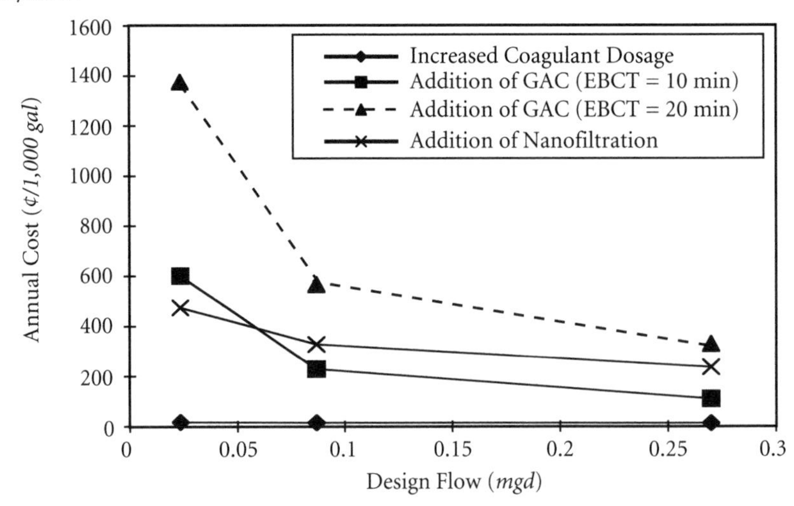

B. Large Systems

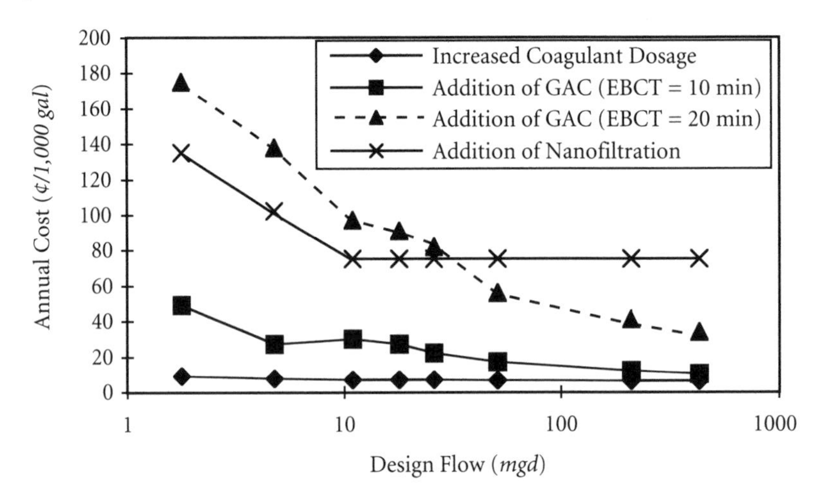

Figure 18-2 Upgrade costs for improved NOM removal

when making decisions on implementing changes to meet DBPs. Further, combinations of technologies may be required to meet different water quality objectives, and objectives beyond DBP control alone may affect implementation choices. A least-cost approach may not always be considered by a system, if other factors are considered beyond DBP compliance.

Finally, it must be noted that the central-tendency costs presented here do not reflect the costs associated with demolition and unique coordination requirements to integrate these technologies into existing facilities. These costs are highly system-specific and could not be universally estimated.

REFERENCES

Allgeier, S.C., and R.S. Summers. 1995. Evaluating NF for DBP Control with the RBSMT. *Jour. AWWA*, 87(3):87.

AWWA. 1998. *Secondary Impacts of Enhanced Coagulation.* Water Industry Technical Action Fund. Denver, Colo.: American Water Works Association.

———. 1997. Compilation and Analysis of Enhanced Coagulation and Enhanced Precipitative Softening Data. Water Industry Technical Action Fund. Prepared by Malcolm Pirnie, Inc.

AWWARF and USEPA. 1989. Design and Use of Granular Activated Carbon - Practical Aspects. In *Proc. Technology Transfer Conference.* Denver, Colo.: American Water Works Association Research Foundation.

Bergman, R.A. 1996. Cost of Membrane Softening in Florida. *Jour. AWWA*, 88(5):32.

Cheng, R., et al. 1995. Enhanced Coagulation: A Preliminary Evaluation. *Jour. AWWA*, 87(2):91.

Chowdhury, Z.K., et al. 1995. Cost of Implementing Enhanced Coagulation and the ICR; A City of Tempe Experience. In *Proc. 1995 AWWA Annual Conference.* Denver, Colo.: American Water Works Association.

Crozes, G., et al. 1995. Enhanced Coagulation: Its Effect on NOM Removal and Chemical Costs. *Jour. AWWA*, 87(1):78.

Shorney, H.L., and S.J. Randtke. 1994. Enhanced Lime Softening for Removal of Disinfection By-Product Precursors. In *Proc. 1994 AWWA Annual Conference.* Denver, Colo.: American Water Works Association.

Summers, R.S. et al. 1995. Bench Scale Evaluation of GAC for NOM Control. *Jour. AWWA*, 87(6):69.

USEPA. 1998a. *Technologies and Costs for the Control of Disinfection By-Products.* Prepared by International Consultants, Inc., and Malcolm Pirnie, Inc.

———. 1998b. *Guidance Manual for Enhanced Coagulation and Enhanced Precipitative Softening.* Prepared by International Consultants, Inc. and Malcolm Pirnie, Inc.

CHAPTER • NINETEEN

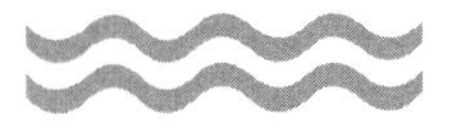

Disinfection and Disinfection By-Products in Europe

RONALD F. PACKHAM

Providing a European perspective on disinfection by-products (DBPs) is difficult because the component 15 countries within the European Union (EU) differ in the way they organize their public water supplies. Also differing are the attitudes and expectations of their populations in regard to drinking water.

Although the inhabitants of many European countries have a high liquid intake, they seem to consume very little water and when they do it is often of the commercially bottled variety (see Table 19-1). In the United Kingdom, consumers have always been led to believe that their drinking water is of excellent quality and on average about 1 L of tap water is consumed per person per day, mostly as drinks such as tea and coffee made with boiling water. Bottled water consumption is rising rapidly, however, partly because it has become fashionable and partly as the result of often ill-founded concerns about drinking water quality.

Table 19-1 Bottled water consumption per annum (1988)

Country	Litres/person
Belgium	73
France	83
Germany	73
Greece	15
Italy	80
Netherlands	11
Spain	30
United Kingdom	3
United States of America	23

Source: British Soft Drinks Association (1992).

Most European countries have highly developed public water supply systems, but there are situations, particularly in rural areas, where there is room for considerable improvement or where pollution risks need to be reduced drastically. On the other hand, in some countries, there is such confidence in the natural quality of local groundwaters or mountain lakes, that the local population resists any form of water treatment, particularly if it might impart a chlorinous taste. In countries with water supplies that are highly dependent on major international rivers, such as the Rhine, water is regarded with greater suspicion and the application of the most advanced methods of treatment is widely regarded as being essential.

Although the European water industry has been closely involved in work on the problem of DBPs and has been instigating actions aimed at minimizing their levels in drinking water, the issue has not caught the attention of the media to the same extent that it has in North America. Water quality has been in the front line of media interest on many occasions, but in Europe, the emphasis has been more on pesticides, nitrate, and lead amongst other things.

Against this background, a European perspective on DBPs will be discussed in relation to the regulatory approach, the compliance with regulations, the application of remedial treatment processes, and the perception of risk.

DRINKING WATER REGULATIONS

Many European countries, including the United Kingdom, had no legally enforceable numerical standards for drinking water prior to the promulgation of the EC Drinking Water Directive in 1980 (European Community 1980). Previously the legal requirement in the United Kingdom was that drinking water had to be "wholesome" and, although there was no specific definition of this term, it was widely held to imply that water not only had to be safe, but also pleasant to use. The drinking water standards of the World Health Organization (WHO 1970) were widely used in the United Kingdom and many other European countries, although they had no legal basis.

The European Commission's first attempt at a drinking water directive was considered as unsatisfactory from scientific and other aspects. Despite all the well-justified criticism, however, there is little doubt that the directive resulted in a move towards the raising of water quality standards across the EC in terms of key microbiological standards, lead, nitrate, and some other chemical parameters. On the matter of DBPs, the directive had little to say, except that the concentration of haloforms should be as low as possible. A "guide level" of 1 μg/L was given, but this had no legal standing.

The 1980 directive is the European water regulation still in force and this has had to be taken into the law of all EU member states. There is nothing to stop member states from adopting their own regulations on water quality provided that these are at least as stringent as those of the directive. This has taken place in some countries, particularly following publication of two editions of WHO *Guidelines for Drinking Water Quality* (WHO 1984, 1993), which include a comprehensive list of guideline values for chemical contaminants, including DBPs.

In the United Kingdom, the Water Supply Water Quality Regulations 1989 (Statutory Instrument No. 1147 1989) adopted a limit of 100 μg/L for the sum of four trihalomethanes (THMs) based on averages of samples collected every three months from within designated water supply zones. In Norway, Italy, Portugal, and some other countries (Aeppli 1997), a limit for chloroform alone of 30 μg/L was adopted following the 1984 WHO *Guidelines for Drinking Water Quality*. In The Netherlands (Kruithof and Evendijk 1995), very stringent limits for THMs were proposed in 1993 based on a lifetime cancer risk of 1 in 10^6 (Statutory Instrument No. 1147 1989). These give limits for individual THMs of around 5 μg/L and imply the phasing out of chlorine as a disinfectant.

At the time of this writing, a new EU directive on drinking water quality was being drafted. The final document (European Community 1998) was promulgated in December 1998 to come into force in December 2003 with a requirement for compliance within five years of that date. In the case of Total THMs, compliance with a parametric value of 100 μg/L is required within 10 years of the directive coming into force. An interim parametric value of 150 μg/L must be complied with between 5 and 10 years. For bromate, a value of 25 μg/L must be complied with between 5 and 10 years, with a value of 10 μg/L after 10 years. The directive sets no limits for haloacetic acids and other DBPs although these may be included in the regulations set by individual member states.

COMPLIANCE

It is impossible at present to get good pan-European data on compliance with current or prospective regulations. In many European countries, the existing management structure governing water supply does not lend itself to the retrieval of nationwide data on performance in relation to EU water regulations; indeed only two member states were able to meet a September 1996 deadline for the provision of information under a Standardized Reporting Directive (EC 1991).

In the United Kingdom, compliance information on all drinking water quality parameters is reported annually and, for THMs, compliance overall is in the high 90 percentiles and is showing year-to-year improvement (Rouse 1996). The United Kingdom has no regulations concerning many DBPs other than THMs, but chloroacetic acids have been surveyed in relation to WHO provisional guideline values of 50 μg/L for dichloroacetic acid and 100 μg/L for trichloroacetic acid. These were found in many waters and there is the possibility of noncompliance with the WHO limits as a result of the treatment of some highly colored waters, but this is not believed to constitute a serious problem (Drinking Water Inspectorate 1998).

The EU is facing a limit for bromate of 10 μg/L and this is causing some concern where ozone is used. Ozone has been used in France, Germany, Switzerland, and other European countries for many years, and much of the technology of ozone production and application is of European origin. The main applications have been for disinfection and taste-and-odor control, but recently ozone has been more widely applied to reduce levels of pesticides to meet the extremely low limits (0.1 μg/L for any pesticide) required under the EU directive (European Community 1980). The presence of bromide in the raw water can lead to the formation of bromate (see chapters 2, 9), sometimes at levels in excess of the proposed EU limit. Although a number of bromate removal methods are being investigated, the preferred option would be prevention of bromate formation if a suitable method could be found.

In the United Kingdom, bromate levels in the treated water from 46 water plants supplying a population of about 13.5 million were reviewed in relation to theoretical limits for bromate of 5, 10, and 25 μg/L (Department of the Environment & Welsh Office 1989). At 25 μg/L (WHO provisional guideline value)(WHO 1993), four plants would be noncompliant on the basis of a maximum admissible concentration (MAC), with only one noncompliant on the basis of a limit based on an annual mean. At 10 μg/L (proposed EU limit), 25 systems (nine of which did not use ozone) would be noncompliant with a MAC, including seven noncompliant on the basis of an annual mean. For a limit of 5 μg/L, drastic changes would be needed in treatment at 36 systems to achieve compliance on the basis of an annual mean.

The presence of bromate in water chlorinated with commercial sodium hypochlorite solution was demonstrated in a study of levels in commercial hypochlorite, raw water, and final water at nine treatment works (Drinking Water Inspectorate 1998). The results (Table 19-2) show bromate concentrations ranging from 50 to 1,150 mg/L in hypochlorite and from 3 to 28 μg/L in treated waters. In a separate study, bromate was found to be absent in treated waters from systems where chlorine gas or on-site electrolytically produced hypochlorite were used. It appears that the

Table 19-2 Bromate in commercial sodium hypochlorite

	Bromate Concentration		
Water System	Hypochlorite (*mg/L*)	Raw Water (*µg/L*)	Final Water (*µg/L*)
A	1,000	<2	3
B	905	<2	9
C	50	<2	<2
D	220	<2	5
E	370	3	8
F	1,100	<2	8
G	800	<2	3
H	1,150	<2	28
I	770	<2	4

Source: Drinking Water Inspectorate (1998).

specifications for commercial hypochlorite should include a strict limit for the bromate concentration.

Chlorite is important only in relation to chlorine dioxide, for which a maximum concentration for the total of ClO_2 + ClO_2^- + ClO_3^- of 500 µg/L in treated water is specified in United Kingdom regulations. Chlorine dioxide is used to a very limited extent in the United Kingdom, but more extensively in continental Europe, particularly in Germany.

Chlorate is an issue in the United Kingdom where there is a growing interest in on-site electrolytic production of hypochlorite. The concern is motivated largely by the health and safety hazards associated with the use, storage, and transportation of liquefied chlorine in bulk. Unfortunately chlorate formation takes place in the electrolytic cell and concentrations of up to 250 µg/L in treated water have been reported (Drinking Water Inspectorate 1998). Table 19-3 gives the results of that study of chlorate levels at seven systems using hypochlorite generated on-site. Levels of up to 0.84 g/L were found in the tank where the hypochlorite was stored for up to 7 days. Levels of up to 166 µg/L chlorate were present in treated waters. There is a current United Kingdom regulation permitting up to 700 µg/L of chlorate in water treated by electrolytic chlorination, but the toxicological basis of this limit is inadequate.

Table 19-3 Chlorate in on-site generated hypochlorite

	Chlorate Concentration		
Water System	Hypochlorite Storage Tank (*g/L*)	Raw Water (*μg/L*)	Final Water (*μg/L*)
E	0.39	<10	20
F	0.21	<10	42
G	0.13	14	61
H	0.32	<10	166
J	0.84	<10	34
K	NA*	<10	24
L	NA	38	38

*Not available.
Source: Drinking Water Inspectorate (1998).

The work with electrolytic hypochlorite production led to the recognition that ordinary hypochlorite solutions could contain high levels of chlorate, particularly after storage, with formation being enhanced by temperature, sunlight, and metal oxides. At present there are no sound limits for chlorate and there is undoubtedly a need for a good toxicological study on the basis of which a limit can be set.

TREATMENT

As in North America, changes in water treatment aimed at reducing DBPs (or more specifically THMs) have focused on reducing their formation by removing precursors and changing disinfection practice rather than on the removal of DBPs once formed. A basic consideration is always that any change made must not result in the disinfection process being compromised.

Precursor removal would normally be by chemical coagulation under conditions as close as possible to the optimum for color removal, although the relatively low pH required for this is difficult to achieve with highly buffered waters. The practice of prechlorination has been reduced or even eliminated and precursor removal is enhanced where oxidative treatment (e.g., ozone) or activated carbon (usually granular activated carbon [GAC]) is used. Where biological filtration (slow sand filtration) is used as an alternative to coagulation, GAC may be incorporated in the filter bed or added as a separate unit process. Ozone and activated carbon are rarely used specifically for DBP control, but are usually a part of the general armory against trace organics, particularly in the treatment of surface-derived waters. Terminal disinfection with chlorine and an appropriate period of retention in contact tanks prior to distribution would be the norm in the United Kingdom. This appears to represent a significant difference between European and US practices, where contact tanks are not an integral component of postfiltration designs.

Apart from the reduced use of prechlorination, other changes in disinfection practice include a reduction in the heavy application of chlorine (e.g., superchlorination–dechlorination) and an increase in the use of chloramine instead of free chlorine. In the United Kingdom, all public water supplies are disinfected and, in virtually all cases, a form of chlorination (including chloramination) is used regardless of whether or not ozone is being applied for other reasons. In continental Europe, chlorine is still used widely, but disinfection with ozone, usually followed by terminal chlorination, has been used for many years in parts of France, Switzerland, and Germany. In The Netherlands, the use of chlorine has been virtually eliminated following the introduction of multiple stages of treatment (e.g., coagulation, sedimentation, filtration, bank filtration and storage, ozonation, and GAC). Ultraviolet (UV) disinfection is practiced on some of the smaller supplies in Europe.

There is no simple universal remedy to situations where high levels of THMs occur. Treatment has to be tailored to deal with the individual circumstances. The same basic principles apply, the most important of which is the maintenance of efficient disinfection.

RISK

Concern about THMs and other DBPs has been a major stimulus to research on drinking water quality over the last 20 years, and as a result of this work, we now know far more about the nature, formation, and interaction of these compounds with the wide range of treatment processes that are available. Still unknown is an evaluation of the true risk to public health that THMs and other DBPs present. This is of vital concern because the presence of these substances in drinking water is by definition an outcome of the essential process of disinfection.

The available tools for estimating the effect on health of trace substances in water are extremely crude. Most epidemiological studies in this field have severe problems with confounding factors and difficulties in assessing exposure over decades of water consumption. The toxicological approach, on which almost all limits and guideline values are based, depends on the extrapolation of data from animal models exposed to vastly higher concentrations than those found in water. This necessitates the application of large safety factors to take into account various uncertainties, including the quality of the data. This tends to result in a stacking up of worst-case scenarios and the definition of limits with a considerable margin of safety. Although it is essential always to err on the side of safety, the actual level of risk becomes important in discussing the "disinfection dilemma," in which the poorly-defined risks associated with DBPs have to be set against the known risks associated with inadequate disinfection.

In most European countries, while DBP levels are minimized, the DBP risk is regarded as minor relative to the risks associated with inadequate disinfection. In the United Kingdom, following a detailed review of epidemiological and toxicological evidence then available, an official committee reported in 1986 that, "we have found no sound reason to conclude that the consumption of the by-products of chlorination, in drinking water which has been treated and chlorinated according to current practices, increases the risk of cancer in humans" (Department of the Environment & Welsh Office 1989). The committee's view that it is prudent to minimize levels of THMs provided that disinfection is not compromised in any way continues, in the author's opinion, to represent the general view in the United Kingdom. It is fortunate that situations are extremely rare where compliance with accepted limits for DBPs presents insurmountable difficulties in relation to disinfection.

Whatever the official view, the increasing sales of home water treatment devices and bottled water undoubtedly reflect a general concern by the public about the risks that may be associated with trace chemicals in drinking water, although the present European concern is probably directed more at agricultural chemicals than DBPs.

SUMMARY

The EU is working towards harmonizing environmental management within member states and, in the field of public water supply, an essential element of this is the adoption of common criteria for assessing the quality of drinking water. A directive on drinking water quality promulgated in 1980 has been updated and includes for the first time mandatory limits for THMs and bromate. Several individual member states have adopted their own national standards for some DBPs following publication of *Guidelines for Drinking Water Quality* by the World Health Organization. Pan-European data on levels of DBPs in public water supplies is unavailable, but such information is published in some countries, including the United Kingdom.

The European approach to the minimization of DBP levels is usually similar to that in North America, i.e., the removal of precursor compounds being a first priority. Reducing the level of prechlorination as far as possible is important in combination with efficient disinfection after filtration. The emphasis on post-filtration disinfection with the necessary installation of chlorination retention tanks is common practice in many European countries and this may represent an important difference between European and American practice. In several countries, ozone has been used as a preferred alternative to chlorine for many years, although even with ozone, chlorine, chloramine, or chlorine dioxide may be used to provide a disinfecting residual. The adoption of a low limit for bromate is likely to present a problem in some situations and for this a satisfactory remedy has still to be found. The lack of any satisfactory toxicology for chlorate, which may occur at fairly high concentrations in hypochlorite solutions, is seen to represent an important research need. There is no doubt within the countries of the EU that, whatever is done to control the level of DBPs, the integrity of the disinfection process must never be compromised.

REFERENCES

Aeppli, J. 1991. *Potable Water Quality Requirements.* International Water Supply Association International Report. London: International Water Supply Association.

British Soft Drinks Association. 1992. Factsheet No. 12 - 11.92. London: British Soft Drinks Association.

Department of the Environment & Welsh Office. 1989. *Guidance on Safeguarding the Quality of Water Supplies.* London: Her Majesty's Stationery Office.

Drinking Water Inspectorate. 1998 (in press). *Disinfection By-products from Drinking Water Treatment.* Marlow, England: Foundation for Water Research.

European Community. 1980. Directive Relating to the Quality of Water Intended for Human Consumption. *Official Journal,* 80/778/EC, No. L229.

———. 1991. Directive on Standardising and Rationalising Reports on the Implementation of Certain Directives Relating to the Environment. *Official Journal,* 91/692/EC, No. L377.

———. 1998. Directive Relating to the Quality of Water Intended for Human Consumption. *Official Journal,* 98/83/EC, No. L330.

Kruithof, J.C., and J.E. Evendijk. 1995. *New WHO Recommendations for Water Quality Standards. Impact on Water Treatment Practices.* National Report of The Netherlands. London: International Water Supply Association.

Rouse, M. 1996. *Drinking Water 1995, a Report By the Chief Inspector, Drinking Water Inspectorate.* Department of the Environment & Welsh Office. London: Her Majesty's Stationery Office.

Statutory Instrument No. 1147. 1989. Water, England and Wales. *The Water Supply (Water Quality) Regulations 1989.* London: Her Majesty's Stationery Office.

WHO. 1970. *European Standards for Drinking Water.* Geneva, Switzerland: World Health Organization.

———. 1984. *Guidelines for Drinking Water Quality.* Geneva, Switzerland: World Health Organization.

———. 1993. *Guidelines for Drinking Water Quality,* 2nd ed. Geneva, Switzerland: World Health Organization.

APPENDIX A

Abbreviations

A	surface area
ANPRM	Advanced Notice for Proposed Rulemaking
AOC	assimilable organic carbon
AOP(s)	advanced oxidation process(es)
AOX	adsorbable organic halogen (halide)
APHA	American Public Health Association
ASTM	American Society of Testing Materials
AWWA	American Water Works Association
AWWARF	American Water Works Association Research Foundation
BAT	best available technology
BDL	below detection limit
BDOC	biodegradable organic carbon
BOM	biodegradable organic material
BV	bed volume
CAM	carbon adsorption method
CBA	cost–benefit analysis
CDC	Centers for Disease Control
CEA	cost-effectiveness analysis
CNCl	cyanogen chloride
CNX	cyanogen halide(s)

CSTR	continuous-flow stirred tank reactor
$C \times T$	concentration of disinfectant times contact time
CUA	cost utility analysis
CUR	carbon useage rate
D(s)	disinfectant(s)
D/DBP(s)	disinfectants/disinfection by-product(s)
DBAA	dibromoacetic acid
DBP(s)	disinfection by-product(s)
DBPFP	disinfection by-product formation potential
DBPP(s)	disinfection by-product precursor(s)
DCAA	dichloroacetic acid
DCAN	dichloroacetonitrile
DOC	dissolved organic carbon
DOX	dissolved organic halogen (halide)
DPD	*N,N*-diethyl-p-phenylenediamine
DWPL	Drinking Water Priority List
EBCT	empty-bed contact time
ECD	electron capture detector
EPA	(US) Environmental Protection Agency
ESWTR	Enhanced Surface Water Treatment Rule
EU	European Union
FACA	Federal Advisory Committee Act
FP	formation potential
GAC	granular activated carbon
GC	gas chromatography
GC/MS	gas chromatography/mass spectrometry
GWDR	Groundwater Disinfection Rule
HAA(s)	haloacetic acid(s)
HAA5	mono-, di-, and trichloroacetic acid, plus mono- and dibromoacetic acid
HAA6	mono-, di-, and trichloroacetic acid, plus mono- and dibromoacetic acid, plus bromochloroacetic acid
HAA9	mono-, di-, and trichloroacetic acid, plus mono-, di-, and tribromoacetic acid, plus bromochloro-, dibromochloro-, and bromodichloroacetic acid
HAN(s)	haloacetonitrile(s)
HK(s)	haloketone(s)
HRT	hydraulic retention residence time
IARC	International Association for Research on Cancer

ICR	Information Collection Rule
IESWTR	Interim Enhanced Surface Water Treatment Rule
ILSI	International Life Sciences Institute
IOC(s)	inorganic chemical(s)
IOP	iron oxide particles
IX	ion exchange
J	transmembrane flux
LLE	liquid–liquid extraction
LTESWTR	Long-Term Enhanced Surface Water Treatment Rule
MAC	maximum allowable concentration
MBAA	monobromoacetic acid
MCAA	monochloroacetic acid
MCL	maximum contaminant level
MCLG	maximum contaminant level goal
MDL	minimum detection level
M/DBP	microbial(s)/disinfection by-product(s) (cluster)
MF	microfiltration
mgd	million gallons per day
MRDL(s)	maximum residual disinfectant level(s)
MRDLG(s)	maximum residual disinfectant level goal(s)
MRL	minimum reporting level
MX	3-chloro-4-(dichloromethyl)-5-hydroxy-2(5H)-furanone
MW	molecular weight
MWCO	molecular weight cutoff
NAS	National Academy of Sciences
NCI	National Cancer Institute
NPDWR	National Primary Drinking Water Regulations
NF	nanofiltration
NIPDWR	National Interim Primary Drinking Water Regulations
NODA	Notice of Data Availability
NOM	natural organic matter
NOMS	National Organics Monitoring Survey
NORS	National Organics Reconnaissance Survey
NPOC	nonpurgeable organic carbon
O&M	operation and maintenance (costs)
OR	odds ratio
OX	organic halogen (halide)

$\Delta\Pi$	osmotic pressure gradient
P	pressure
PAC	powdered activated carbon
PAC-UF	powdered activated carbon - ultrafiltration
POE	point of entry
POU	point of use
RMCL	recommended maximum contaminant level
RO	reverse osmosis
Q	flow rate
R	feedwater recovery
R_m	hydraulic resistance
RR	relative risk
RSSCT	rapid small-scale column test
SBA	strong-base anion (exchange)
SDS	simulated distribution system
SDWA	Safe Drinking Water Act
SOC(s)	synthetic organic chemical(s)
SPW	State Project Water
SUVA	specific ultraviolet absorbance
SWTR	Surface Water Treatment Rule
TCA	trichloroacetone
TCAA	trichloroacetic acid
TCAN	trichloroacetonitrile
TCR	total coliform rule
TDS	total dissolved solids
THM(s)	trihalomethane(s)
THMFP	trihalomethane formation potential
TOBr	total organic bromide
TOC	total organic carbon
TOCl	total organic chloride
TOX	total organic halogen (halide)
TOXFP	total organic halide formation potential
TTHM(s)	total trihalomethane(s): chloroform, bromodichloromethane, dibromochloromethane, and bromoform
TWG(s)	technical working group(s)
UF	ultrafiltration
UFC	uniform formation conditions
USEPA	US Environmental Protection Agency

USPHS	US Public Health Service
UV	ultraviolet absorbance
UV-254	ultraviolet absorbance at 254 nm
VOC(s)	volatile organic chemical(s)
WBA	weak-base anion (exchange)
WHO	World Health Organization
WIDB	Water Industry Data Base

Index

NOTE: *f.* indicates a figure, *t.* indicates a table.

A

B

C

D

E

H

I

J

K

L

M

N

U

V

W

X